The Handbook of Environmental Chemistry

Volume 114

Founding Editor: Otto Hutzinger

Series Editors: Damià Barceló • Andrey G. Kostianoy

In over four decades, *The Handbook of Environmental Chemistry* has established itself as the premier reference source, providing sound and solid knowledge about environmental topics from a chemical perspective. Written by leading experts with practical experience in the field, the series continues to be essential reading for environmental scientists as well as for environmental managers and decision-makers in industry, government, agencies and public-interest groups.

Two distinguished Series Editors, internationally renowned volume editors as well as a prestigious Editorial Board safeguard publication of volumes according to high scientific standards.

Presenting a wide spectrum of viewpoints and approaches in topical volumes, the scope of the series covers topics such as

- local and global changes of natural environment and climate
- anthropogenic impact on the environment
- water, air and soil pollution
- remediation and waste characterization
- environmental contaminants
- biogeochemistry and geoecology
- chemical reactions and processes
- chemical and biological transformations as well as physical transport of chemicals in the environment
- environmental modeling

A particular focus of the series lies on methodological advances in environmental analytical chemistry.

The Handbook of Environmental Chemistry is available both in print and online via https://link.springer.com/bookseries/698. Articles are published online as soon as they have been reviewed and approved for publication.

Meeting the needs of the scientific community, publication of volumes in subseries has been discontinued to achieve a broader scope for the series as a whole.

Emerging Pollutants in Sewage Sludge and Soils

Volume Editors: Avelino Núñez-Delgado · Manuel Arias-Estévez

With contributions by

B. Abishek Kumar · M. B. Almendro-Candel · E. Alonso · E. Álvarez-Rodríguez · S. Álvarez-Torrellas · V. Andreu · I. Aparicio · D. Arenas-Lago · M. Arias-Estévez · A. K. Aseperi · J. Barker · R. Busquets · F. M. Caccamo · M. Carballa · M. Carnevale Miino · J. S. Carsella · R. Cela-Dablanca · P. C. W. Cheung · R. B. Chirayath · M. C. Collivignarelli · D. Fernández-Calviño · M. J. Fernández-Sanjurjo · A. Ferraro · L. Gonzalez-Gil · P. S. Hooda · A. Huang · V. Ismael Águeda Maté · M. Iyer · J. C. Jiménez · C. A. Kinney · J. M. Lema · I. G. Lucas · J. Martín · C. Mejías · B. H. Mendioroz · A. M. Montes · J. Mutz · A. P. Nair · J. Navarro-Pedreño · A. Núñez-Delgado · I. Ortiz · A. Panico · Y. Picó · H. Pirasteh-Anosheh · F. Pirozzi · D. Polo · M. Race · J. G. Rodríguez · A. Rodríguez-Seijo · J. L. Romalde · M.-F. San-Román · V. Santás-Miguel · J. L. Santos · S. Schröder · D. Spasiano · M. D. Subramaniam · G. Trancone · B. Vellingiri · D. Vitale · D. Zhi · Y. Zhou · A. A. Zorpas · T. Zuniga

Springer

Editors
Avelino Núñez-Delgado
Department of Soil Science and Agricultural Chemistry
University of Santiago de Compostela
Lugo, Spain

Manuel Arias-Estévez
Department of Soil Science and Agricultural Chemistry
University of Vigo
Ourense, Spain

ISSN 1867-979X ISSN 1616-864X (electronic)
The Handbook of Environmental Chemistry
ISBN 978-3-031-07608-4 ISBN 978-3-031-07609-1 (eBook)
https://doi.org/10.1007/978-3-031-07609-1

This Springer imprint is published by the registered company Springer Nature Switzerland AG
The registered company address is: Gewerbestrasse 11, 6330 Cham, Switzerland

Series Preface

With remarkable vision, Prof. Otto Hutzinger initiated *The Handbook of Environmental Chemistry* in 1980 and became the founding Editor-in-Chief. At that time, environmental chemistry was an emerging field, aiming at a complete description of the Earth's environment, encompassing the physical, chemical, biological, and geological transformations of chemical substances occurring on a local as well as a global scale. Environmental chemistry was intended to provide an account of the impact of man's activities on the natural environment by describing observed changes.

While a considerable amount of knowledge has been accumulated over the last four decades, as reflected in the more than 150 volumes of *The Handbook of Environmental Chemistry*, there are still many scientific and policy challenges ahead due to the complexity and interdisciplinary nature of the field. The series will therefore continue to provide compilations of current knowledge. Contributions are written by leading experts with practical experience in their fields. *The Handbook of Environmental Chemistry* grows with the increases in our scientific understanding, and provides a valuable source not only for scientists but also for environmental managers and decision-makers. Today, the series covers a broad range of environmental topics from a chemical perspective, including methodological advances in environmental analytical chemistry.

In recent years, there has been a growing tendency to include subject matter of societal relevance in the broad view of environmental chemistry. Topics include life cycle analysis, environmental management, sustainable development, and socio-economic, legal and even political problems, among others. While these topics are of great importance for the development and acceptance of *The Handbook of Environmental Chemistry*, the publisher and Editors-in-Chief have decided to keep the handbook essentially a source of information on "hard sciences" with a particular emphasis on chemistry, but also covering biology, geology, hydrology and engineering as applied to environmental sciences.

The volumes of the series are written at an advanced level, addressing the needs of both researchers and graduate students, as well as of people outside the field of "pure" chemistry, including those in industry, business, government, research establishments, and public interest groups. It would be very satisfying to see these volumes used as a basis for graduate courses in environmental chemistry. With its high standards of scientific quality and clarity, *The Handbook of Environmental Chemistry* provides a solid basis from which scientists can share their knowledge on the different aspects of environmental problems, presenting a wide spectrum of viewpoints and approaches.

The Handbook of Environmental Chemistry is available both in print and online via https://link.springer.com/bookseries/698. Articles are published online as soon as they have been approved for publication. Authors, Volume Editors and Editors-in-Chief are rewarded by the broad acceptance of *The Handbook of Environmental Chemistry* by the scientific community, from whom suggestions for new topics to the Editors-in-Chief are always very welcome.

Damià Barceló
Andrey G. Kostianoy
Series Editors

Preface

The editors of this book have been working for years on various kinds of emerging pollutants. Some of their recent papers, books, and book chapters have focused on environmental pollution due to antibiotics, but they have also worked previously on heavy metals, pesticides, polycyclic aromatic hydrocarbons, some pathogenic microorganisms, and other contaminants. Taking into account current environmental and public health issues caused by emerging pollutants, as well as risks of future eventual hazards, the editors thought that a book on emerging contaminants derived from both abiotic and biotic origins would be highly interesting.

Now, the book is ready, including high-quality chapters from top researchers around the world (Spain, Italy, India, U.K., China, and USA). Specifically, one chapter deals with "Biotic and Abiotic Contamination due to Emerging Pollutants in Sewage Sludge and Soils: A Country-Based Perspective", another one is focused on "Pharmaceuticals and Their Metabolites in Sewage Sludge and Soils: Distribution and Environmental Risk Assessment", a third chapter deals with "Microplastics in Soils as a Source of Pollution and Environmental Risk", while a fourth chapter has the title "Environmental Transmission of Human Pathogens Associated with SARS-CoV-2 and the Effect on Soil and Aquatic Ecosystem: A Biological Way of Management". The other chapters have the following titles: "Innovative Treatment Processes for Emerging Contaminants Removal from Sewage Sludge", "Emerging Pollutants That Can Be Transformed into PCDD/Fs", "An Innovative Technology to Minimize Biological Sludge Production and Improve Its Quality in a Circular Economy Perspective", "Fate of Neonicotinoids in the Environment: Why Bees Are Threatened", "Current Progress of Microplastics in Sewage Sludge", "Revision of the Most Harmful Organic Compounds Present in Sewage and Sludge", "Identifying Emerging Pollutants Using Non-target or Wide-Screening Liquid Chromatography-Mass Spectrometry", "Fate of Emerging Pollutants During Anaerobic Digestion of Sewage Sludge", "Pharmaceuticals and Personal Care Products as Contaminants of Emerging Concern in Sewage Sludge and Soils and the Role of Transformation Products in Their Fate and Environmental Impact", "Current Methodology for Extraction, Separation, Identification and

Quantification of Microplastics in Terrestrial Systems", and "Emerging Viruses in Sewage Sludge and Soils".

The editors think that the result is a very interesting book including high-quality chapters, and they would like to thank all authors contributing to it, as well as to the reviewers and the staff of Springer-Nature involved in the task. All the people that have participated in the elaboration of the book hope that it will be clearly useful for researchers and the overall society, now and for years, as it covers in a rigorous and scientific manner current and eventual future environmental and public health issues of enormous concern.

The scientific editors of the Book: Dr. Avelino Núñez-Delgado (University of Santiago de Compostela, Spain) and Dr. Manuel Arias-Estévez (University of Vigo, Spain).

Lugo, Spain — Avelino Núñez-Delgado
Ourense, Spain — Manuel Arias-Estévez

Contents

Biotic and Abiotic Contamination Due to Emerging Pollutants in Sewage Sludge and Soils: A Country-Based Perspective

Raquel Cela-Dablanca, Esperanza Álvarez-Rodríguez, María J. Fernández-Sanjurjo, Manuel Arias-Estévez, Vanesa Santás-Miguel, David Fernández-Calviño, and Avelino Núñez-Delgado

Contents

Abstract This book chapter deals with emerging pollutants in sewage sludge and soils, showing a country-based perspective for contaminants from both biotic and abiotic origins included in the category. Using scientific searching tools, a view on the evolution of the number of studies performed in this field can be obtained by years and by countries. As regards the countries showing higher number of results

R. Cela-Dablanca, E. Álvarez-Rodríguez, M. J. Fernández-Sanjurjo, and A. Núñez-Delgado (✉)
Department of Soil Science and Agricultural Chemistry, Engineering Polytechnic School, University of Santiago de Compostela, Lugo, Spain
e-mail: raquel.dablanca@usc.es; esperanza.alvarez@usc.es; mf.sanjurjo@usc.es; avelino.nunez@usc.es

M. Arias-Estévez, V. Santás-Miguel, and D. Fernández-Calviño
Soil Science and Agricultural Chemistry, Faculty of Sciences, University of Vigo, Ourense, Spain
e-mail: mastevez@uvigo.es; davidfc@uvigo.es

Avelino Núñez-Delgado and Manuel Arias-Estévez (eds.), *Emerging Pollutants in Sewage Sludge and Soils*, Hdb Env Chem (2023) 114: 1–18, DOI 10.1007/698_2022_846,

Published online: 12 February 2022

for publications on the subject, the searching tool Web of Science (WOS) indicates that: (a) for the searching string "emerging pollutants sewage sludge," the countries on the top are China, Spain, and the USA; (b) for the searching string "emerging pollutants soil," the top countries are China, the USA, and Spain. For both searching strings, these three countries have numbers of publications constituting more than 10% of the total of the world production (as per WOS). Overall, the total number of results found by the scientific searching tools used (Google Scholar, Scopus, and Web of Science) indicates that the field of research is still growing, and it is expected to continue in this way, as it is clear that emerging pollutants are of real concern for human health and the overall environment.

Keywords Abiotic contaminants, Biotic contaminants, Emerging contaminants, Geographic level, Sludge pollution, Soil pollution

1 Introduction

As this chapter focuses on emerging pollutants from abiotic and biotic origins, having a definition for the specific concepts under study would be a clear need to start a review on this matter. With this in mind, and regarding emerging pollutants from abiotic origin, as indicated by Dulio et al. [1], since around 2008 "contaminants of emerging concern (CECs) became a common term for chemicals that are currently not regulated (not submitted to a routine monitoring and/or emission control regime) but may be under scrutiny for future regulation." These authors also comment that "in addition, it is now common knowledge that the contaminants of interest are not necessarily newly developed chemicals: most CECs are substances that have entered the environment for years, even decades, but their presence has only recently begun to be investigated." They also report that "most known CECs include industrial compounds, pharmaceuticals, personal-care products, biocides, and plant protection products, but the list of compounds is constantly growing."

As regards biotic emerging contaminants (emerging pathogens), Watkins [2] indicates that "in 1992, an expert committee reported on emerging infections and defined them as new, reemerging, or drug-resistant infections whose incidence in humans has increased within the past two decades or whose incidence threatens to increase in the near future."

Below (Fig. 1) we included images corresponding to some areas affected by various kinds of contaminants containing emerging pollutants, where (or with which) the authors of the chapter have worked or are still working on.

As one of the main focuses of this chapter is placed on publications dealing with emerging pollutants present in sewage sludge, Fig. 2 shows some examples of materials, facilities, and areas related to the treatment and spread of sewage sludge where the authors of this chapter have worked or are still working on.

Fig. 1 Images of materials and areas including various kinds of emerging pollutants. Above: accumulation of materials from oil spills; below: cattle slurry pit, on the left, and farm generating boiler manure and slurry, on the right (original from authors, not previously published)

Bearing in mind that the second main focus of the chapter is situated on publications regarding the presence of emerging pollutants in soils (including mine soils and some degraded areas), Fig. 3 shows some examples of areas affected by these issues, with some of them being studied by the authors of this chapter.

This chapter will focus on these main aspects (publications on emerging pollutants present in sewage sludge and soils), but special attention will be paid to the consideration of a country-based perspective, searching for eventual differences provided by scientific searching tools in this regard.

Fig. 2 Images of raw sewage sludge and a sewage sludge treatment facility (above), and sewage sludge derived material spread on a soil (below) (original from authors, not previously published)

Within the wide field of research related to emerging pollutants, it is also relevant to bear in mind that a variety of alternatives have been considered to inactivate, retain, and/or remove these emerging contaminants present in environmental compartments. Some of them are based on the use of sorbents, including bio-sorbents, by-products, agricultural and forest waste materials, compost, or biochar, raw or modified, as promising low-cost and potentially effective materials [3–16].

Fig. 3 Images of soil over an area where emerging pollutants were buried and confined (above), an open mining area, and a mine spoil area after the first steps corresponding to restoration tasks (below) (original from authors, not previously published)

Figures 4 and 5 show some examples in this regard, corresponding to materials and areas where the authors of this chapter have worked or are still working on.

As previously indicated, in most of the materials and areas shown in the images included in Figs. 1, 2, 3, 4, and 5, the authors of this chapter made research works in the past or are currently working on, indicating that for the members of this team of research it can be considered a field of high relevance, both for their past and current works, and probably for the future.

Fig. 4 Images of sorbent materials and by-products used to be mixed with sewage sludge and other wastes containing emerging pollutants, including by-products from mussel shell processing (above), individual vegetable remains and mixtures of them (below) (original from authors, not previously published)

2 Emerging Pollutants in Sewage Sludge

To have an overall view, and then a country-based perspective, we used various scientific searching tools, with keywords and searching strings related to the subject.

Specifically, searching for "emerging pollutants sewage sludge" on Google Scholar (GS), a total of 57,400 results were provided, while the number was reduced to 3,920 when the searching string was modified to "emerging-pollutants sewage-sludge." On the opposite side, Scopus provided just 441 results for "emerging pollutants sewage sludge."

In an intermediate position, Web of Science (WOS) found a total of 1,593 results when searching for "emerging pollutants sewage sludge."

WOS allows to differentiate results by country, with data in this regard presented in Table 1.

Table 2 shows the results provided by WOS corresponding to the number of publications by year, starting in 1975, date of the first publication on the subject reported by this searching tool.

Fig. 5 Images of plants growing on different kinds of waste materials treated with sorbents and by-products in the laboratory (above), and an example of test areas corresponding to restoration tasks performed in a mine spoil area (below) (original from authors, not previously published)

2.1 Emerging Pollutants of Abiotic Origin in Sewage Sludge

The results shown by the searching tools regarding emerging pollutants of abiotic origin in sewage sludge are the following. With the searching string "abiotic emerging-pollutants sewage-sludge," the total number of results found by GS was 968, while changing the string to "abiotic emerging-pollutants sewage sludge" the number found was 1,840. Scopus reported just 3 results for these strings. Searching

Table 1 Results by country corresponding to the searching string "emerging pollutants sewage sludge." There are other various countries with less than 10 records that are not included in the table. Source: WOS

Countries/regions	Number of records	Percentage of results
China	364 + 198	22.85 + 12.43 = 35.28
Spain	245	15.38
USA	232	14.56
UK + England + Scotland	61 + 56 + 18	3.83 + 3.52 + 1.13 = 8.48
India	105	6.59
Italy	96	6.03
Canada	82	5.15
Australia	76	4.77
Brazil	72	4.52
France	63	3.96
South Korea	54	3.39
Germany	52	3.26
Portugal	50	3.14
Greece	42	2.64
Mexico	35	2.20
Poland	31	1.95
Singapore	27	1.70
Malaysia	26	1.63
Netherlands	26	1.63
South Africa	26	1.63
Sweden	26	1.63
Finland	24	1.51
Japan	24	1.51
Switzerland	23	1.44
Iran	22	1.38
Turkey	20	1.26
Saudi Arabia	19	1.19
Belgium	18	1.13
Norway	17	1.07
Denmark	16	1.00
Pakistan	16	1.00
Argentina	15	0.94
Colombia	14	0.88
Taiwan	14	0.88
Czech Republic	12	0.75
New Zealand	11	0.69

for "abiotic emerging pollutants sewage sludge," WOS indicated that the number of publications was 27.

Table 2 Results by years corresponding to the searching string "emerging pollutants sewage sludge." Source: WOS

Publication year	Number of records	Percentage of results
2020	198	12.43
2019	314	19.71
2018	214	13.43
2017	193	12.12
2016	156	9.79
2015	114	7.16
2014	73	4.58
2013	76	4.77
2012	60	3.77
2011	40	2.51
2010	36	2.26
2009	22	1.38
2008	10	0.63
2007	10	0.63
2006	9	0.57
2005	5	0.31
2004	4	0.25
2003	4	0.25
2002	4	0.25
1978	1	0.06
1975	1	0.06

As regards a country-based perspective, the results provided by WOS show 11 publications for China, 6 for the UK, 4 for Spain, Australia, and Canada, and 3 for the USA and France, with all other countries reported having just 1 publication.

The authors of this chapter have previously published on emerging pollutants of abiotic origin present in sewage sludge. Some examples are the publications by Pousada-Ferradás et al. [12], Núñez-Delgado et al. [17], and Núñez-Delgado et al. [18].

Considering the results provided by GS, sorted by relevance, the paper situated first for the search "abiotic emerging-pollutants sewage-sludge" is the one by Mailler et al. [19], followed by the one by Núñez-Delgado et al. [18], and by Bletsou et al. [20].

2.2 *Emerging Pollutants of Biotic Origin in Sewage Sludge*

The results shown by the searching tools regarding emerging pollutants of biotic origin in sewage sludge are the following. With the searching string "biotic emerging-pollutants sewage-sludge," the total number of results found by GS was

1,770, while changing the string to "biotic emerging-pollutants sewage sludge" the result was 3,500. Scopus reported between 2 and 3 results for these two strings. Searching for "biotic emerging pollutants sewage sludge" WOS indicated that the number of results was 13.

As regards a country-based perspective, the results provided by WOS indicate that 3 were from China, 2 from Canada, South Korea, and U.K., and the other countries reported had just 1 publication.

The authors of this chapter have also previously published on emerging pollutants of biotic origin in sewage sludge. Some examples are the papers by Pousada-Ferradás et al. [12], and by Pousada-Ferradás et al. [21].

Considering the results provided by GS, sorted by relevance, the paper situated first for the search "biotic emerging-pollutants sewage-sludge" is the one by Trapp and Eggen [22], followed by Bletsou et al. [20] and by Carmona and Picó [23].

3 Emerging Pollutants in Soils

When GS was used to search for "emerging pollutants soil," the total number of results was 228,000, which was reduced to 13,000 when the string was changed to "emerging-pollutants soil." Scopus gave a total of 1,341 results for "emerging pollutants soil."

For the string "emerging pollutants soil," WOS provided a total of 9,772 results.

Table 3 shows the results provided by WOS presented by country.

Table 4 shows the results provided by WOS presented by year.

3.1 Emerging Pollutants of Abiotic Origin in Soils

The results shown by the searching tools regarding emerging pollutants of abiotic origin in soils are the following. With the searching string "abiotic emerging-pollutants soil," the total number of results found by GS was 2,360, while Scopus reported 3 results (however, the number was 43 when the string was changed to "abiotic emerging pollutants soil"). Searching for "abiotic emerging pollutants soil," WOS indicated that the number of results was 698.

As regards a country-based perspective, the results provided by WOS indicate that USA was placed first, followed by China, U.K., Canada, Spain, Italy, Germany, France, Portugal, and then other countries with less than 30 publications.

Considered by years, the number of results is continuously increasing (from 2 in 1999 to 120 in 2020).

The authors of this chapter have previously published on emerging pollutants of abiotic origin in soils. Some examples are the papers by Conde-Cid et al. [24], Conde-Cid et al. [25], Alvarez-Esmorís et al. [26], Cela-Dablanca et al. [27], and Santás-Miguel et al. [28].

Table 3 Results by country corresponding to the searching string "emerging pollutants soil." There are other various countries with less than 10 records that are not included in the table. Source: WOS

Countries/regions	Number of records	Percentage of results
People's Republic of China	2,143 + 372	21.93 + 3.81 = 25.74
USA	2,291	23.45
Spain	1,083	11.08
UK + England + Scotland	381 + 494 + 101	3.90 + 5.06 + 1.03 = 9.05
Canada	651	6.66
India	600	6.14
Italy	565	5.78
Germany	560	5.73
France	436	4.46
Australia	395	4.04
Brazil	374	3.83
South Korea	262	2.68
Portugal	239	2.45
Sweden	213	2.18
Japan	206	2.11
Netherlands	202	2.07
Switzerland	182	1.86
Denmark	148	1.52
Belgium	145	1.48
Greece	143	1.46
Norway	137	1.40
Mexico	136	1.39
Taiwan	136	1.39
Poland	133	1.36
Pakistan	116	1.19
Singapore	113	1.16
Czech Republic	111	1.14
Iran	107	1.10
South Africa	95	0.97
Saudi Arabia	94	0.96
Malaysia	87	0.89
Turkey	87	0.89
Argentina	80	0.82
Finland	77	0.79
New Zealand	77	0.79
Austria	70	0.72
Colombia	63	0.65
Israel	61	0.62
Ireland	55	0.56
Vietnam	55	0.56
Chile	53	0.54

(continued)

Table 3 (continued)

Countries/regions	Number of records	Percentage of results
Egypt	52	0.53
Wales	43	0.44
Serbia	42	0.43
Romania	39	0.40
Russia	37	0.38
Thailand	37	0.38
Nigeria	35	0.36
Tunisia	35	0.36
Croatia	33	0.34
Slovenia	31	0.32
Cyprus	30	0.31
Bangladesh	29	0.30
Slovakia	29	0.30
United Arab Emirates	27	0.28
Sri Lanka	26	0.27
Hungary	23	0.24
North Ireland	22	0.23
Philippines	22	0.23
Algeria	21	0.22
Indonesia	20	0.21
Kenya	18	0.18
Lebanon	14	0.14
Ghana	13	0.13
Estonia	12	0.12
Luxembourg	12	0.12
Ecuador	11	0.11
Ethiopia	11	0.11
Costa Rica	10	0.10
Lithuania	10	0.10
Morocco	10	0.10
Qatar	10	0.10
Ukraine	10	0.10

Considering the results provided by GS, sorted by relevance, the paper situated first for the search "abiotic emerging-pollutants soil" is the one by Bletsou et al. [20], followed by Conde-Cid et al. [29] and by Gavrilescu et al. [30].

Table 4 Results by years corresponding to the searching string "emerging pollutants soil." Source: WOS

Publication year	Number of records	Percentage of results
2020	1,613	16.51
2019	1,387	14.19
2018	965	9.88
2017	765	7.83
2016	745	7.62
2015	630	6.45
2014	533	5.45
2013	470	4.81
2012	428	4.38
2011	286	2.93
2010	290	2.97
2009	199	2.04
2008	167	1.71
2007	109	1.12
2006	127	1.30
2005	100	1.02
2004	79	0.81
2003	72	0.74
2002	59	0.60
2001	37	0.38
2000	37	0.38
1999	29	0.30
1998	25	0.26
1997	13	0.13
1996	14	0.14
1995	19	0.19
1994	7	0.07
1993	7	0.07
1992	11	0.11
1991	10	0.10
1990	11	0.11
1989	5	0.05
1988	5	0.05
1987	2	0.02
1986	6	0.06
1985	3	0.03
1984	9	0.092
1983	1	0.01
1982	1	0.01
1981	2	0.02
1980	5	0.05
1979	4	0.04

(continued)

Table 4 (continued)

Publication year	Number of records	Percentage of results
1978	1	0.01
1977	1	0.01
1976	1	0.01
1975	2	0.02

3.2 *Emerging Pollutants of Biotic Origin in Soils*

The results shown by the searching tools regarding emerging pollutants of biotic origin in soils are the following. With the searching string "biotic emerging-pollutants soil," the total number of results found by GS was 4,780, while Scopus reported 3 results (however, changing the string to "biotic emerging pollutants soil" the number was 24). Searching for "biotic emerging pollutants soil," WOS indicated that the number of results was 115.

As regards a country-based perspective, the results provided by WOS indicate that the ranking is: USA, China, Spain, Canada, France, Germany, Italy, Australia, Greece, India, and Sweden, and then other countries having less than 5 publications.

Regarding the number of results by year, it is increasing, with the top placed in 2020.

The authors of this chapter have also published previously on emerging pollutants of biotic origin in soils. Some examples are the publications by Núñez-Delgado [31], Anand et al. [32], Conde-Cid et al. [33], and Conde-Cid et al. [34].

Considering the results provided by GS, sorted by relevance, the paper situated first for the search "biotic emerging-pollutants soil" is the one by Geissen et al. [35], followed by Bletsou et al. [20] and by Carmona and Picó [23].

4 Overall Situation and Perspectives

In this chapter we reviewed publications on emerging pollutants present in sewage sludge and soils, differencing between those from biotic and those from abiotic origins, and showing a country-based perspective for these publications. Using classical searching tools for academic and research works, a high number of results were displayed, being clearly different for each of the three tools considered (Google Scholar, Scopus, and Web of Science). Based on results from WOS, a tool which allows to see differences among countries where the published research was carried out, China, USA and Spain are the three geographic areas where a higher percentage of publications has been performed on the subject. Regarding the number of publications, the trend is to increase, and it can be expected that they will still grow for years, taking into account that emerging pollutants are globally considered a real concern for human health and the overall environment. It is clear that more research on the subject will be needed for the coming years, due to increasing threats related to chemicals and pathogens reaching environmental compartments.

Acknowledgements *Funding:* This work was supported by the Spanish Ministry of Science, Innovation and Universities [grant numbers RTI2018-099574-B-C21 and RTI2018-099574-B-C22]. It also received funds from the European Regional Development Fund (ERDF) (FEDER in Spain), being a complement to the previous grants, without additional grant number.

The sponsors had not involvement in study design; in the collection, analysis, and interpretation of data; in the writing of the report, and in the decision to submit the article for publication.

D. Fernández-Calviño holds a Ramón y Cajal contract (RYC-2016-20411) financed by the Spanish Ministry of Economy, Industry, and Competitiveness.

References

1. Dulio V, van Bavel B, Brorström-Lundén E, Harmsen J, Hollender J, Schlabach M, Slobodnik J, Kevin Thomas K, Jan Koschorreck J (2018) Emerging pollutants in the EU: 10 years of NORMAN in support of environmental policies and regulations. Environ Sci Eur 30:5. https://doi.org/10.1186/s12302-018-0135-3
2. Watkins K (2018) Emerging infectious diseases: a review. Curr Emerg Hosp Med Rep 6(3): 86–93. https://doi.org/10.1007/s40138-018-0162-9
3. Ahmed W, Mehmood S, Núñez-Delgado A, Qaswar M, Ali S, Ying H, Liu Z, Mahmood M, Chen DY (2021) Fabrication, characterization and U(VI) sorption properties of a novel biochar derived from *Tribulus terrestris* via two different approaches. Sci Total Environ 780:146617. https://doi.org/10.1016/j.scitotenv.2021.146617
4. Bashir MA, Naveed M, Ahmad Z, Gao B, Mustafa A, Núñez-Delgado A (2020) Combined application of biochar and sulfur regulated growth, physiological, antioxidant responses and Cr removal capacity of maize (*Zea mays* L.) in tannery polluted soils. J Environ Manage. https://doi.org/10.1016/j.jenvman.2019.110051
5. Cela-Dablanca R, Conde-Cid M, Ferreira-Coelho G, Arias-Estévez M, Fernández-Calviño D, Núñez-Delgado A, Fernández-Sanjurjo MJ, Álvarez-Rodríguez E (2021) Adsorption of tetracycline and sulfadiazine onto three different bioadsorbents in binary competitive systems. Processes 9:28. https://doi.org/10.3390/pr9010028
6. Chen B, Yuan M, Liu H (2011) Removal of polycyclic aromatic hydrocarbons from aqueous solution using plant residue materials as a biosorbent. J Hazard Mater 188(1–3):436–442. https://doi.org/10.1016/j.jhazmat.2011.01.114
7. Conde-Cid M, Cela-Dablanca R, Ferreira-Coelho G, Fernández-Calviño D, Núñez-Delgado A, Fernández-Sanjurjo MJ, Arias-Estévez M, Álvarez-Rodríguez E (2021) Sulfadiazine, sulfamethazine and sulfachloropyridazine removal using three different porous materials: pine bark, "oak ash" and mussel shell. Environ Res 195:110814. https://doi.org/10.1016/j.envres.2021.110814
8. Estévez-Schwarz I, Seoane S, Núñez A, López-Mosquera ME (2009) Characterization and evaluation of compost utilized as ornamental plant substrate. Compost Sci Util 17(4):210–219. https://doi.org/10.1080/1065657X.2009.10702426
9. Estévez-Schwarz I, Seoane S, Núñez-Delgado A, López-Mosquera ME (2012) Production and characterization of compost made from garden and other waste. Pol J Environ Stud 21(4): 855–864
10. Fernández-Calviño D, Cutillas-Barreiro L, Nóvoa-Muñoz JC, Díaz-Raviña M, Fernández-Sanjurjo MJ, Álvarez-Rodríguez E, Núñez-Delgado A, Arias-Estévez M, Rousk J (2018) Using pine bark and mussel shell amendments to reclaim microbial functions in a Cu polluted acid mine soil. Appl Soil Ecol 127:102–111. https://doi.org/10.1016/j.apsoil.2018.03.010

11. Núñez-Delgado A, Álvarez-Rodríguez E, Fernández-Sanjurjo MJ, Nóvoa-Muñoz JC, Arias-Estévez M, Fernández-Calviño D (2015) Perspectives on the use of by-products to treat soil and water pollution. Microporous Mesoporous Mater 210:199–201. https://doi.org/10.1016/j.micromeso.2015.02.001
12. Pousada-Ferradás Y, Seoane-Labandeira S, Mora-Gutiérrez A, Núñez-Delgado A (2012) Risk of water pollution due to ash–sludge mixtures: column trials. Int J Environ Sci Technol 9:21–29. https://doi.org/10.1007/s13762-011-0014-6
13. Santás-Miguel V, Fernández-Sanjurjo MJ, Núñez-Delgado A, Álvarez-Rodríguez E, Díaz-Raviña M, Arias-Estévez M, Fernández-Calviño D (2020) Use of biomass ash to reduce toxicity affecting soil bacterial community growth due to tetracycline antibiotics. J Environ Manage 269:110838. https://doi.org/10.1016/j.jenvman.2020.110838
14. Santás-Miguel V, Fernández-Sanjurjo MJ, Núñez-Delgado A, Álvarez-Rodríguez E, Díaz-Raviña M, Arias-Estévez M, Fernández-Calviño D (2021) Use of waste materials to prevent tetracycline antibiotics toxicity on the growth of soil bacterial communities. Environ Res 193: 110404. https://doi.org/10.1016/j.envres.2020.110404
15. Seco-Reigosa N, Bermúdez-Couso A, Garrido-Rodríguez B, Arias-Estévez M, Fernández-Sanjurjo MJ, Álvarez-Rodríguez E, Núñez-Delgado A (2013) As(V) retention on soils and forest by-products and other waste materials. Environ Sci Pollut Res 20:6574–6583. https://doi.org/10.1007/s11356-013-1730-x
16. Wang S, Zhang W, Jia F, Fu H, Liu T, Zhang X, Liu B, Núñez-Delgado A, Han N (2021) Novel Ag_3PO_4/boron-carbon-nitrogen photocatalyst for highly efficient degradation of organic pollutants under visible-light irradiation. J Environ Manage 292:112763. https://doi.org/10.1016/j.jenvman.2021.112763
17. Núñez-Delgado A, Sueiro-Blanco P, Seoane-Labandeira S, Cela-Torrijos R (2018) Polycyclic aromatic hydrocarbons concentrations in a waste from fuel oil spill and its mixture with other materials: time-course evolution. J Clean Prod 172:1910–1917. https://doi.org/10.1016/j.jclepro.2017.11.239
18. Núñez-Delgado A, Pousada-Ferradás Y, Álvarez-Rodríguez E, Fernández-Sanjurjo MJ, Conde-Cid M, Nóvoa-Muñoz JD, Arias-Estévez M (2019) Chapter 1. Effects of microbiological and non-microbiological treatments of sewage sludge on antibiotics as emerging pollutants present in wastewater: a review. In: Shah MP, Rodríguez-Couto S (eds) Microbial wastewater treatment. Elsevier, pp 1–17. ISBN: 978-0-12-816809-7. https://doi.org/10.1016/B978-0-12-816809-7.00001-4
19. Mailler R, Gasperi J, Chebbo G, Rocher V (2014) Priority and emerging pollutants in sewage sludge and fate during sludge treatment. Waste Manag 34(7):1217–1226. https://doi.org/10.1016/j.wasman.2014.03.028
20. Bletsou AA, Jeon J, Hollender J, Archontaki E, Thomaidis NS (2015) Targeted and non-targeted liquid chromatography-mass spectrometric workflows for identification of transformation products of emerging pollutants in the aquatic environment. TrAC Trends Anal Chem 66:32–44. https://doi.org/10.1016/j.trac.2014.11.009
21. Pousada-Ferradás Y, Seoane-Labandeira S, Blanco M, Núñez-Delgado A (2014) The effect of aging on element plant availability and bacterial counts of mixtures of wood ash and sewage sludge. Maderas-Cienc Tecnol 13(3):307–318. https://doi.org/10.4067/S0718-221X2011000300006
22. Trapp S, Eggen T (2013) Simulation of the plant uptake of organophosphates and other emerging pollutants for greenhouse experiments and field conditions. Environ Sci Pollut Res 20:4018–4029. https://doi.org/10.1007/s11356-012-1337-7
23. Carmona E, Picó Y (2018) The use of chromatographic methods coupled to mass spectrometry for the study of emerging pollutants in the environment. Crit Rev Anal Chem 48(4):305–316. https://doi.org/10.1080/10408347.2018.1430555
24. Conde-Cid M, Núñez-Delgado A, Fernández-Sanjurjo MJ, Álvarez-Rodríguez E, Fernández-Calviño D, Arias-Estévez M (2020) Tetracycline and sulfonamide antibiotics in soils: presence, fate and environmental risks. Processes 8:1479. https://doi.org/10.3390/pr8111479

25. Conde-Cid M, Ferreira-Coelho G, Fernández-Calviño D, Núñez-Delgado A, Fernández-Sanjurjo MJ, Arias-Estévez M, Álvarez-Rodríguez E (2020) Single and simultaneous adsorption of three sulfonamides in agricultural soils: effects of pH and organic matter content. Sci Total Environ 744:140872. https://doi.org/10.1016/j.scitotenv.2020.140872
26. Alvarez-Esmorís C, Conde-Cid M, Fernández-Sanjurjo MJ, Núñez-Delgado A, Álvarez-Rodríguez E, Arias-Estévez M (2021) Environmental relevance of adsorption of doxycycline, enrofloxacin, and sulfamethoxypyridazine before and after the removal of organic matter from soils. J Environ Manage 287:112354. https://doi.org/10.1016/j.jenvman.2021.112354
27. Cela-Dablanca R, Nebot C, López LR, Ferández-Calviño D, Arias-Estévez M, Núñez-Delgado A, Álvarez-Rodríguez E, Fernández-Sanjurjo MJ (2021) Retention of the antibiotic cefuroxime onto agricultural and forest soils. Appl Sci 11:4663. https://doi.org/10.3390/app11104663
28. Santás-Miguel V, Díaz-Raviña M, Martín A, García-Campos E, Barreiro A, Núñez-Delgado A, Álvarez-Rodríguez E, Arias-Estévez M, Fernández-Calviño D (2021) Soil enzymatic activities and microbial community structure in soils polluted with tetracycline antibiotics. Agronomy 11: 906. https://doi.org/10.3390/agronomy11050906
29. Conde-Cid M, Fernández-Calviño D, Nóvoa-Muñoz JC, Arias-Estévez M, Díaz-Raviña M, Fernández-Sanjurjo MJ, Núñez-Delgado A, Álvarez-Rodríguez E (2018) Biotic and abiotic dissipation of tetracyclines using simulated sunlight and in the dark. Sci Total Environ 635: 1520–1529. https://doi.org/10.1016/j.scitotenv.2018.04.233
30. Gavrilescu M, Demnerová K, Aamand J, Agathos S, Fava F (2015) Emerging pollutants in the environment: present and future challenges in biomonitoring, ecological risks and bioremediation. N Biotechnol 32(1):147–156. https://doi.org/10.1016/j.nbt.2014.01.001
31. Núñez-Delgado A (2020) SARS-CoV-2 in soils. Environ Res 190:110045. https://doi.org/10.1016/j.envres.2020.110045
32. Anand U, Bianco F, Suresh S, Tripathi V, Núñez-Delgado A, Race M (2021) SARS-CoV-2 and other viruses in soil: an environmental outlook. Environ Res 198:111297. https://doi.org/10.1016/j.envres.2021.111297
33. Conde-Cid M, Arias-Estévez M, Núñez-Delgado A (2021) How to study SARS-CoV-2 in soils? Environ Res 198:110464. https://doi.org/10.1016/j.envres.2020.110464
34. Conde-Cid M, Arias-Estévez M, Núñez-Delgado A (2021) SARS-CoV-2 and other pathogens could be determined in liquid samples from soils. Environ Pollut 273:116445. https://doi.org/10.1016/j.envpol.2021.116445
35. Geissen V, Mol H, Klumpp E, Umlauf G, Nadal M, van der Ploeg M, van de Zee SEATM, Ritsema CJ (2015) Emerging pollutants in the environment: a challenge for water resource management. Int Soil Water Conserv Res 3(1):57–65. https://doi.org/10.1016/j.iswcr.2015.03.002

Pharmaceuticals and Their Metabolites in Sewage Sludge and Soils: Distribution and Environmental Risk Assessment

Juan Luis Santos, Julia Martín, Carmen Mejías, Irene Aparicio, and Esteban Alonso

Contents

Abstract Pharmaceutical compounds (PhCs) are continually discharged to sewer systems through human excreta. In wastewater treatment plants, these compounds are partially removed by biodegradation or retention onto the sludge generated during wastewater treatment. As a result, they can end up in the aquatic environment, through the discharge of wastewater effluents to the receiving waters, or to the soil, through the application of the sludge as organic amended, or by the irrigation with recycled water. Moreover, these compounds are partially metabolized after their consumption, and, as a result, PhCs and their metabolites are present in the environment. This chapter summarizes recent research on the occurrence of PhCs and their metabolites in sewage sludge stabilization processes and on sludge-amended soils. Recent studies have shown that antibiotics, non-steroidal anti-inflammatory drugs, antidepressants, and antidiabetics are the most abundant PhCs found in sludge matrices. Overall, attenuation of PhCs concentrations occurs during sludge stabilization, and particularly during anaerobic digestion and composting. The potential ecotoxicological risk associated with the presence of PhCs in amended soils is

J. L. Santos (✉), J. Martín, C. Mejías, I. Aparicio, and E. Alonso
Departamento de Química Analítica, Escuela Politécnica Superior, Universidad de Sevilla, Seville, Spain
e-mail: jlsantos@us.es

Avelino Núñez-Delgado and Manuel Arias-Estévez (eds.),
Emerging Pollutants in Sewage Sludge and Soils,
Hdb Env Chem (2023) 114: 19–36, DOI 10.1007/698_2022_847,

Published online: 12 February 2022

medium-low for most PhCs. The most critical compounds found in sludge-amended soils are ciprofloxacin, 17α-ethinylestradiol, and 17β-estradiol.

Keywords Environmental risk, Metabolites, Occurrence, Sludge stabilization treatments, Sludge-amended soil

1 Introduction

In the last years, numerous studies have described the presence of pharmaceutical compounds (PhCs) in the environment [1]. These compounds are continually discharged to the sewer system through human excreta. In wastewater treatment plants (WWTPs), these compounds are partially removed by biodegradation or retention onto the sludge generated during wastewater treatment. As a result, they can end up in the aquatic environment, through the discharge of wastewater effluents to the receiving waters, or to the soil, through the application of the sludge as organic amended, or by irrigation with recycled water [2]. Moreover, veterinary pharmaceuticals used in livestock are excreted by the animals and end up in soils via grazing livestock or manure used as agricultural fertilizer [3]. Among PhCs frequently detected are anti-inflammatories like acetaminophen, ibuprofen, naproxen, or diclofenac [1, 4, 5]; antibiotics as sulfamethoxazole, trimethoprim, norfloxacin, or sulfonamides [4]; or antiepileptics as carbamazepine [6]. Moreover, these compounds are partially metabolized after their consumption [1, 2, 7]. Consequently, both pharmaceuticals and their metabolites have been detected not only in their sources, wastewater, and sludge [8, 9], but also in their main fates, surface waters [10], and soil [11, 12].

The amount of sewage sludge generated in WWTPs has increased strikingly in recent years. In the European Union, the most usual final destiny of these sludges is their use as organic amended in soil [5, 13]. For example, it is estimated that around 40% of the sludge produced in 2021 will be used as a source of organic matter and nutrients for agricultural purposes [5, 14], although different application rates are used among the Member States of the EU [14, 15]. The main stabilization processes applied to the sludge previously to their application onto the soil are anaerobic and aerobic digestion, composting, and, particularly in little municipalities, low-cost wastewater treatments, as lagooning [5, 14, 16]. However, several studies have described the persistence of PhCs along these treatments [2, 16–19].

In this chapter, a discussion is carried out on the main studies reported in the last years about the distribution of the most recurrent PhCs and their metabolites in sludges stabilization processes, and in soils amended with these sludges. Moreover, the risk associated with the presence of these compounds in sludge applied to the soil is discussed too.

2 Occurrence of PhCs and Metabolites Alongside Sludge Stabilization Treatments

The most usual sludge stabilization treatments are their digestion under anaerobic (in high-populated cities) and aerobic conditions (in low populated municipalities). In both processes, the sludge separated from influent wastewater in the primary sedimentation tank is concentrated in a gravity thickener (primary sludge) and mixed with the secondary sludge obtained from the secondary settler and concentrated in the flotation thickener. Mixed sludge is digested under anaerobic (in anaerobic stabilization plants) or aerobic (in aerobic stabilization plants) conditions. The final product of these treatments is usually subjected to composting processes carried out in the open air, forming piles in which the aeration is thermally controlled by turning [2].

In small municipalities, low-cost wastewater treatment is usually applied to wastewater treatment and sludge stabilization. The most usual treatment is lagooning, in which the sludge stabilization takes place at the bottom of a lagoon under anaerobic conditions [2].

In this chapter, the main findings included in more than 50 works published in the last 20 years about the presence and distribution of PhCs and their metabolites in sludge are evaluated (Fig. 1). About 70% of the published data have focused on compounds measured in the final product of the stabilization treatment (digested, dehydrated, or composted sludge). The studies about the distribution of metabolites in sludge stabilization treatments are even scarcer. Anti-inflammatories (36 papers, mainly about diclofenac, ibuprofen, naproxen, and ketoprofen), antibiotics (33 papers, mainly about ciprofloxacin, ofloxacin, enrofloxacin, norfloxacin, sulfamethoxazole, sulfamethazine, tetracycline, oxytetracycline, and trimethoprim), and antiepileptics (33 papers, mainly about carbamazepine) are the most studied therapeutic groups in sludge, followed by beta-blockers (22 papers, mainly about propranolol, metoprolol, atenolol), antilipemics (19 papers, mainly about bezafibrate and gemfibrozil), and stimulants (18 papers, mainly about caffeine).

Considering metabolites, the most studies have been focused on those derived from anti-inflammatories, antilipemics, or antiepileptics (mainly the metabolites of carbamazepine). On the contrary, the studies of the metabolites of antibiotics are very sparse.

Moreover, most of these studies evaluate the distribution on sludge stabilization treatments for only a few metabolites, as the ones from diclofenac [2], ibuprofen [2], carbamazepine [2, 15, 20], or sulfamethoxazole [20] (Fig. 1). This could be due to the complexity of the analysis of these samples and to the lack of commercial analytical standards that allow their determination.

Figures 2 and 3 show the concentrations of 180 PhCs, measured in sludge samples worldwide in the last 20 years, grouped by therapeutic group (data collected from [5]). Measured concentrations ranged from ng or μg per kilogram to even mg per kilogram, depending on the consumption, physicochemical properties of the compounds, and the characteristics of the sludge.

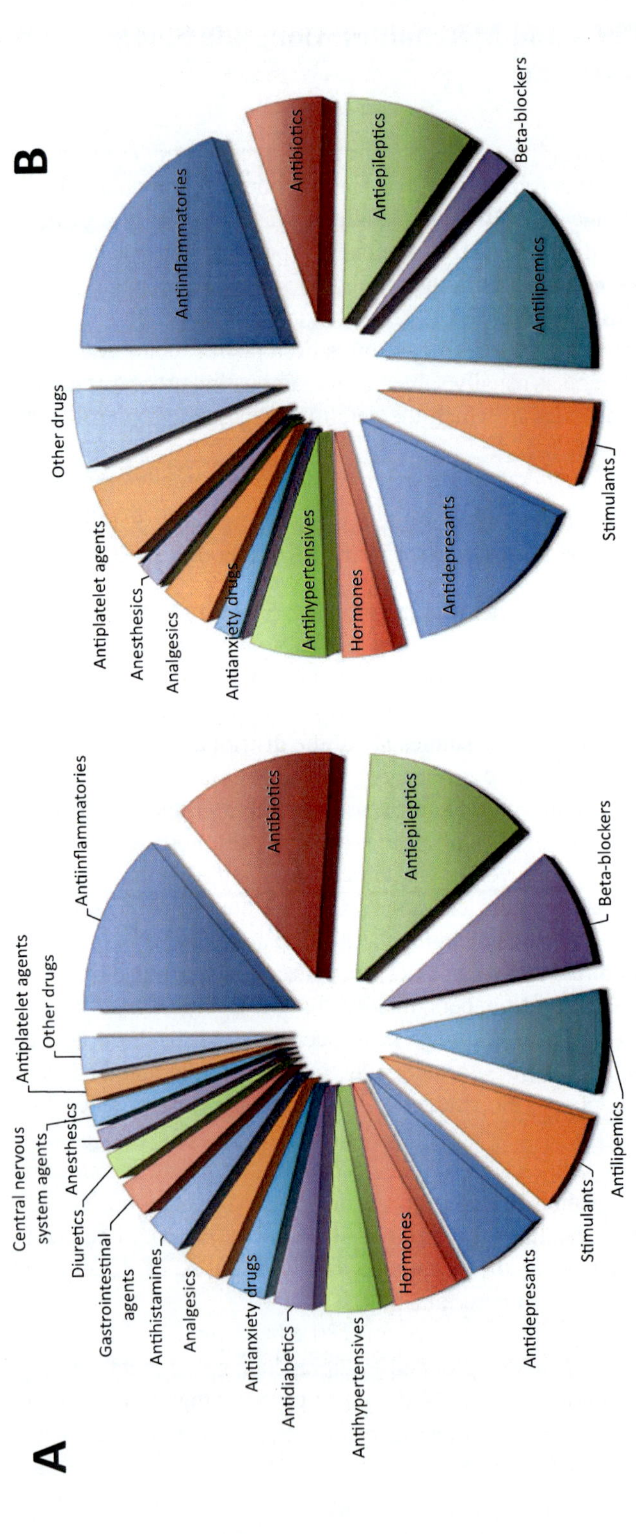

Fig. 1 Therapeutic groups studied in sludge in the last 20 years (Data from [5])

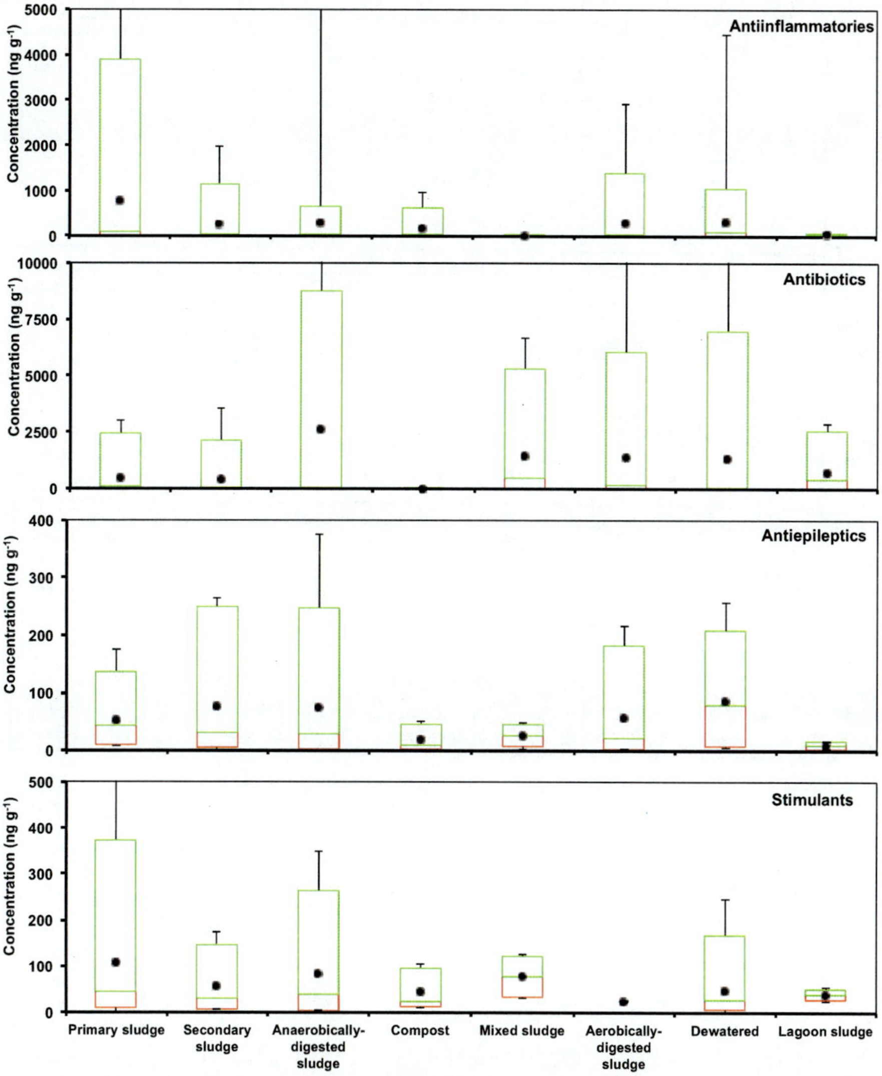

Fig. 2 Concentration of antibiotics, anti-inflammatories, antiepileptics, and stimulants measured alongside sludge stabilization processes. Raw sludge was considered as mixed sludge (Data collected from [5])

The higher concentrations in fresh sludge (primary, secondary sludge, and mixed) have been described in the case of antibiotics (in mixed sludge) and anti-inflammatories (mainly in primary sludge), followed by stimulants (mainly caffeine). The compounds more frequently measured and at the highest concentrations in these sludges are fluoroquinolones [14, 15, 20–22] as ciprofloxacin, norfloxacin, and ofloxacin (for example, concentrations of ciprofloxacin up to 12,858 ng g^{-1} were measured in raw sludge from France [22]), the anti-inflammatories diclofenac [2, 20, 22–26] (up to 7,020 ng g^{-1} measured in Germany) and ibuprofen [2, 24, 25]

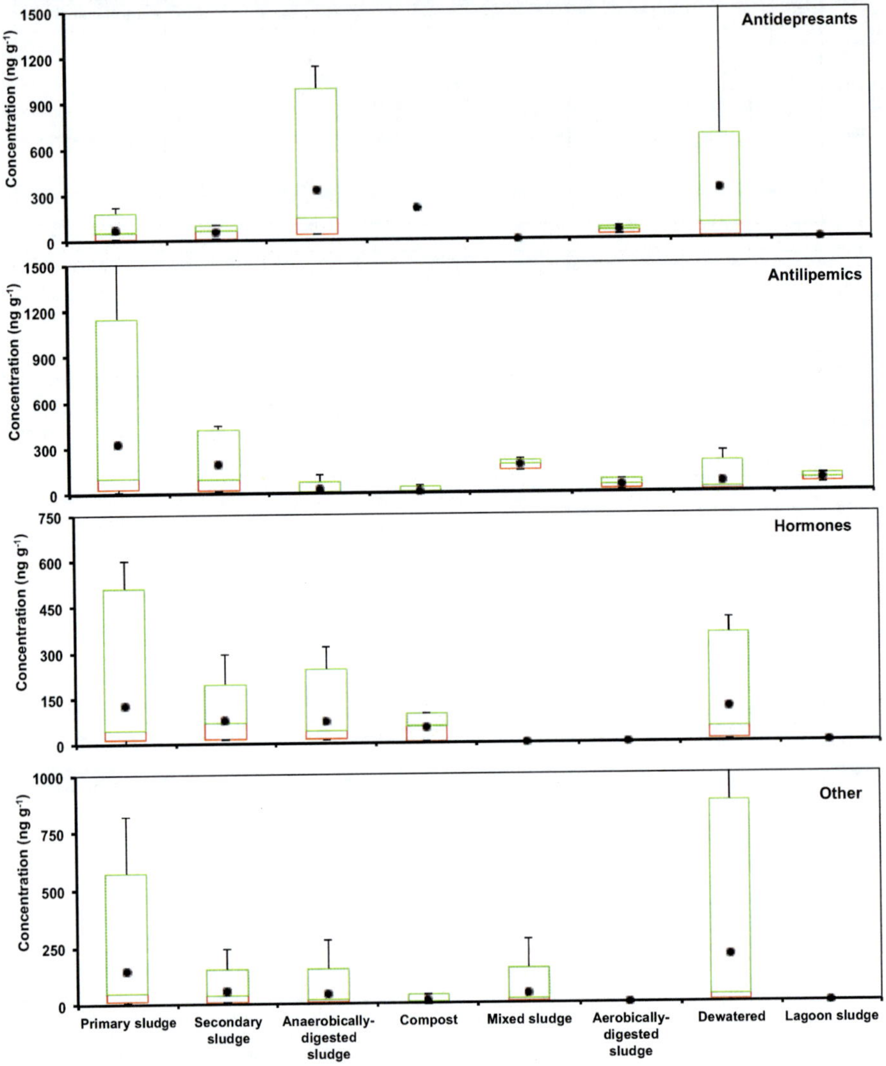

Fig. 3 Concentration of antidepressants, hormones, and other pharmaceuticals measured alongside sludge stabilization processes. Raw sludge was considered as mixed sludge (Data collected from [5])

(ranging from 11.1 to 4,105 ng g^{-1}), the antilipemic gemfibrozil (concentrations up to 2,026 ng g^{-1} have been measured in primary sludge from Spain [27]), the stimulant caffeine (up to 2,828 ng g^{-1} measured in Canada [28]) or hormones [22, 26, 28–31] (up to 599 ng g^{-1} dw and 421 ng g^{-1} dw in the case of 17-α-ethynylestradiol and estrone, respectively [31, 32]).

Other compounds, as sulfonamide or macrolide antibiotics, anti-inflammatory drugs as acetaminophen, naproxen or ketoprofen [5, 8, 27–31, 33–35], antidepressants as carbamazepine [19, 29, 31, 33, 36] or beta-blockers as propranolol and

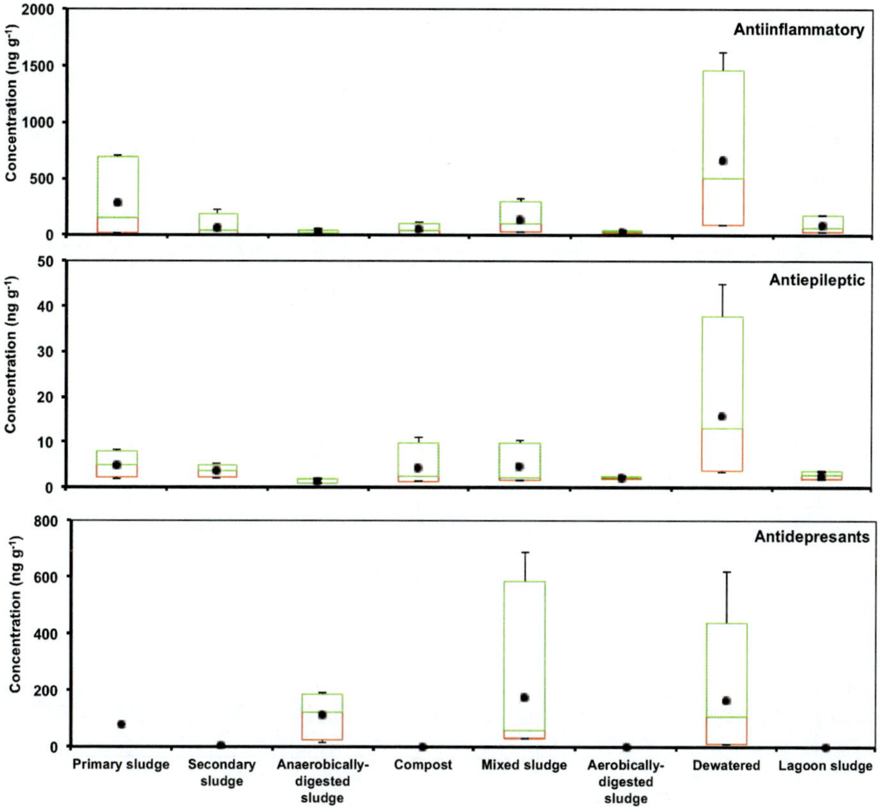

Fig. 4 Concentration of metabolites of anti-inflammatories, antiepileptics, and antidepressants measured alongside sludge stabilization processes. Raw sludge was considered as mixed sludge (Data collected from [5])

atenolol [8, 19, 20, 22, 27, 29, 31, 33] have been frequently detected, although at lower concentrations.

Considering pharmaceutical metabolites, the highest concentrations have been measured in the case of those derived from salicylic acid (concentrations up to 931 ng g^{-1} have been detected in sludge stabilization treatments from the south of Spain [8, 33]) and ibuprofen (up to 204 and 100 ng g^{-1}, in the case of carboxyibuprofen and 2-hydroxy ibuprofen, respectively [2]). Other compounds measured in fresh sludge (although at lower concentrations) are metabolites of antilipemics as clofibric acid [14, 26, 30], metabolite of clofibrate; N-desmethylcitalopram [37, 38] and norsertraline [19, 22, 37–39], metabolites of antidepressants, and paraxanthine (PX) [2], metabolite of caffeine. These metabolites have usually been found at lower concentrations than their parent compounds in fresh sludges (Figs. 2, 3, 4, and 5). This contrasts with the results obtained in aqueous environment, where some metabolites, as 2-hydroxyibuprofen (2OH-IBU) and carboxyibuprofen (CBX-IBU), have been measured at higher

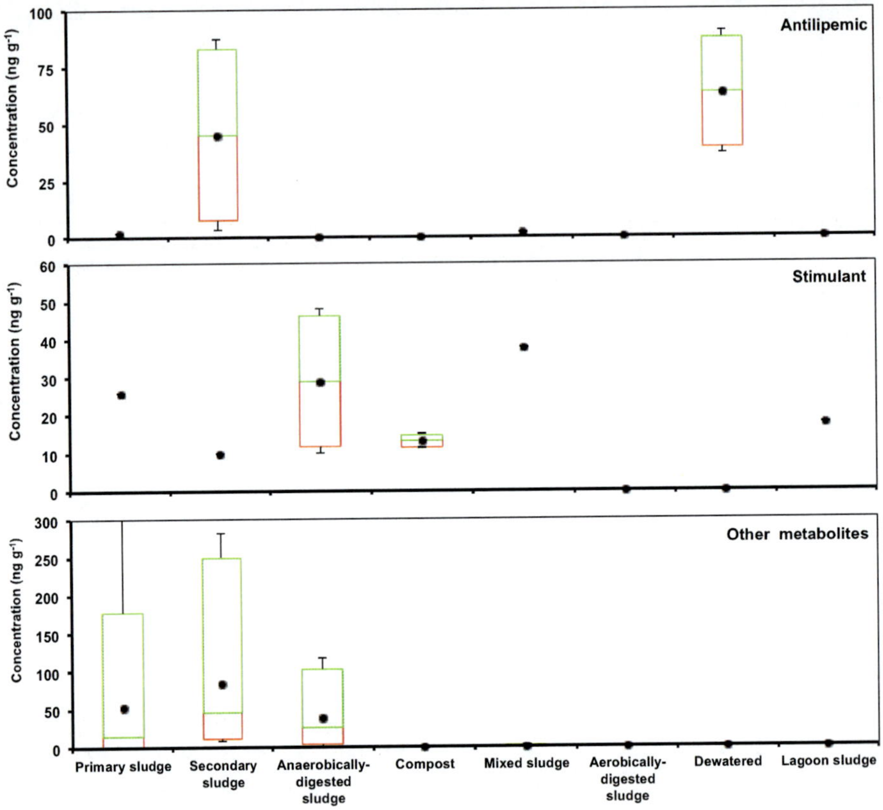

Fig. 5 Concentration of metabolites of antilipemics, stimulants, and other PhCs measured alongside sludge stabilization processes. Raw sludge was considered as mixed sludge (Data collected from [5])

concentrations than their parent compounds (specifically in wastewater samples). However, this could be explained by the high number of studies reporting the concentration of PhCs in sludge in relation to those reporting concentrations of their metabolites. On the contrary, Malvar et al. [2] found in sludge samples concentrations of 10-hydroxycarbamazepine (10OH-CBZ), 4-hydroxydiclofenac (4OH-DIC), 2OH-IBU and CBX-IBU higher than those measured for their parent compounds, as it was described for water samples. This shows the concentrations in influent wastewater as the main factor governing the concentration of these compounds in fresh sludge, in spite of the different sorption capacity of these compounds onto the sludge [5]. Moreover, the concentration of PhCs and metabolites in secondary and mixed sludges could be affected by the biodegradation of these compounds in secondary wastewater treatment. For example, recently Malvar et al. [2] related the concentrations of 10OH-CBZ, 2OH-IBU and CBX-IBU measured in secondary sludge (higher than those measured in primary sludge) with the biodegradation of ibuprofen during secondary treatment.

The PhCs and metabolites more frequently studied in fresh sludge and found at highest concentrations were the compounds most studied and detected in treated sludge: anti-inflammatory drugs, antibiotics, and antiepileptics, and, in less extension, antidepressants. Data published in the literature showed higher concentrations of antibiotics in treated sludge than those measured in fresh sludges, especially in the case of fluoroquinolone antibiotics, with concentrations up to 12,858 ng g^{-1} (ciprofloxacin), 6,049 ng g^{-1} (norfloxacin) and 6,712 ng g^{-1} (ofloxacin). The same behavior could be observed in the case of antiepileptics (mainly carbamazepine), which shows the high persistence of these compounds in the digestion processes. Other compounds, as anti-inflammatories, show a decrease of the concentrations measured in digested sludges with respect to those measured in fresh sludge (Fig. 2). In spite of that, concentrations up to 7,020 and 4,105 ng g^{-1} have been measured for diclofenac and ibuprofen, respectively. In the case of other PhCs, as stimulants, antidepressants or hormones, their behavior in sludge stabilization treatments seems to depend on the digestion process. Most of them showed lower concentrations in aerobic processes than those measured in sludge treated under anaerobic conditions. The differences in concentration pattern between aerobic and anaerobic stabilization processes have been previously described in the case of pharmaceuticals as azithromycin, irbesartan, sertraline, which are more frequently detected in aerobically digested sludge [15]. Other studies [40–42] showed a higher mitigation of the concentration of clarithromycin and azithromycin under anaerobic conditions, while caffeine showed a high persistence to the anaerobic treatment [8, 42], whereas under aerobic condition was widely removed. Moreover, several works have described the importance, not only of the anaerobic or aerobic conditions, but also of other parameters, such as temperature or treatment time. For example, 60% of diclofenac and diazepam were removed under mesophilic anaerobic conditions (38°C), while under thermophilic conditions (55°C) only 38% of diclofenac and 73% of diazepam were removed [43]. Other works have even showed an improvement in the removal of these compounds by the combination of the two conditions. For example, removals up to 90% have been measured in the case of diclofenac, oxazepam, ofloxacin, or propranolol using combined anaerobic and aerobic conditions [44, 45]. Regarding composting, in general the concentrations measured after the composting process are lower than those measured in digested sludge for the most of the PhCs. Some works showed the photodegradation of hydrosoluble PhCs, as well as mineralization and even the dilution due to the mixture with other products, as potential ways of removal [46].

Considering PhC metabolites, only one study has been reported in the literature [42], which shows the importance of performing a further investigation on this issue. The results obtained in this work described that the distribution of PhC metabolites depend on the compound and the conditions applied to the stabilization. Compounds as PX or the metabolites of ibuprofen showed high decrease of their concentrations alongside the sludge stabilization treatments, while other, as the metabolites of carbamazepine and diclofenac, were highly persistent to all treatments evaluated. Moreover, transformations between metabolites and parent compounds were only observed in the case of ibuprofen.

3 Distribution of PhCs and Metabolites in Soil

One of the main disposal options for the sludge generated during sludge stabilization treatments is its application to the soil as organic amended. As a result, organic contaminants in the sludge, among them PhCs and their metabolites, end up into the soil. In a study carried out in the Slovak Republic [15], it was estimated a total input load of up to 120 kg/year of fexofenadine disposed into the soil through the application of sewage sludge, together with other PhCs, including antihistamines, antidepressants, or antihypertensives. PhCs frequently detected and measured at highest concentrations in soil are antibiotics, as fluoroquinolones (up to 550 ng g^{-1}), tetracyclines (tetracycline and oxytetracycline, up to 63.8 and 101 ng g^{-1}, respectively) and sulfamethoxazole (47.9 ng g^{-1}) [47, 48], as well as other compounds as diclofenac [11, 49], ibuprofen [11], carbamazepine [11, 50], or caffeine [11].

Considering PhC metabolites, only a few data have been reported in the literature about their presence in soils. García-Galán et al. [51] evaluated the presence of several sulfonamide antibiotics and their metabolites in soils collected in rural areas from Catalonia (North East of Spain). Among the measured compounds were acetylsulfamethoxazole (up to 1.38 ng g^{-1}) and acetylsulfapiridine (up to 0.77 ng g^{-1}). Other metabolites as CBX-IBU [52] and 4OH-DIC [35] have been measured in soils at concentrations up to 46.1 and 3.3 ng g^{-1}, respectively.

In the soil, PhCs and their metabolites may undergo different routes, such as sorption/desorption processes [53, 54], transport by leaching [55], or degradation/transformation [2, 11]. The concentrations of PhCs and metabolites in the soil depend, in addition to the sludge application rates and frequency, on several factors, as soil properties, physicochemical characteristic of the compounds, precipitations or even land relief. Highly mobile compounds could contaminate surface water through runoff or groundwater by leaching. Highly adsorbed compounds could be accumulated into the soil [5]. In this way, several works have showed the mobility of ibuprofen, acetaminophen, or sulfamethoxazole [56, 57], which have been measured in leachates from sludge-amended soils, while other compounds as carbamazepine, diclofenac, trimethoprim, or propranolol showed high retention in the soil matrix [6, 56, 58]. Regarding PhCs metabolites, several works have shown, in laboratory experiments, the adsorption behavior of these compounds, mainly for carbamazepine [54, 55, 59]. These studies show different soil retention of PhC metabolites compared to those observed in the case of their parent compounds, what could be due to the different physicochemical characteristics of the metabolites. For example, Paz et al. [54] showed how the relative charge densities for metabolites of carbamazepine, due to the electronegative oxygen atoms, could contribute to the different adsorption behavior of these compounds.

Considering the degradation of these compounds in soils, only a few studies have assessed its dissipation in the edaphic environment. Some compounds as norfloxacin, ciprofloxacin or azithromycin have shown high persistence [60], while other as sulfamethoxazole, diclofenac or caffeine showed a high dissipation in a few days [42]. Until now, only one study has been reported in the literature about the dissipation of metabolites of PhCs in soil [42]. In this work, batch experiments carried out with three different soils spiked with these substances showed differences between metabolites and parents compounds regarding their persistence in soil. For example, carbamazepine and epoxy-carbamazepine showed high persistence, while the metabolites 3OH-carbamazepine and 10OH-CBZ showed a rapid dissipation (up to 20 days). Other compounds, as caffeine and its metabolite PX and sulfamethoxazole and its metabolite acetylsulfamethoxazole, showed persistence between 20 and 60 days, depending on the soil characteristics. Only in the case of ibuprofen, transformation of ibuprofen in its metabolite 2OH-IBU could be considered.

4 Environmental Risk Assessment of Pharmaceuticals and Metabolites in Soil

The most common approach used to assess the environmental risk caused by the presence of PhCs and metabolites in soil is based on the European Medicines Agency Guideline [61] through the use of the risk quotient (RQ). The RQ is the relation between the measured (MEC) or predicted environmental concentrations (PEC) and the predicted no-effect concentrations (PNEC). To calculate the PNEC values, it is needed to use the lowest acute and chronic toxicity data in fish, *Daphnia magna,* or algae organisms and dividing them with an assessment factor of 1,000 or 100, respectively, to consider the worst-case scenario. Because of the lack of ecotoxicological data for terrestrial organisms, many authors usually take the values in aquatic organisms to estimate the $PNEC_{soil}$ through the equilibrium partition approach: $PNEC_{soil} = K_d \times PNEC_{water}$, as suggested by the European Commission [8, 22, 33, 62]. Recently, Mejías et al. [5] recompiled the toxicity and K_{dsoil} data reported in the literature. This work draws attention as regards the lack of available data in the particular case of metabolites.

Once the RQ is estimated, the criteria proposed by Hernando et al. [63] is applied to evaluate the risk, considering low risk for RQ < 0.1, medium risk for RQ between 0.1–1 and a high risk when RQ > 1. Table 1 shows the minimum and maximum RQ values calculated based on the lowest and highest concentration levels found for PhCs and metabolites in digested sludge or compost, and the K_d compiled from the literature. The PEC values in soil were estimated according to the EC-TGD [62] by the equation:

$$PEC_{soil} = C_{sludge} \times APPL_{sludge}/DEPTH_{soil} \times RHO_{soil} \quad (1)$$

where C_{sludge} is the concentration measured in digested sludge or compost; $APPL_{sludge}$ is the dry-sludge application rate (0.5 kg m^{-2} year); $DEPTH_{soil}$ is the mixing soil depth (0.20 m), and RHO_{soil} is the bulk density of wet soil (1,700 kg m^{-3}).

Table 1 Risk quotient (minimum and maximum values) in soil amended due to PhCss and metabolites measured in digested sludge or compost in worldwide ($PNEC_{water}$ and K_d data taken from Mejías et al. [5])

Pharmaceuticals and metabolites	$PNEC_{soil}$ (ng g^{-1})	PEC_{soil} (min) (ng g^{-1})	PEC_{soil} (max) (ng g^{-1})	RQ (min)	RQ (max)
Ciprofloxacin	2.14	0.110	15.43	5.2E−02	7.2E+00
17α-ethinylestradiol	0.10	0.018	0.460	1.8E−01	4.6E+00
17ß-estradiol	0.03	0.062	0.062	1.9E+00	1.9E+00
Sulfamethoxazole	1.20	0.006	0.978	4.8E−03	8.1E−01
Sertraline	1.58	0.085	0.925	5.4E−02	5.8E−01
Tylosin	4.35	2.229	2.229	5.1E−01	5.1E−01
Carbamazepine	48.9	0.005	16.25	1.0E−04	3.3E−01
Diclofenac	34.2	0.016	6.233	4.6E−04	1.8E−01
Estrone	1.23	0.015	0.201	1.2E−02	1.6E−01
Fluoxetine	3.22	0.025	0.251	7.9E−03	7.8E−02
Atenolol	37.5	0.006	2.426	1.6E−04	6.5E−02
Ofloxacin	246	0.078	14.03	3.2E−04	5.7E−02
Oxytetracycline	213	0.002	10.83	7.6E−06	5.1E−02
Ketoprofen	140	0.012	6.553	8.4E−05	4.7E−02
Ibuprofen	46.5	0.071	1.686	1.5E−03	3.6E−02
Caffeine	10.3	0.011	0.365	1.1E−03	3.5E−02
Erythromycin	4.08	0.001	0.120	2.7E−04	2.9E−02
Tetracycline	146	0.006	3.493	4.3E−05	2.4E−02
Clarithromycin	6.72	0.008	0.147	1.1E−03	2.2E−02
Propranolol	40.6	0.015	0.545	3.7E−04	1.3E−02
Estriol	95.8	0.081	0.597	8.4E−04	6.2E−03
Naproxen	28.8	0.015	0.175	5.1E−04	6.1E−03
Simvastatin	68.9	0.382	0.382	5.6E−03	5.6E−03
Sulfamethazine	38.3	0.038	0.204	1.0E−03	5.3E−03
Trimethoprim	7.28	0.001	0.038	1.9E−04	5.3E−03
Metoprolol	146	0.009	0.588	6.3E−05	4.0E−03
Gemfibrozil	64.7	0.012	0.223	1.9E−04	3.5E−03
Acetaminophen	294	0.033	0.391	1.1E−04	1.3E−03
Sulfapyridine	173	0.036	0.178	2.1E−04	1.0E−03
Valsartan	365	0.038	0.221	1.0E−04	6.0E−04
Clofibric acid	113	0.012	0.054	1.1E−04	4.8E−04
Metformin	13,427	0.095	6.146	7.1E−06	4.6E−04
Bezafibrate	84.0	0.019	0.038	2.3E−04	4.6E−04
Salycilic acid	7380	0.027	2.390	3.6E−06	3.2E−04
Sulfathiazole	418	0.107	0.113	2.6E−04	2.7E−04
Irbesartan	273	0.071	0.072	2.6E−04	2.6E−04
Codeine	240	0.009	0.024	3.9E−05	9.8E−05
Norfloxacin	86,865	0.040	8.006	4.6E−07	9.2E−05
Azithromycin	40,964	0.032	1.232	7.9E−07	3.0E−05
Telmisartan	45,214	0.235	0.853	5.2E−06	1.9E−05

Although the estimated results showed an overall low risk for most of the compounds assessed, the evaluation of the environmental risks of ciprofloxacin (RQ 7.2) and estrogens (17α-ethinylestradiol (RQ 4.6) and 17β-estradiol (RQ 1.9)) should not be underestimated when the highest concentrations found are used.

Similar results have been reported by other authors [8, 22, 33, 64, 65]. Martín et al. [8, 33] indicated that 17β-estradiol, 17α-ethinylestradiol, ibuprofen, gemfibrozil, and sulfamethoxazole are among the most hazardous pharmaceuticals when performing a risk assessment in digested sludge and compost. Nevertheless, an important decrease in RQ was perceived when sludge is amended on soils, being 17β-estradiol the only substance showing some potential toxic effects. This practice caused a "dilution" effect, resulting RQ < 0.10. More recently, Gros et al. [65] and Bastos et al. [64] reported RQ > 1 for fluoroquinolones causing risk for soil life and crops. The scientific data available on the potential environmental consequences of sludge amendment to soils have increased the studies on the crop uptake of pharmaceuticals in last years [66]; however, there are gaps in the literature regarding the food chain and the risk to human health.

Finally, it is important to note that the use of highly sensitive organisms such as *H. attenuate* or *B. calyciflorus* in the study could have a high implication on the ecotoxicological risk assessment. In a near future, more research is needed to draw firm conclusions in (1) terrestrial organisms and type of crops and endpoints; (2) including metabolites; and (3) conducting these studies at lower environmentally relevant concentrations.

5 Conclusions and Future Trends

Many works have been published in the last 20 years about pharmaceutical compounds in the environment. Their distribution has been evaluated in their main sources, wastewater, and their main fates, surface waters. However, the studies reporting their distribution in sludge and, especially, their occurrence in sludge stabilization treatments are scarce. The few studies reported in the literature show that the decrease of the concentration of PhC and their metabolites in the sludge stabilization treatments depends not only on the compound but also on the conditions and the process applied to the sludge stabilization. In spite of these studies, there is a lack of information about the behavior of PhCs, and especially their metabolites (and not identified compounds), on both, wastewater and sludge treatment technologies.

Regarding the distribution of PhC and metabolites in soils, their adsorption onto the edaphic matrix as well as their degradation depends on the physicochemical characteristics of the compounds and the properties of the soil. However, for most of these pollutants, the mechanisms governing the occurrence of these compounds, the influence of their characteristics, and the properties of the soil are unknown.

On this basis, in-depth studies are necessary to elucidate the behavior of pharmaceuticals, and especially their degradation products, in different sludge stabilization treatments, with different technologies, as well as there is a need to evaluate advanced technologies that allow their complete removal, especially in the case of sludges that are going to be applied to the soil. Moreover, further studies evaluating, not only the distribution in soil/water systems, but also their degradation in soils with different characteristics, are necessary.

Concerning the potential environmental risk of PhCs and their metabolites, in spite of the toxicological studies that are being carried out, more toxicological data, especially in the case of pharmaceutical metabolites, are necessary, in order to achieve a complete evaluation of environmental risks due to pharmaceutical compounds.

Acknowledgements This work was financially supported by the Ministerio de Ciencia e Innovación from Spanish Government (Project No. PID2020-117641RB-I00 and CTM2017-82778-R).

References

1. Ibáñez M, Bijlsma L, Pitarch E, López FJ, Hernández F (2021) Occurrence of pharmaceutical metabolites and transformation products in the aquatic environment of the Mediterranean area. Trends Environ Anal Chem 29:00118. https://doi.org/10.1016/j.teac.2021.e00118
2. Malvar JL, Santos JL, Martín J, Aparicio I, Alonso E (2020) Occurrence of the main metabolites of pharmaceuticals and personal care products in sludge stabilization treatments. Waste Manag 116:22–30. https://doi.org/10.1016/j.wasman.2020.07.051
3. Kemper N (2008) Veterinary antibiotics in the aquatic and terrestrial environment. Ecol Indic 8(1):1–13. https://doi.org/10.1016/j.ecolind.2007.06.002
4. Chaturvedi P, Shukla P, Giri BS, Chowdhary P, Chandra R, Gupta P, Pandey A (2021) Prevalence and hazardous impact of pharmaceutical and personal care products and antibiotics in environment: a review on emerging contaminants. Environ Res 194:110664. https://doi.org/10.1016/j.envres.2020.110664
5. Mejías C, Martín J, Santos JL, Aparicio I, Alonso E (2021) Occurrence of pharmaceuticals and their metabolites in sewage sludge and soil: a review on their distribution and environmental risk assessment. Trends Environ Anal Chem 30:e00125. https://doi.org/10.1016/j.teac.2021.e00125
6. Shao Y, Yang K, Jia R, Tian C, Zhu Y (2018) Degradation of triclosan and carbamazepine in two agricultural and garden soils with different textures amended with composted sewage sludge. Int J Environ Res Public Health 15(11):1–14. https://doi.org/10.3390/ijerph15112557
7. Brezina E, Prasse C, Meyer J, Mückter H, Ternes TA (2017) Investigation and risk evaluation of the occurrence of carbamazepine, oxcarbazepine, their human metabolites and transformation products in the urban water cycle. Environ Pollut 225:261–269. https://doi.org/10.1016/j.envpol.2016.10.106
8. Martín J, Camacho-Muñoz D, Santos JL, Aparicio I, Alonso E (2012) Occurrence of pharmaceutical compounds in wastewater and sludge from wastewater treatment plants: removal and ecotoxicological impact of wastewater discharges and sludge disposal. J Hazard Mater 239–240. https://doi.org/10.1016/j.jhazmat.2012.04.068
9. Martín J, Santos JL, Aparicio I, Alonso E (2015) Pharmaceutically active compounds in sludge stabilization treatments: anaerobic and aerobic digestion, wastewater stabilization ponds and composting. Sci Total Environ 503–504. https://doi.org/10.1016/j.scitotenv.2014.05.089

10. Biswas P, Vellanki BP (2021) Occurrence of emerging contaminants in highly anthropogenically influenced river Yamuna in India. Sci Total Environ 782:146741. https://doi.org/10.1016/j.scitotenv.2021.146741
11. Biel-Maeso M, Corada-Fernández C, Lara-Martín PA (2018) Monitoring the occurrence of pharmaceuticals in soils irrigated with reclaimed wastewater. Environ Pollut 235:312–321. https://doi.org/10.1016/j.envpol.2017.12.085
12. Chen C, Li J, Chen P, Ding R, Zhang P, Li X (2014) Occurrence of antibiotics and antibiotic resistances in soils from wastewater irrigation areas in Beijing and Tianjin, China. Environ Pollut 193:94–101. https://doi.org/10.1016/j.envpol.2014.06.005
13. Clarke BO, Smith SR (2011) Review of "emerging" organic contaminants in biosolids and assessment of international research priorities for the agricultural use of biosolids. Environ Int 37(1):226–247. https://doi.org/10.1016/j.envint.2010.06.004
14. Fijalkowski K, Rorat A, Grobelak A, Kacprzak MJ (2017) The presence of contaminations in sewage sludge – the current situation. J Environ Manage 203:1126–1136. https://doi.org/10.1016/j.jenvman.2017.05.068
15. Ivanová L, Mackuľak T, Grabic R, Golovko O, Koba O, Staňová AV, Szabová P, Grenčíková A, Bodík I (2018) Pharmaceuticals and illicit drugs – a new threat to the application of sewage sludge in agriculture. Sci Total Environ 634:606–615. https://doi.org/10.1016/j.scitotenv.2018.04.001
16. Patureau D, Mailler R, Delgenes N, Danel A, Vulliet E, Deshayes S, Moilleron R, Rocher V, Gasperi J (2021) Fate of emerging and priority micropollutants during the sewage sludge treatment – part 2: mass balances of organic contaminants on sludge treatments are challenging. Waste Manag 125:122–131. https://doi.org/10.1016/j.wasman.2021.02.034
17. Mailler R, Gasperi J, Patureau D, Vulliet E, Delgenes N, Danel A, Deshayes S, Eudes V, Guerin S, Moilleron R, Chebbo G, Rocher V (2017) Fate of emerging and priority micropollutants during the sewage sludge treatment: case study of Paris conurbation. Part 1: Contamination of the different types of sewage sludge. Waste Manag 59:379–393. https://doi.org/10.1016/j.wasman.2016.11.010
18. Shi X, Leong KY, Ng HY (2017) Anaerobic treatment of pharmaceutical wastewater: a critical review. Bioresour Technol 245:1238–1244. https://doi.org/10.1016/j.biortech.2017.08.150
19. Stasinakis AS (2012) Review on the fate of emerging contaminants during sludge anaerobic digestion. Bioresour Technol 121:432–440. https://doi.org/10.1016/j.biortech.2012.06.074
20. Golovko O, Örn S, Sörengård M, Frieberg K, Nassazzi W (2021) Occurrence and removal of chemicals of emerging concern in wastewater treatment plants and their impact on receiving water systems. Sci Total Environ 754:142122. https://doi.org/10.1016/j.scitotenv.2020.142122
21. Martínez-Alcalá IM, Guillén-Navarro JM, Fernández-López C (2017) Pharmaceutical biological degradation, sorption and mass balance determination in a conventional activated-sludge wastewater treatment plant from Murcia, Spain. Chem Eng J 316:332–340. https://doi.org/10.1016/j.cej.2017.01.048
22. Verlicchi P, Zambello E (2015) Pharmaceuticals and personal care products in untreated and treated sewage sludge: occurrence and environmental risk in the case of application on soil – a critical review. Sci Total Environ 538:750–767. https://doi.org/10.1016/j.scitotenv.2015.08.108
23. Cerqueira MBR, Soares KL, Caldas SS, Primel EG (2018) Sample as solid support in MSPD: a new possibility for determination of pharmaceuticals, personal care and degradation products in sewage sludge. Chemosphere 211:875–883. https://doi.org/10.1016/j.chemosphere.2018.07.165
24. Pérez-Lemus N, López-Serna R, Pérez-Elvira SI, Barrado E (2019) Analytical methodologies for the determination of pharmaceuticals and personal care products (PPCPs) in sewage sludge: a critical review. Anal Chim Acta 1083:19–40. https://doi.org/10.1016/j.aca.2019.06.044
25. Saleh A, Larsson E, Yamini Y, Jönsson JÅ (2011) Hollow fiber liquid phase microextraction as a preconcentration and clean-up step after pressurized hot water extraction for the determination of non-steroidal anti-inflammatory drugs in sewage sludge. J Chromatogr A 1218:1331–1339. https://doi.org/10.1016/j.chroma.2011.01.011

26. Zuloaga O, Navarro P, Bizkarguenaga E, Iparraguirre A, Vallejo A, Olivares M, Prieto A (2012) Overview of extraction, clean-up and detection techniques for the determination of organic pollutants in sewage sludge: a review. Anal Chim Acta 736:7–29. https://doi.org/10.1016/j.aca.2012.05.016
27. Radjenović J, Jelic A, Petrovic M, Barceló D (2009) Determination of pharmaceuticals in sewage sludge by pressurized liquid extraction (PLE) coupled to liquid chromatography-tandem mass spectrometry (LC-MS/MS). Anal Bioanal Chem 393:1685–1695. https://doi.org/10.1007/s00216-009-2604-4
28. Chen Y, Yu G, Cao Q, Zhang H, Lin Q, Hong Y (2013) Occurrence and environmental implications of pharmaceuticals in Chinese municipal sewage sludge. Chemosphere 93:1765–1772. https://doi.org/10.1016/j.chemosphere.2013.06.007
29. Cucina M, Ricci A, Zadra C, Pezzolla D, Tacconi C, Sordi S, Gigliotti G (2019) Benefits and risks of long-term recycling of pharmaceutical sewage sludge on agricultural soil. Sci Total Environ 695:133762. https://doi.org/10.1016/j.scitotenv.2019.133762
30. Urra J, Alkorta I, Mijangos I, Epelde L, Garbisu C (2019) Application of sewage sludge to agricultural soil increases the abundance of antibiotic resistance genes without altering the composition of prokaryotic communities. Sci Total Environ 647:1410–1420. https://doi.org/10.1016/j.scitotenv.2018.08.092
31. Wei Z, Li W, Zhao D, Seo Y, Spinney R, Dionysiou DD, Wang Y, Zeng W, Xiao R (2019) Electrophilicity index as a critical indicator for the biodegradation of the pharmaceuticals in aerobic activated sludge processes. Water Res 160:10–17. https://doi.org/10.1016/j.watres.2019.05.057
32. Herrero P, Borrull F, Pocurull E, Marcé RM (2014) A quick, easy, cheap, effective, rugged and safe extraction method followed by liquid chromatography- (Orbitrap) high resolution mass spectrometry to determine benzotriazole, benzothiazole and benzenesulfonamide derivates in sewage sludge. J Chromatogr A 1339:34–41. https://doi.org/10.1016/j.chroma.2014.02.081
33. Martín J, Camacho-Muñoz MD, Santos JL, Aparicio I, Alonso E (2012) Distribution and temporal evolution of pharmaceutically active compounds alongside sewage sludge treatment. Risk assessment of sludge application onto soils. J Environ Manage 102:18–25. https://doi.org/10.1016/j.jenvman.2012.02.020
34. Peysson W, Vulliet E (2013) Determination of 136 pharmaceuticals and hormones in sewage sludge using quick, easy, cheap, effective, rugged and safe extraction followed by analysis with liquid chromatography-time-of-flight-mass spectrometry. J Chromatogr A 1290:46–61. https://doi.org/10.1016/j.chroma.2013.03.057
35. Rossini D, Ciofi L, Ancillotti C, Checchini L, Bruzzoniti MC, Rivoira L, Fibbi D, Orlandini S, Del Bubba M (2016) Innovative combination of QuEChERS extraction with on-line solid-phase extract purification and pre-concentration, followed by liquid chromatography-tandem mass spectrometry for the determination of non-steroidal anti-inflammatory drugs and their metabolites in sewage sludge. Anal Chim Acta 935:269–281. https://doi.org/10.1016/j.aca.2016.06.023
36. Brown AK, Ackerman J, Cicek N, Wong CS (2020) In situ kinetics of human pharmaceutical conjugates and the impact of transformation, deconjugation, and sorption on persistence in wastewater batch bioreactors. Environ Pollut 265:114852. https://doi.org/10.1016/j.envpol.2020.114852
37. Miao XS, Yang JJ, Metcalfe CD (2005) Carbamazepine and its metabolites in wastewater and in biosolids in a municipal wastewater treatment plant. Environ Sci Technol 39:7469–7475. https://doi.org/10.1021/es050261e
38. Nieto A, Borrull F, Pocurull E, Marcé RM (2007) Pressurized liquid extraction of pharmaceuticals from sewage-sludge. J Sep Sci 30:979–984. https://doi.org/10.1002/jssc.200600360
39. Berthod L, Roberts G, Sharpe A, Whitley DC, Greenwood R, Mills GA (2017) Effect of sewage sludge type on the partitioning behaviour of pharmaceuticals: a meta-analysis. Environ Sci Water Res Technol 2:154–163. https://doi.org/10.1039/c5ew00171d

40. Alenzi A, Hunter C, Spencer J, Roberts J, Craft J, Pahl O, Escudero A (2021) Pharmaceuticals effect and removal, at environmentally relevant concentrations, from sewage sludge during anaerobic digestion. Bioresour Technol 319:124102. https://doi.org/10.1016/j.biortech.2020.124102
41. Joss A, Andersen H, Ternes T, Richle PR, Siegrist H (2004) Removal of estrogens in municipal wastewater treatment under aerobic and anaerobic conditions: consequences for plant optimization. Environ Sci Tech 38:3047–3055. https://doi.org/10.1021/es0351488
42. Malvar JL, Santos JL, Martín J, Aparicio I, Alonso E (2021) Occurrence of the main metabolites of the most recurrent pharmaceuticals and personal care products in Mediterranean soils. J Environ Manage 278:111584. https://doi.org/10.1016/j.jenvman.2020.111584
43. Carballa M, Omil F, Lema JM, Llompart M, García-Jares C, Rodríguez I, Gómez M, Ternes T (2004) Behavior of pharmaceuticals, cosmetics and hormones in a sewage treatment plant. Water Res 38(12):2918–2926. https://doi.org/10.1016/j.watres.2004.03.029
44. Ahmad M, Eskicioglu C (2019) Fate of sterols, polycyclic aromatic hydrocarbons, pharmaceuticals, ammonia and solids in single-stage anaerobic and sequential anaerobic/aerobic/anoxic sludge digestion. Waste Manag 93:72–82. https://doi.org/10.1016/j.wasman.2019.05.018
45. González-Salgado I, Cavaille L, Dubos S, Mengelle E, Kim C, Bounouba MP, Pommier S, Bessiere Y (2020) Combining thermophilic aerobic reactor (TAR) with mesophilic anaerobic digestion (MAD) improves the degradation of pharmaceutical compounds. Water Res 182: 116033. https://doi.org/10.1016/j.watres.2020.116033
46. Lai HT, Lin JJ (2009) Degradation of oxolinic acid and flumequine in aquaculture pond waters and sediments. Chemosphere 75(4):462–468. https://doi.org/10.1016/j.chemosphere.2008.12.060
47. Arun S, Kumar RM, Ruppa J, Mukhopadhyay M, Ilango K, Chakraborty P (2020) Occurrence, sources and risk assessment of fluoroquinolones in dumpsite soil and sewage sludge from Chennai, India. Environ Toxicol Pharmacol 79. https://doi.org/10.1016/j.etap.2020.103410
48. Song L, Li L, Yang S, Lan J, He H, McElmurry SP, Zhao Y (2016) Sulfamethoxazole, tetracycline and oxytetracycline and related antibiotic resistance genes in a large-scale landfill, China. Sci Total Environ 551–552(266):9–15. https://doi.org/10.1016/j.scitotenv.2016.02.007
49. Sathishkumar P, Meena RAA, Palanisami T, Ashokkumar V, Palvannan T, Gu FL (2020) Occurrence, interactive effects and ecological risk of diclofenac in environmental compartments and biota – a review. Sci Total Environ 698:134057. https://doi.org/10.1016/j.scitotenv.2019.134057
50. Ben Mordechay E, Tarchitzky J, Chen Y, Shenker M, Chefetz B (2018) Composted biosolids and treated wastewater as sources of pharmaceuticals and personal care products for plant uptake: a case study with carbamazepine. Environ Pollut 232:164–172. https://doi.org/10.1016/j.envpol.2017.09.029
51. García-Galán MJ, Díaz-Cruz S, Barceló D (2013) Multiresidue trace analysis of sulfonamide antibiotics and their metabolites in soils and sewage sludge by pressurized liquid extraction followed by liquid chromatography-electrospray-quadrupole linear ion trap mass spectrometry. J Chromatogr A 1275:32–40. https://doi.org/10.1016/j.chroma.2012.12.004
52. Bragança I, Plácido A, Paíga P, Domingues VF, Delerue-Matos C (2012) QuEChERS: a new sample preparation approach for the determination of ibuprofen and its metabolites in soils. Sci Total Environ 433:281–289. https://doi.org/10.1016/j.scitotenv.2012.06.035
53. Kočárek M, Kodešová R, Vondráčková L, Golovko O, Fer M, Klement A, Nikodem A, Jakšíka O, Grabic R (2016) Simultaneous sorption of four ionizable pharmaceuticals in different horizons of three soil types. Environ Pollut 218:563–573. https://doi.org/10.1016/j.envpol.2016.07.039
54. Paz A, Tadmor G, Malchi T, Blotevolgel J, Borch T, Polubesova T, Chefetz B (2016) Fate of carbamazepine, its metabolites, and lamotrigine in soils irrigated with reclaimed wastewater: sorption, leaching and plant uptake. Chemosphere 160:22–29. https://doi.org/10.1016/j.chemosphere.2016.06.048

55. Wojslawski J, Białk-Bielińska A, Stepnowski P, Dołżonek J (2019) Leaching behavior of pharmaceuticals and their metabolites in the soil environment. Chemosphere 231:269–275. https://doi.org/10.1016/j.chemosphere.2019.05.031
56. Barron L, Havel J, Purcell M, Szpak M, Kelleher B, Paull B (2009) Predicting sorption of pharmaceuticals and personal care products onto soil and digested sludge using artificial neural networks. Analyst 124:663–670. https://doi.org/10.1039/b817822d
57. Lachassagne D, Soubrand M, Casellas M, Gonzalez-Ospina A, Dagot C (2015) Impact of sludge stabilization processes and sludge origin (urban or hospital) on the mobility of pharmaceutical compounds following sludge landspreading in laboratory soil-column experiments. Environ Sci Pollut Res 22:17135–17150. https://doi.org/10.1007/s11356-015-4918-4
58. Chefetz B, Mualem T, Ben-Ari J (2008) Sorption and mobility of pharmaceutical compounds in soil irrigated with reclaimed wastewater. Chemosphere 73:1335–1343. https://doi.org/10.1016/j.chemosphere.2008.06.070
59. Fenet H, Mathieu O, Mahjoub O, Li Z, Hillaire-Buys D, Casellasa C, Gomeza E (2012) Carbamazepine, carbamazepine epoxide and dihydroxycarbamazepine sorption to soil and occurrence in a wastewater reuse site in Tunisia. Chemosphere 88(1):49–54. https://doi.org/10.1016/j.chemosphere.2012.02.050
60. Walters E, McClellan K, Halden RU (2010) Occurrence and loss over three years of 72 pharmaceuticals and personal care products from biosolids – soil mixtures in outdoor mesocosms. Water Res 44:6011–6020. https://doi.org/10.1016/j.watres.2010.07.051
61. EMEA (2006) Guideline on the environmental risk assessment of medicinal products for human use, EMEA/CHMP/SWP/4447/00 corr 21, European Medicine Agency, Committee for Medicinal Products for Human Use (CHMP)
62. EC-TGD (2003) Technical guidance document on risk assessment, part II, EUR 20418 EN/ 2, European Commission, Joint Research Center
63. Hernando MD, Mezcua M, Fernandez-Alba AR, Barceló D (2006) Environmental risk assessment of pharmaceutical residues in wastewater effluents, surface waters and sediments. Talanta 69:334–342. https://doi.org/10.1016/j.talanta.2005.09.037
64. Bastos MC, Soubrand M, Le Guet T, Le Floch É, Joussein E, Baudu M, Casellas M (2020) Occurrence, fate and environmental risk assessment of pharmaceutical compounds in soils amended with organic wastes. Geoderma 375:114498. https://doi.org/10.1016/j.geoderma.2020.114498
65. Gros M, Mas-Pla J, Boy-Roura M, Geli I, Domingo F, Petrovic M (2019) Veterinary pharmaceuticals and antibiotics in manure and slurry and their fate in amended agricultural soils: findings from an experimental field site. Sci Total Environ 654:1337–1349. https://doi.org/10.1016/j.scitotenv.2018.11.061
66. Klampfl CW (2019) Metabolization of pharmaceuticals by plants after uptake from water and soil: a review. Trends Anal Chem 111:13–26. https://doi.org/10.1016/j.trac.2018.11.042

Microplastics in Soils as a Source of Pollution and Environmental Risk

Aida Moya Montes, Jose Navarro-Pedreño, María Belén Almendro-Candel, Ignacio Gómez Lucas, and Antonis A. Zorpas

Contents

Abstract This chapter gives an overview of the relations between microplastics and soils based on a deep review of the latest research, mainly centered in the last 10 years. Plastics are widely used materials and their degradation produces plastics of minor size (<5 mm), microplastics, that are persistent and appear distributed all over the world. A high interest is related to the presence of plastics in soils due to possible environmental pollution, considering that soil is a basic resource associated with the food chain. The main objectives of this chapter are to review the main sources of plastics and microplastics, their environmental concern, and the methods used to determine their presence in soils. As conclusions, it can be stated that further future research needs to be done focusing on microplastics in soils, and that there is

A. M. Montes (✉), J. Navarro-Pedreño (✉), M. B. Almendro-Candel, and I. G. Lucas
Department of Agrochemistry and Environment, University Miguel Hernández of Elche, Alicante, Spain
e-mail: jonavar@umh.es

A. A. Zorpas
Laboratory of Chemical Engineering and Engineering Sustainability, Environmental Conservation and Management, Faculty of Pure and Applied Sciences, Open University of Cyprus, Nicosia, Cyprus

Avelino Núñez-Delgado and Manuel Arias-Estévez (eds.),
Emerging Pollutants in Sewage Sludge and Soils,
Hdb Env Chem (2023) 114: 37–60, DOI 10.1007/698_2022_848,

Published online: 16 April 2022

an urgent need of developing standardized methods for sampling, extraction, and detection of plastic when analyzing soil samples.

Keywords Analytical methods, Environmental pollution, Persistence, Plastics, Soil contamination

1 Introduction

For several decades, environmental awareness has increased on topics such as resource scarcity, climate change, waste production, coastal erosion, etc., which in the end have come into attention of several research activities that directly or intricately affect the civil society [1–4]. Such is the case of plastics and their increasing presence in the environment.

Plastics have been obtained for more than a century, with an exponential increment in production after the second half of the twentieth century. Considering the period between 1950 and 2019, plastic production increased from 1.5 to 368 million metric tons [5–7]. It is estimated that the continuation of the current production and waste management trends will result in about 12 billion tons of plastic waste in natural environments by 2050 [8]. According to Laskar and Kumar [9], in the last 10 years we have produced and used more plastics than in the last 100 years. Hence, the amount of plastics entering the sea is near 13 million tons per year.

Nowadays, enormous amounts of plastic wastes are emitted and released into the environment, mainly in soils and water bodies. The origin of the plastic industry dates back to 1862, when the metallurgical chemist Alexander Parkes exhibited the first objects made from parkesine, a semi-synthetic plastic based on cellulose nitrate. The first synthetic plastic produced in commercial quantities, bakelite, was invented at the beginning of the twentieth century [10]. A scarcity of resources and the need to improve existing technologies after World War II were the driving forces behind the development of new synthetic materials, such as plastics [11, 12]. Plastics are one of the most advanced materials in terms of their range of properties and applications. The attractive qualities of these materials make an annual global demand over 400 million tons and an expected annual production rate of 1,100 tons by 2050 [13]. However, in recent years, these materials are suffering great opposition from various public and private sectors [14] due to their persistence in the environment and their consequences for biodiversity, the food chain, and human health.

Plastics, due to multiple attributes, are used in a wide range of industrial and consumer applications such as (1) water bottles, dispensing containers, biscuit trays produced from PET; (2) shampoo bottles, milk bottles, freezer bags, ice cream containers produced from HDPE; (3) bags, trays, containers, food packaging film produced from LDPE; (4) potato chip bags, microwave dishes, ice cream tubs, bottle caps, single-use face masks produced from PP; (5) cutlery, plates, cups produced from PS; (6) protective packaging, hot drying cups produced from EPS; (7) hub caps

(ABS), optical fibers (PBT), eyeglasses lenses, roofing sheets (PC), touch screens (PMMA), cable coating in telecommunications (PTFE), and many others in aerospace, medical implants, surgical devices, membranes, valves and seals, protective coatings, etc. [15, 16].

Their relatively low density gives these materials the advantage of light weight, and most of them have excellent electrical and thermal insulation properties. They can be manufactured in such a way that they conduct electricity when necessary [12]. Even more, they can be resistant to corrosion by many substances (as acids, water, and air exposure), making them durable and suitable for their use in harsh environments. In addition, its property balance can be modified by adding reinforcing fillers, plasticizers, colors, flame retardants, foaming agents, etc., to meet the demands of the specific application [12].

Plastics can be developed with almost any combination of properties to suit a multitude of applications on user demand. As a consequence of these versatile properties (resistant to water or fire, hardness, density), plastics can be used for packaging, in construction materials, transportation, electronics, agriculture, healthcare, sports, energy, etc.

Plastics improve daily life in countless ways, but they also become waste and microscopic pollution. The evidence is unmistakable, showing that plastic is the world's most durable, insidious, and intimate waste, and there is no doubt that we live in the age of plastic [17, 18].

Currently, plastic waste is relatively well defined as a group of mixed substances composed of natural, synthetic, or highly modified polymers, solid and insoluble at 20°C [19]. Synthetic plastics are derived from crude oil, natural gas, or coal. Bio-based plastics are mainly derived from renewable products such as carbohydrates, fats, vegetable oils, bacteria, and other biological substances. Because of the easy manufacturing methods involved in crude oil processing, most of the plastics currently on the market are synthetic [20]. Furthermore, the availability of bio-based feedstocks is expected to increase in the near future, providing an alternative to the use of fossil fuels for production. However, being bio-based does not necessarily imply that the plastic is biodegradable; in fact, resins such as bio-PE (bio-polyethylene) or bio-PET (bio-polyethylene terephthalate) are developed to match the properties of conventional ones, allowing the same useful life and applications [21]. A plastic is considered environmentally friendly when it is decomposed and degraded in a short period of time without giving hazardous wastes. A simplified diagram showing the production of common artificial plastics is shown in Fig. 1.

1.1 General Framework of Plastic Composition and Uses

The common term plastic is used to define a subcategory of a wider class of materials called polymers. Polymers are molecules that have an architecture comprising multiple repeating units derived from low molar mass molecules called monomers

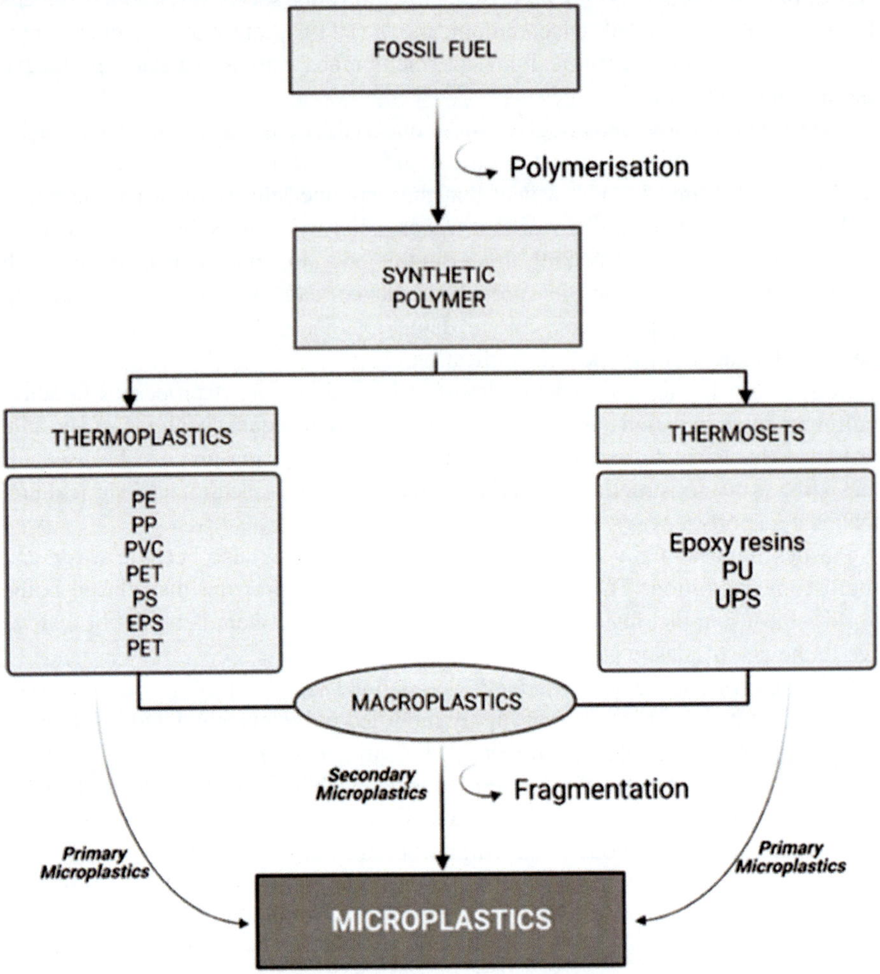

Fig. 1 Production of common synthetic plastics. All plastics can undergo fragmentation with environmental exposure and degradation. Source: own elaboration based on GESAMP [21]

[22], therefore plastic polymers are large molecules and have high molecular weights. Polymers can be formed by identical units (homopolymers), as well as structures of different monomers in various possible sequences (copolymers), the latter although less frequent also present the potential to create microscopic residues in the environment [21]. Homopolymers have a unique molecular structure, consequently, to chemically identify a homopolymer the structure of the repeating unit is specified, i.e., the structure $(\text{-CH2-CH2-CH2-})_n$ represents polyethylene (PE). The longer the molecular chains, the greater the van der Waals forces between them, obtaining exceptional mechanical properties such as strength and fracture toughness [23]. The physical properties of plastic, such as stiffness, flexibility, and elasticity,

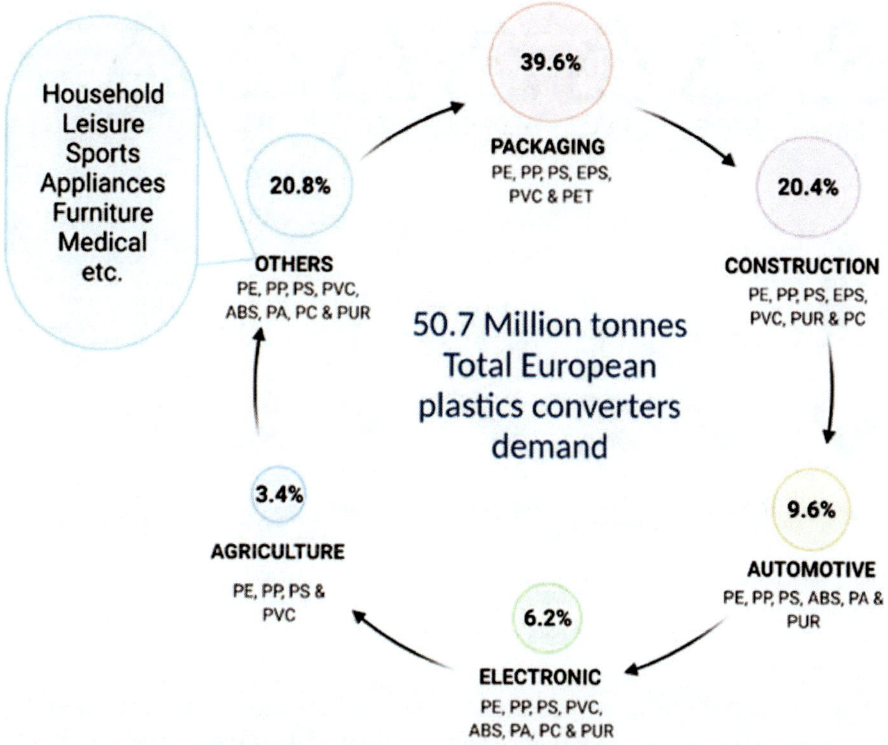

Fig. 2 European plastic demand with main applications and selected polymers. Source: own elaboration based on Plastics Europe [16]

are influenced by the molecular weight distribution of the polymer and the organization of the chains [24].

There are millions of uses for plastic, and hundreds, if not thousands, of different types of plastics. The most abundant plastics in the environment are the following: polyethylene (PE, high and low density), polypropylene (PP), polyvinyl chloride (PVC), polystyrene (PS, including expanded EPS), polyurethane (PUR), polycarbonate (PC), polyamide (PA), polyethylene terephthalate (PET), and acrylonitrile butadiene styrene (ABS) [20, 25]. The applications of the polymers mentioned above are shown in Fig. 2.

In order to improve some specific properties almost all plastic products contain additives, such as UV stabilizers, antioxidants, plasticizers, colorants, etc. Besides, they may also contain unintentionally added substances such as residues from the recycling process, oligomers or initiators of the polymerization process. Those sum another complication when analyzing the "real" environmental exposures and the risks associated with them [22]. Additives can be an important source of pollution during the degradation process of a plastic [26].

The European Commission has published several regulations to specify what types and abundance of plastic contaminants exist, labeled forms to classify them,

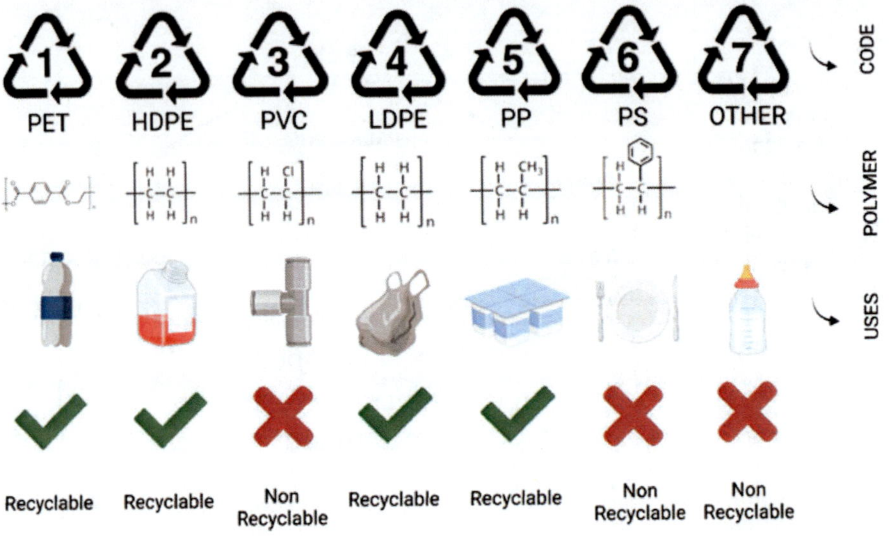

Fig. 3 Identification code for the main types of polymers, their chemical structure, applications and recyclable properties. The Mobius loop with a number at the center indicates the type of plastic the product is made from. Source: own elaboration based on European Directive 94/62/EC on packaging and packaging waste [27]

actuations to understand the hazards that these contaminants suppose, and measures to reduce plastic waste in the environment. One of the first directives is the 94/62/EC on packaging and packaging waste, which specifies the symbols that may be indicated on a product to specify the kind of plastic that the packaging is made from [27]. Figure 3 shows the symbols that identify the type of plastic polymers, some common uses, and their recyclable potential [27].

Regarding the requirements of the European Commission to establish harmonized specifications for reducing the impact of certain products that can be a hazard to the environment, the 2019/904 Directive was published [28]. Based on the marking specifications by Single-Use Plastic Directive 2019/904, on 22th December 2020, the Regulation 2020/2151 published by the European Commission gives some specifications applicable since 3rd July 2021 [29]. This directive lays down marking instructions for certain single-use plastic products that habitually are inappropriately disposed. The specifications are focused on limiting their use through raising awareness in consumers, establishing design requirements to connect the material and the possible hazard, introducing labeling requirements to inform the European population about the plastic content in commonly used products, and to settle waste management and clean obligations for producers. Figure 4 shows the marking specification for single-use plastic products that are applicable from 3rd of July 2021 according to 2019/904 directive.

The single-use products addressed by the Directive are sanitary towels, tampons, tampon applicators, wet wipes, tobacco products, cups for beverages, cotton bud sticks, stirrers, cutlery, plates, straws, balloons, sticks for balloons, food containers

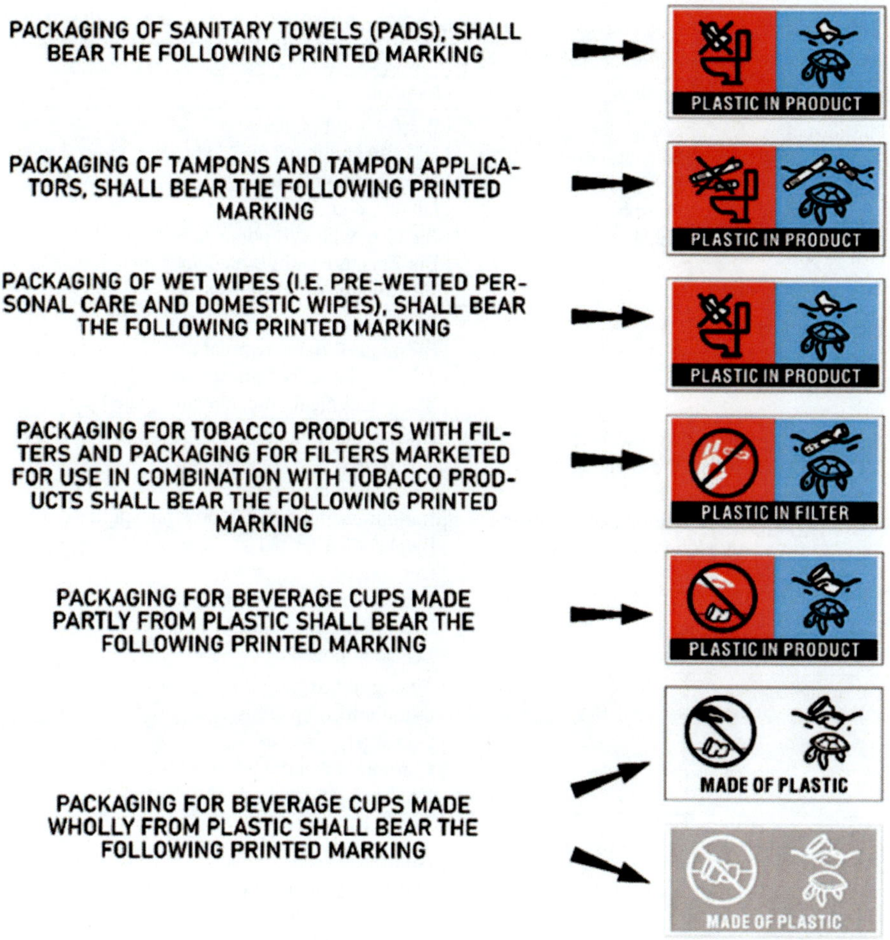

Fig. 4 Marking instructions for certain single-use plastic. Own elaboration based on European Directive 2019/904 [29]

and plastic bags, where sustainable alternatives are easily accessible. Also, some single-use plastic products are totally banned from 3rd of July 2021 like cutlery, straws, plates, stirrers, cotton bud sticks, and sticks for balloons. The ban is also applicable to products made of EPS like food and beverage containers, and all oxo-degradable plastic products.

Most of the single-use plastics that are manufactured contain additives or other substances to improve physical, chemical, and even biological properties of the polymers, which can function as a main component of their structures. That is the case of the oxo-degradable plastics that contain additives to lead the oxidation and fragmentation of the product into micro-fragments that are harmful for the environment, as referred in the EU Directive 2019/904 of the European Parliament [28].

Table 1 Common plastic additives used in the manufacture of plastic products. Adapted from the report of the European Union about impacts of plastics and microplastics use [31]

Function	Additive	Examples
Plasticizers	Esters Aliphatic esters Polyesters Phosphates Phthalates	Benzyl butyl phthalate Di-isoheptylphthalate Di-isobutyl phthalate Dibutyl phthalate Bis (2-ethylhexyl) phthalate Bis(2-methoxyethyl) phthalate Tris(2-chloroethyl) phosphate
Flame retardants		Boric acid Brominated flame retardants Tris(2-chloroethyl) phosphate Short- and medium-chain chlorinated paraffins
Stabilizers	Antioxidants Preservatives Absorbers Biological stabilizers UV stabilizers	Arsenic compounds Triclosan Organic tin compounds Barium-cadmium-zinc-epoxyphosphite Bisphenol A (BPA) Octylphenol Nonylphenol compounds Cadmium compounds Lead compounds
Colorants		Titanium dioxide Cadmium compounds Cobalt (II) diacetate Chromium compounds Lead compounds
Curing agents		Formaldehyde
Antistats		Amines Quaternary ammonium compounds Organic phosphates Polyoxyethylene glycol esters
Processing aids	Lubricants Flow controls Other aids	Calcium, zinc, and lead stearates Fatty esters Amides Petroleum and polyethylene waxes

These additives are not always covalently bonded to the matrix of the plastic, which means that are more susceptible of being released into the environment, increasing this risk with the gradual degradation of the materials. There is also scientific evidence that the occurrence of chemical releasing of additives like plasticizers, antioxidants, and antimicrobial agents into the environment can lead to a broad range of toxicological responses, including oxidative stress in cells and sub-lethal responses in marine organisms [30]. In Table 1, some examples of the most common additives and their main functions are shown.

2 Microplastics

Microplastics include fibers (mostly synthetic) and waste from plastics. Synthetic fibers are extremely harmful to any marine life [9], as the sea fish very often consume microplastics, directly affecting in that way the food chain and the health of a variety of living beings through the bloodstream. Anything sized less than 5 mm among them is recognized as one of the largest threats to marine ecosystem. Major sources for microplastics in urban areas and coastal environments are agricultural greenhouses and single plastic bags [6, 32, 33].

Plastic microfibers (<5 mm) and nanofibers (<100 nm) have been identified in ecosystems in all regions of the globe and have been estimated to comprise up to 35% of primary microplastics in marine environments, being a major proportion of microplastics on coastal shorelines, and are considered to persist for decades in soils treated with sludge from wastewater treatment plants [34].

Essentially, plastics are characterized by plasticity, a term correlated to the deformation of materials in response to applied forces. The technical report for plastics vocabulary of the International Organization for Standardization [35] defines plastic as: "a material that contains as an essential ingredient a high polymer and which, at some stage in its processing into finished products, can be shaped by flow," with the notation that elastomeric materials are not considered plastics even though they can be shaped by flow.

In the current literature the term "microplastic" is used flexibly including a surprising range of particle sizes, from 5 mm to a few microns in diameter. Therefore, it is useful to establish a sub-classification of sizes included within the category of microplastics. A practical classification was proposed by GESAMP [21]. However, also a European classification was proposed in 2013 [23, 36], and both are included in Table 2. Examples of the differences in the categorization of plastics according to their size in the scientific literature and in institutional reports were given by Hartmann et al. [43]. In Table 2, an example of size categorization of plastics is shown.

The formation of microplastics is a process that depends on many factors and on the type of plastic. However, it is assumed that high temperatures accelerate the aging of plastics and promote their fragmentation to microplastics [44], although this

Table 2 Categorization of plastics by size according to various sources in the scientific literature and in institutional reports. It should be noted that this does not represent a comprehensive overview of all size classes

Nanoplastics	Microplastics	Mesoplastics	Macroplastics	Source
	<5 mm			NOAA [37]
1–100 nm				Nature [38]
	<5 mm	5–25 mm	>2.5 cm	MSFD EU [39]
<1 μm	1–1,000 μm	1–25 mm	2.5–100 cm	European Commission [40]
1–100 nm	< 5 mm			EFSA [41]
<100 nm	< 5 mm			European Commission [42]

temperature is related to their thermal properties [45]. This means that in warm climates, for instance in Mediterranean environments, the fragmentation of plastics could be faster than in cold climate zones and the presence of microplastics would be relevant.

Plastic size continuously decreases after degradation by physicochemical (chemical, thermal, and photochemical alteration) and biological processes [45]. However, despite the processes by which plastics are susceptible to degradation, many times these processes are extremely slow and may even stop in certain conditions. Furthermore, some plastics based on aromatic polyesters, such as PET, are virtually impossible to biodegrade biologically under any circumstances [42]. In such case, the persistence of the plastic along time is too high and can derive into environmental risks.

It should be noted that environmental sustainability concerns have promoted the development of different types of plastics with a higher propensity to degradation, trying to reduce the ecological footprint, reaching materials such as oxo-degradable, biodegradable, bioplastics, and compostable plastics. However, bioplastics, although made from bio-based rather than petrochemical feedstocks, are not necessarily more biodegradable than conventional ones, and oxo-degradable ones are not more biodegradable either, they simply break down faster into microplastics and, in fact, worsen and increase microplastic pollution [42]. For this reason, oxo-degradable plastics are banned in the European Union under the so-called Single-Use Plastics Directive [28].

In order to obtain biodegradable plastics that truly reduce the ecological footprint, it is important to ensure that they are environmentally benign and that when decomposing they do not produce toxicity or release toxic additives [46]. Through the European regulation [28] criteria are established that will ensure that the polymer is capable of decomposing physically and biologically under natural conditions in the soil system and aquatic environments throughout the European Union, so that it only decomposes into carbon dioxide, biomass, and water. This law aims to ensure that the use of polymers does not lead to the accumulation of plastics in the environment, mainly in soils and water.

It can be possible to differentiate between primary and secondary microplastics. Primary microplastics are those directly manufactured with a diameter equal to or less than 5 mm. These microplastics include polyethylene (PE), polypropylene (PP), and polystyrene (PS) particles. These are used in various fields such as cosmetic, medical, and technological [47]. Due to the harmful effects to the environment, the sale of cosmetics that include microplastics has been banned in several countries such as Canada or the United States of America [48]. Secondary microplastics are those derived from the decomposition of larger plastic waste. This is because, over time, the physical, biological, and chemical processes that plastics undergo can reduce their structural integrity and result in their fragmentation into various forms such as granules, fibers, and fragments [47, 49]. Typical densities of some plastics are shown in Table 3.

Table 3 Polymer densities of main plastic polymers. The densities are the average of one or more experimental values. Own elaboration based on Crow Polymer Physics [50]

Polymer	Density (g/cm^3)
Polyethylene	0.854
Polypropylene	0.861
Polystyrene	1.052
Polyamides (nylon)	1–1.078
Polyvinylchloride	1.388
Polycarbonate	1.206
Polyethylene terephthalate	1.017
Polychloroprene (neoprene)	1.243

3 Pathways of Occurrence in Soils

In recent years, the scientific community has recognized plastic pollution as a major environmental threat, as referred by the Committee of the European Parliament [31]. Soil scientists revealed a new angle of the problem by exposing the presence of microplastics in terrestrial environments. It is necessary to focus on assessing the magnitude of this, such as the potential effects and frequency of occurrence of these substances. For this purpose, more field studies are needed to allow the quantification and study of the adverse effects on the terrestrial environment [51]. It is estimated that the abundance of microplastics in terrestrial environments is between 4 and 23 times higher than that in the ocean, so it is believed that soils could be larger plastic reservoirs [42]. On the other hand, soil is a heterogeneous matrix, due to different soil characteristics. These characteristics could be the ones causing different physical fractions to respond atypically under the interference of the external environment and different responses and interactions between soil particles and microplastics.

Due to the disparity in characteristics between the aquatic and terrestrial environments, the physicochemical and biological processes in the soil ecosystem also differ. It is precisely because of the heterogeneous properties of soils that publications on the distribution, movement, and biological effects of microplastics are still limited compared to those in aquatic systems. Clearly, relevant impact mechanisms are worth to be investigated given their expected importance to understand ecosystem impacts [52].

There are a multitude of pathways by which microplastics can enter the soil system, through fertilizers, irrigation water, use of plastic mulch, greenhouse coverings, application of sewage sludge, atmospheric deposition, irrigation with treated wastewater, and other sources [17, 53, 54]. In addition, the input and output of microplastics, as well as transport through soils, can be influenced by runoff, infiltration, weather conditions, and even living organisms, so it is necessary to evaluate various spatial-temporal scales to reach a conclusion on which pollution sources are predominant in a given area [55].

In particular, wastewater treatment plants (WWTPs) are an important pathway for the occurrence of microplastics in the environment, as many times treated waters are used for crop irrigation, mainly in warm areas where water scarcity is common. It is

Fig. 5 (1) Mulching applied in crop production; (2) soil profile after mulching, fragmentation of plastic is perceptible at different depths; (3) microplastic fibers observed in a soil sample

important to note that seasonal variations influence the total concentration of microplastics [56].

In the same way, sewage sludge is of concern as it is commonly used in agriculture and is rich in synthetic fibers and other plastic polymers. Microplastics derived from household and industrial processes are mainly conveyed to municipal wastewater treatment plants. In these plants, microplastic particles are removed from the water being retained in the sludge [57]. Europe uses about a 50% of the sewage sludge as a fertilizer in agriculture. Due to their application, European agricultural soils are of particular concern because of the presence of contaminants resulting from their use. On the other hand, the application of sewage sludge in agricultural areas is strictly regulated for the presence of some organic compounds and heavy metals [33, 58–60], however, the presence of plastic elements is not yet regulated [51, 53]. Due to the existing concern about microplastic concentration in sewage sludge, new studies are emerging to quantify the concentration and removal efficiency in several wastewater treatment plants [61, 62].

In addition, the use of plastic films (mulching) is another major cause of microplastic contamination in soils. These plastics are used to increase soil temperature, control weeds, and improve water absorption efficiency. These attractive properties influence their extensive use by farmers (Fig. 5), however, these plastic films suffer degradation in the soil by thermal processes, photodegradation, oxidation, hydrolysis, and mechanical degradation, contributing to the emission of microscopic plastic contaminant particles [52, 63].

In soil, plastic as contaminant is part of a complex mixture with organic matter and minerals, since the surface charge of plastics is commonly negative and interacts with positively charged particles or ions presented in soil solution and/or matrix. For this reason, plastic particles in soils can affect several properties such as some extractable ions (cation exchange capacity), dissolved organic matter, water holding capacity, aggregate stability, and bulk density. These interactions, soil-plastics, evidence the influence on the entire chemical and physical environment of the soil, the habitat of organisms and thus the living beings present in the soil [17].

According to the research published to date, poor information on the presence of microplastics associated with soils and land use have been reported, although it seems that it is increasing in recent years. This lack of enough results weakens the possibilities for ambitious research questions related to toxicology and/or mitigation on a larger scale. More studies providing data on microplastics in different contexts are urgently needed to get an idea of the real magnitude of the problem [51].

As an important consideration, the main sources influencing plastic content in agricultural soils are irrigation with wastewater, the use of some amendments as composted sewage sludge, and the use of plastics for crop cultivation. However, as it occurs in soils of urban areas, diffuse pollution form plastics derived from abandonment wastes should be also considered.

4 Environmental Risks

In view of the presence and degradation processes of microplastics in the environment, related environmental risks are of high concern. Moreover, from 5 years until now, the research about microplastics, environmental impact and risk, and soil pollution has been increasing [10, 14, 63–65].

Nowadays there are no standardized methods for testing the effects of microplastics on the environment, although plastic pollution has various effects on organisms and their habitats, depending on the size and type of plastic, and the level of biological organization [66]. However, Loizia et al. [5], as well as Voukali et al. [67], indicate common methods regarding the assessment of microplastic in coastal areas. According to Loizia et al. [5] and Voukali et al. [67], to float microplastics, a known quantity of homogenized sand sample (100 ± 1 g) was mixed with a known quantity of hypersaline solution (250 ± 0.01 ml in 10% v/v NaCl) for 7 min. After air drying, the samples were passed through separated sieves. The sieving samples were classified as macro-plastics if the diameter was between 2.5 and 50 cm, meso-plastics if the diameter was between 0.5 and 2.5 cm, and microplastics if the diameter was less than 0.5 cm. With this method, when samples are decanted into high-density solutions, plastic particles float on the surface of the solution. However, the denser samples materials remain at the bottom of the solution gradient. It is important to note that, with this method, there is no estimation of any light density plastics, i.e. PP or PE, but only general concentrations.

Microplastics have become iconic symbols of anthropogenic waste and environmental pollution. In fact, plastics themselves may not be the most toxic pollutants in the environment, but they are consistently increasing in natural environments worldwide. Plastics are produced, used, and disposed in terrestrial or continental systems where they start to interact with all other environmental compartments [68].

As a kind of emerging pollutant, it is receiving increasing attention from researchers, because microplastics can potentially impact terrestrial ecosystems either directly or through the toxic and endocrine-disrupting substances added during plastics manufacturing [69]. Furthermore, it has been shown that the dispersion of plastic particles is concomitant with that of their associated microorganisms and mobile genetic elements, including antibiotic resistance genes, pathogenicity islands, and various metabolic pathways, which create new challenges for scientists [54]. This makes evident that quantification of microplastics in different ecosystems is critical, because of their associated potential risks [42].

Since plastics are generated and often disposed or handled in terrestrial areas, it is precisely in this environment that they are commonly dispersed. Once plastic particles reach the soil and accumulate to some extent, they can affect soil properties, soil functionality, as well as biodiversity, causing potential damage to soil ecosystems [70]. Experimental studies in soil matrix involving biota are limited, but those performed show that microplastics exposure can negatively impact organism's health and behavior. For example, Kim and An [71] showed that soil microplastics can affect the movement of springtail species, and also some studies have evidenced that the presence of microplastics may alter the physical properties as well as biophysical function and structure of soils, and consequently influence the interaction between soil and plants and the soil microorganism community, therefore causing worrying effects on crops, soil biota, and the trophic chain [51, 72].

Plastics can be effective transport mediums for organism and other organic matter due to their potential for surface adhesion and the low density of certain types of plastics. In addition, microplastics can transport contaminants (i.e., pesticides) in the soil matrix; in fact, it has been shown that the presence of microplastics in soils decreases the retention capacity of pesticides and other pollutants by 4–15%. In addition, there are studies that show that plastic particles can act as an efficient adsorbent for heavy metals and persistent organic pollutants [36, 53], contributing to make them more available to migrate through the soil profile to groundwater reservoirs, where they can cause contamination of drinking water sources [53].

The ability of microplastics to sequester and transport pollutants in the environment is influenced by the properties of the plastic particle (polarity, point of zero charge, type of polymer, degree of crystallinity, surface area, functional groups, surface topography, etc.), properties of the contaminants (i.e., hydrophobicity), and matrix properties (salinity, temperature, pH), and there is scientific evidence that hydrophobic compounds are the most likely to be adsorbed by plastic particles. In fact, enriched microplastics are a major source of pollution, as they allow contaminants to travel long distances and consequently affect the distribution of pollutants in the environment [53]. Furthermore, microplastics can also release harmful additives such as phthalates and flame retardants, which are endocrine disruptors and consequently affect organisms present in the soil in a negative way [36].

Humans are also exposed to microplastic pollution through the ingestion and inhalation ways. The toxicity of plastic particles in humans is still being unclear due to the lack of research in the area [66]. In spite of this fact, The World Health Organization (WHO), The Food and Agriculture Organization of the United Nations (FAO), and The European Food Safety Authority (EFSA) recently carried out assessments of human exposure to microplastics and its sorbed chemicals, using conservative worst-case scenarios to estimate the levels of these pollutants [41, 73, 74].

Usually, the size and concentrations of microplastics used in ecotoxicological research do not reflect the concentrations or sizes of plastic particles found in the environment. Microplastics effects studies are often performed using concentrations that are much higher than those reported in the environment, or very small sized, for which limited data of occurrence exists [22]. The Environment and Climate Change of Canada suggests that researchers studying the effects of plastic particles pollution should use plastics of similar size, shape, and composition to those found in the environment [66]. There is consequently a need to develop standard methods to evaluate the potential hazard effects associated with the presence of plastics and toxicity in the environment.

In light of the possible direct and indirect interactions of microplastics with abiotic and biotic aspects of ecosystems, plastics may well represent an important driver of global climate change. It is possible that many terrestrial species are already or will soon be exposed to levels of pollution capable of changing the baselines of basic functional ecosystem processes globally. Moreover, some species, particularly those with short generation times, may already be under evolutionary pressure from this new anthropogenic stressor [75]. For all these reasons, and considering that there is a lack of studies on microplastics in soil, further research is needed to fully understand the interactive effects that plastic pollution will cause on all the compartments of the soil ecosystem, including the potential uptake into food crops and pasture fauna.

5 Introduction to Analysis of Microplastics in Soils

Representative sampling is one of the most important steps of all methodology, as non-representative sampling will cause unreliable data, even though the subsequent analysis would be fully reliable. Regarding the first step in soil studies, soil sampling is generally more challenging than sampling water. Sampling soil matrixes can require significantly more effort and resources and depends on the area and soil distribution. The procedure for taking soil samples when quantifying microplastics will vary according to the purpose of the study. Aspects to consider would be the depth (vertical profile or surface samples), choosing the best location for taking the samples, and the sample size [21].

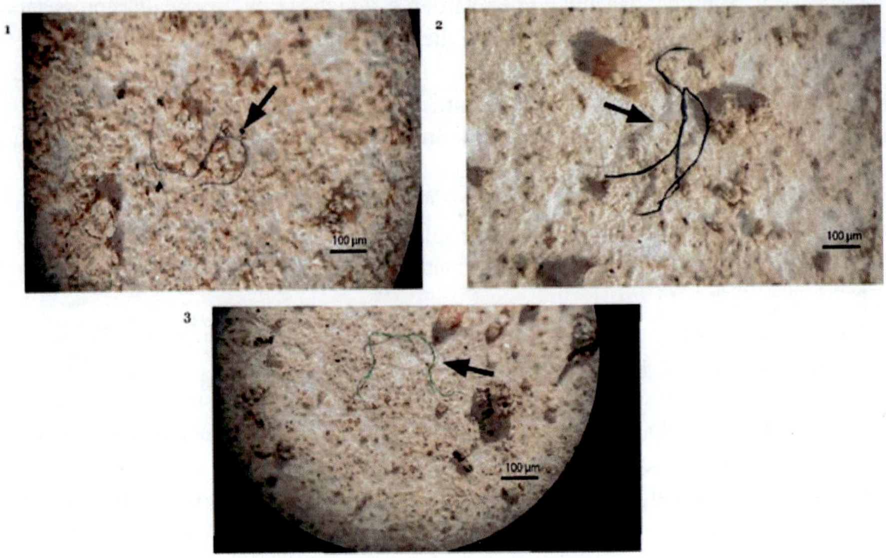

Fig. 6 Images of plastic particles in urban soil matrices. It is noticed that fibers of different origin (color) are presented in the soil

The sampling depth should always be defined and documented. As soils are a three-dimensional body, if the deposition of microplastics on the surface and an undisturbed soil are in the focus of the study, then bulk sampling of the first centimeters of the surface would be sufficient. Contrarily, if the varying depths of the soil contaminated by microplastics are to be determined, core sampling would be required [64, 76, 77]. Usually, the most probable depth for the presence of plastic particles is the upper 10 cm of the soil surface [78, 79]. After selecting the most suitable sampling approach according to the research objective, the number of sampling points at each site must also be accounted.

Once the samples are in the laboratory, the analysis is a multi-step process that may include: sample preparation, such as air drying, homogenization, and sieving. After this process, microplastics need to be separated from other organic and inorganic particles prior to be quantified, weighed, and the polymer type identified if possible (Fig. 6).

Initial separation from soil particles can be achieved by density using concentrated/saturated solutions like NaCl, $ZnCl_2$, NaI, among others. Organic particles present in the sample may interfere with microscopic identification, so these particles would be removed previously by using hydrogen peroxide and/or sulfuric acid [21]. Since there is not a standardized method for the extraction and identification of microplastic particles in soil samples, one of the existing methods that can be selected is the following: after having passed the sample through a 2 mm sieve (or 5 mm, depending on the plastic size under investigation), an amount of soil (i.e., 5 or 10 g) is introduced into a high-density saline solution and allowed to settle for at least 8 h in order to float microplastic particles [63]; ultrasonic stirring can help

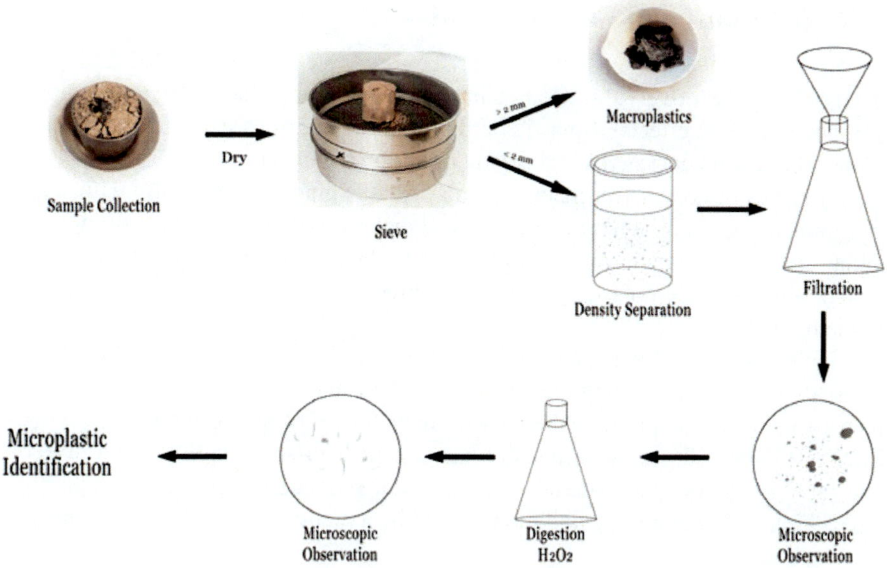

Fig. 7 Schematic diagram in analytical procedures for microplastics present in soil samples

during the separation and floating process; then, the dilution supernatant is filtered, and the sediment retained in the filter is subjected to organic matter digestion; after being digested, the resulting sediment is floated and filtered again, and this procedure is performed to obtain microplastic particles (Fig. 7); this step should be run at least three times; finally, all filters with retained materials are observed through a microscope and plastic polymers identified by using several methods described below.

It becomes really challenging to distinguish plastic particles from non-plastic particles with decreasing size using optical microscopic examination alone. To confirm the presence of plastics and provide the polymer composition, several spectroscopy techniques can be useful [77]. In fact, to confirm the identification of plastics, and their synthetic polymer types, spectroscopy is required.

Microscopic Fourier transformed infra-red (FTIR) and Raman spectroscopy are the most useful methodologies to identify microplastics. These techniques rely on light transmission and wavelengths, when the particle is smaller than those wavelengths, none polymer IR spectrum can be detected, which means that none Infrared spectrum from a polymer can be detected. In laboratory experiments, fluorescently labeled microplastics can be used to evaluate their behavior. This can be useful in laboratory experiments but does not solve the problem when measuring microplastics in environmental samples. Development of new analytical techniques is required to extract, isolate, and identify microplastics in the soil environment [21, 64, 77].

6 Conclusions and Future Trends

In this chapter, an overview about reliable and persistent facts, the morphology, presence, fate, and effects of plastic particles has been discussed. It is relevant that studies about soil ecosystems and plastic pollution have just recently started, due to characteristics of soil matrix making it more challenging in comparison with water environments. Overall, it has been evidenced that, in contrast to marine microplastic pollution, the problematic of plastics in terrestrial environments is far poorly investigated, although microplastics in soils are directly connected to human and environment health.

To improve the problematic of soil ecotoxicity derived from plastic pollution, it is necessary to begin decreasing the amount of plastic at the source, as well as to avoid the use of plastic, and promote the reuse or recycling of plastic materials before final safe environmentally sound deposition/management. Also, to promote and investigate about biodegradable materials that match the properties of the common plastics is of concern. Reducing soil pollution should be a main target in all the administration's agendas.

Future research about microplastic pollution should consider different scenarios such as sources, ways of transfer, fate, and ecotoxicological effects. As one of the big supplies of plastic in terrestrial environments is through sewage sludge, measures to prevent environmental pollution at this source should be conducted.

The last point to highlight in this section is the urgency of standardized methods for sampling, extraction, and detection of plastic when analyzing soil samples, as a basic need to properly conduct reliable research in this field.

References

1. Boring S, Laibach N, Wustmans M (2020) Innovation types in the bioeconomy. J Clean Prod 266:121939. https://doi.org/10.1016/j.jclepro.2020.121939
2. Zorpas AA (2020) Strategy development in the framework of waste management. Sci Total Environ 716:137088. https://doi.org/10.1016/j.scitotenv.2020.137088
3. Zorpas AA, Lasaridi K, Abeliotis K, Voukkali I, Loizia P, Fitiri L, Chroni C, Bikaki N (2014) Waste prevention campaign regarding the waste framework directive. Fresen Environ Bull 23(11a):2876–2883
4. Zorpas AA, Lasaridi K, Mihaela Pociovălişteanu D, Loizia P (2018) Household compost monitoring and evaluation in the framework of waste prevention strategy. J Clean Prod 172: 3567–3577. https://doi.org/10.1016/j.jclepro.2017.03.155
5. Loizia P, Voukkali I, Chatziparaskeva G, Navarro-Pedreño J, Zorpas AA (2021) Measuring the level of environmental performance on coastal environment before and during Covid-19 pandemic. A case study from Cyprus. Sustainability 13:2485. https://doi.org/10.3390/su13052485
6. Sourkouni G, Kalogirou C, Moritz P, Gödde A, Pandis KP, Höfft O, Vouyiouka S, Zorpas AA, Argirusis C (2021) Study on the influence of advanced treatment processes on the surface properties of polylactic acid for a bio-based circular economy for plastics. Ultrason Sonochem 10(76):105627. https://doi.org/10.1016/j.ultsonch.2021.105627

7. Tiseo I (2021) Global plastic production 1950–2019. In: Statista. https://www.statista.com/statistics/282732/global-production-of-plastics-since-1950/. Accessed 1 July 2021
8. Geyer R, Jambeck JR, Law KL (2017) Production, use, and fate of all plastics ever made. Sci Adv 3(7):e1700782. https://doi.org/10.1126/sciadv.1700782
9. Laskar N, Kumar U (2019) Plastics and microplastics: a threat to environment. Environ Technol Innov 14:100352. https://doi.org/10.1016/j.eti.2019.100352
10. Callapez ME (2021) History of the beginnings of the plastics industry in Portugal. ChemTexts 7(3):17. https://doi.org/10.1007/s40828-021-00134-1
11. Gilbert M (2016) Plastics materials: introduction and historical development. In: Brydson's plastics materials. 8th edn. Elsevier, pp 2–18. https://doi.org/10.1016/B978-0-323-35824-8.00001-3
12. Plastics Europe (2021) History of plastics. https://www.plasticseurope.org/en/about-plastics/what-are-plastics/history. Accessed 8 May 2021
13. Lear G, Kingsbury JM, Franchini S, Gambarini V, Maday SDM, Wallbank JA, Weaver L, Pantos O (2021) Plastics and the microbiome: impacts and solutions. Environ Microbiomes 16: 1–19. https://doi.org/10.1186/s40793-020-00371-w
14. Skoczinski P, Krause L, Raschka A, Dammer L, Carus M (2021) Current status and future development of plastics: solutions for a circular economy and limitations of environmental degradation. Methods in enzymology, Academic Press Inc., pp 1–26. https://doi.org/10.1016/bs.mie.2020.11.001
15. Plastics Europe (2019) Plastics the Facts, 2019. An analysis of European plastics production, demand and waste data. https://www.plasticseurope.org/application/files/9715/7129/9584/FINAL_web_version_Plastics_the_facts2019_14102019.pdf. Accessed 8 July 2021
16. Plastics Europe (2020) Plastics the facts, 2020. An analysis of European plastics production, demand and waste data. https://www.plasticseurope.org/application/files/3416/2270/7211/Plastics_the_facts-WEB-2020_versionJun21_final.pdf. Accessed 15 Jun 2021
17. Blöcker L, Watson C, Wichern F (2020) Living in the plastic age – different short-term microbial response to microplastics addition to arable soils with contrasting soil organic matter content and farm management legacy. Environ Pollut 267:115468. https://doi.org/10.1016/j.envpol.2020.115468
18. ORB (2017) Invisibles – Multimedia | Orb. https://orbmedia.org/stories/Invisibles_plastics. Accessed 1 July 2021
19. Zhou Y, Wang J, Zou M, Jia Z, Zhou S, Li Y (2020) Microplastics in soils: a review of methods, occurrence, fate, transport, ecological and environmental risks. Sci Total Environ 748:141368. https://doi.org/10.1016/j.scitotenv.2020.141368
20. British Plastic Federation (2021) How is Plastic Made? https://www.bpf.co.uk/plastipedia/how-is-plastic-made.aspx. Accessed 8 May 2021
21. GESAMP 2015 Sources, fate and effects of microplastics in the marine environment: a global assessment" (Kershaw, P. J., ed.). (IMO/FAO/UNESCO-IOC/UNIDO/WMO/IAEA/UN/UNEP/UNDP Joint Group of Experts on the Scientific Aspects of Marine Environmental Protection). Rep. Stud. GESAMP No. 90, 96 p
22. SAPEA (2019) Science advice for policy by European Academies 2019. A scientific perspective on microplastics in nature and society. SAPEA, Berlin, https://doi.org/10.26356/microplastics
23. Andrady AL (2017) The plastic in microplastics: a review. Mar Pollut Bull 119:12–22. https://doi.org/10.1016/j.marpolbul.2017.01.082
24. Sperling LH (2005) Introduction to physical polymer science.4th edn. Wiley, Bethlehem
25. Environment and Climate Change Canada (2019) Canada's Plastics Science Agenda. https://www.canada.ca/en/environment-climate-change/services/sciencetechnology.html. Accessed 21 May 2021
26. Hahladakis JN, Velis CA, Weber R, Iacovidou E, Purnell P (2018) An overview of chemical additives present in plastics: migration, release, fate and environmental impact during their use,

disposal and recycling. J Hazard Mater 344:179–199. https://doi.org/10.1016/j.jhazmat.2017.10.014
27. EU (2018) European Directive (EU) 94/62/EC packaging and packaging waste. European Parliament, Council of the European Union. (Text with EEA relevance) PE/12/2018/REV/2. https://eur-lex.europa.eu/legalcontent/EN/TXT/?uri=CELEX %3A01994L0062-20180704. Accessed 16 Jun 2021
28. EU (2019) European Directive (EU) 2019/904 of the European Parliament and of the Council of 5 June 2019 on the reduction of the impact of certain plastic products on the environment (Text with EEA relevance) PE/11/2019/REV/1. https://eur-lex.europa.eu/eli/dir/2019/904/oj. Accessed 16 Jun 2021
29. EU (2020) European Parliament (2020) Commission Implementing Regulation (EU) 2020/2151 of 17 December 2020 laying down rules on harmonised marking specifications on single-use plastic products listed in Part D of the Annex to Directive (EU) 2019/904 of the European Parliament and of the Council on the reduction of the impact of certain plastic products on the environment (Text with EEA relevance) C/2020/9199. https://eur-lex.europa.eu/legalcontent/EN/TXT/P DF/?uri=CELEX:32020R2151&from=EN. Accessed 16 Jun 2021
30. Capolupo M, Sørensen L, Jayasena KDR, Booth AM, Fabbri E (2020) Chemical composition and ecotoxicity of plastic and car tire rubber leachates to aquatic organisms. Water Res 169:115270. https://doi.org/10.1016/j.watres.2019.115270
31. Pinto Costa J da Rocha-santos T, Duarte AC (2020) The environmental impacts of plastics and micro-plastics use, waste and pollution: EU and national measures. Aveiro. https://www.europarl.europa.eu/RegData/etudes/STUD/2020/658279/IPOL_STU(2020)658279_EN.pdf
32. Zorpas AA, Navarro-Pedreño J, Panagiotakis I, Dermatas D (2021) Steps forward to adopt a circular economy strategy by the tourism industry. Waste Manag Res:1–3. https://doi.org/10.1177/0734242X211029087
33. Zorpas AA, Navarro-Pedreño J, Almendro Candel MB (2021) Heavy metals overview in the framework of wastewater, sewage sludge, and soil treatment and removal. The study-case of zeolite. Arab J Geosci 14:1098. https://doi.org/10.1007/s12517-021-07443-2
34. Henry B, Laitala K, Klepp IG (2019) Microfibres from apparel and home textiles: prospects for including microplastics in environmental sustainability assessment. Sci Total Environ 652:483–494. https://doi.org/10.1016/j.scitotenv.2018.10.166
35. ISO 472:2013/AMD 1:2018 (2018) International standards organisation – 472:2013/Amd 1:2018 – plastics — vocabulary — amendment 1: additional items. In: ICS: 01.040.83 rubber and plastics industries (Vocabularies) 83.080.01 Plastics in general. https://www.iso.org/standard/45082.html. Accessed 1 July 2021
36. Lin J, Xu XM, Yue BY, Xu XP, Liu JZ, Zhu Q, Wang JH (2020) Multidecadal records of microplastic accumulation in the coastal sediments of the East China Sea. Chemosphere:128658. https://doi.org/10.1016/j.chemosphere.2020.128658
37. Arthur C, Baker J, Bamford H (2009) Proceedings of the International Research Workshop on Microplastic Marine Debris I OR&R's Marine Debris Program. In: NOAA Technical Memorandum. Sept 9–11, 2008; National Oceanic and Atmospheric Administration: 2009. https://marinedebris.noaa.gov/proceedings-international-research-workshop-microplastic-marine-debris. Accessed 1 July 2021
38. Stamm H (2011) Risk factors: nanomaterials should be defined. Nature 476(7361):399. https://doi.org/10.1038/476399c
39. Hanke G, Galgani F, Werner S, Oosterbaan L, Nilsson P, Fleet D, Kinsey S, Thompson R, Palatinus A, Van Franeker J, Vlachogianni T, Scoullos M, Veiga J, Matiddi M, Alcaro L, Maes T, Korpinen S, Budziak A, Leslie H, Gago J, Liebezeit G (2013) MSDF guidance on monitoring of Marine litter in European Seas . EUR 26113. Publications Office of the European Union, Luxembourg, JRC83985
40. Rauscher H, Roebben G, Boix Sanfeliu A, Emons H, Gibson P, Koeber R, Linsinger T, Rasmussen K, Riego Sintes J, Sokull-Kluettgen B, Stamm H (2015) Towards a review of the EC recommendation for a definition of the term "nanomaterial": part 3: scientific-technical

evaluation of options to clarify the definition and to facilitate its implementation. EUR 27240. Publications Office of the European Union, Luxembourg, JRC95675. https://publications.jrc.ec.europa.eu/repository/handle/JRC95675. Accessed 1 July 2021

41. EFSA (2016) Presence of microplastics and nanoplastics in food, with particular focus on seafood. EFSA J 14(6). https://doi.org/10.2903/j.efsa.2016.4501
42. European Commission (2019) Environmental and Health Risks of Microplastic Pollution. http://europa.eu. Accessed 10 May 2021
43. Hartmann NB, Hu T, Thompson RC, Hassello M, Verschoor A, Daugaard AE, Rist S, Karlsson T, Brennholt N, Cole M, Herrling MP, Hess MC, Ivleva NP, Lusher AL, Wagner M (2019) Are we speaking the same language? Recommendations for a definition and categorization framework for plastic debris. https://doi.org/10.1021/acs.est.8b05297
44. Hu L, Fu J, Wang S, Xiang Y, Pan X (2020) Microplastics generated under simulated fire scenarios: characteristics, antimony leaching, and toxicity. Environ Pollut 269:115905. https://doi.org/10.1016/j.envpol.2020.115905
45. Zhang K, Hamidian AH, Tubic A, Zhang Y, Fang JKH, Wu C, Lam PKS (2021) Understanding plastic degradation and microplastic formation in the environment: a review. Environ Pollut 274:116554. https://doi.org/10.1016/j.envpol.2021.116554
46. UNEP (2018) Exploring the potential for adopting alternative materials to reduce marine plastic litter. https://www.unep.org/resources/report/exploring-potential-adopting-alternative-materials-reduce-marine-plastic-litter. Accessed 12 Jun 2021
47. Akdogan Z, Guven B (2019) Microplastics in the environment: a critical review of current understanding and identification of future research needs. Environ Pollut 254:113011. https://doi.org/10.1016/j.envpol.2019.113011
48. Ballent A, Corcoran PL, Madden O, Helm PA, Longstaffe FJ (2016) Sources and sinks of microplastics in Canadian Lake Ontario nearshore, tributary and beach sediments. Mar Pollut Bull 110(1):383–395. https://doi.org/10.1016/j.marpolbul.2016.06.037
49. Cole M, Lindeque P, Halsband C, Galloway TS (2011) Microplastics as contaminants in the marine environment: a review. Mar Pollut Bull 62:2588–2597. https://doi.org/10.1016/j.marpolbul.2011.09.025
50. Crow Polymer Physics (2021) Polymer properties database. https://polymerdatabase.com/polymer%20physics/Polymer%20Density.html. Accessed 21 Jun 2021
51. Corradini F, Casado F, Leiva V, Huerta-Lwanga E, Geissen V (2021) Microplastics occurrence and frequency in soils under different land uses on a regional scale. Sci Total Environ 752: 141917. https://doi.org/10.1016/j.scitotenv.2020.141917
52. Yu H, Fan P, Hou J, Dang Q, Cui D, Xi B, Tan W (2020) Inhibitory effect of microplastics on soil extracellular enzymatic activities by changing soil properties and direct adsorption: An investigation at the aggregate-fraction level. Environ Pollut 267:115544. https://doi.org/10.1016/j.envpol.2020.115544
53. Šunta U, Prosenc F, Trebše P, Bulc TG, Kralj MB (2020) Adsorption of acetamiprid, chlorantraniliprole and flubendiamide on different type of microplastics present in alluvial soil. Chemosphere 261:127762. https://doi.org/10.1016/j.chemosphere.2020.127762
54. Wu RT, Cai YF, Xing SC, Yang YW, Mi JD, di Liao X (2020) A novel method for extraction of polypropylene microplastics in swine manure. Environ Sci Pollut Res:1–10. https://doi.org/10.1007/s11356-020-11111-5
55. Zhang S, Wang J, Yan P, Hao X, Xu B, Wang W, Aurangzeib M (2020) Non-biodegradable microplastics in soils: a brief review and challenge. J Hazard Mater 409:124525. https://doi.org/10.1016/j.jhazmat.2020.124525
56. Ben-David EA, Habibi M, Haddad E, Hasanin M, Angel DL, Booth AM, Sabbah I (2021) Microplastic distributions in a domestic wastewater treatment plant: removal efficiency, seasonal variation and influence of sampling technique. Sci Total Environ 752:141880. https://doi.org/10.1016/j.scitotenv.2020.141880

57. Stubenrauch J, Ekardt F (2020) Plastic pollution in soils: governance approaches to foster soil health and closed nutrient cycles. Environments 7(5):38. https://doi.org/10.3390/environments7050038
58. Inglezakis JV, Doula KM, Aggelatou V, Zorpas AA (2010) Removal of iron and manganese from underground water by use of natural minerals in batch mode treatment. Desalin Water Treat 18(1–3):341–346. https://doi.org/10.5004/dwt.2010.1102
59. Zorpas AA, Vlyssides GA, Loizidou M (1999) Dewater anaerobically stabilized primary sewage sludge composting. Metal leach ability and uptake by natural Clinoptilolite. Commun Soil Sci Plant Anal 30(11/12):1603–1614. https://doi.org/10.1080/00103629909370312
60. Zorpas AA, Coumi C, Drtil M, Voukkali I (2011) Municipal sewage sludge characteristics and waste water treatment plant effectiveness under warm climate conditions. Desalin Water Treat 36:319–333. https://doi.org/10.5004/DWT.2011.2773
61. Carr SA, Liu J, Tesoro AG (2016) Transport and fate of microplastic particles in wastewater treatment plants. Water Res 91:174–182. https://doi.org/10.1016/j.watres.2016.01.002
62. Ois Clayer F, Jartun M, Buenaventura NT, Guerrero J-L, Lusher A (2021) Bypass of booming inputs of urban and sludge-derived microplastics in a large Nordic Lake. Environ Sci Technol 55:7949–7958. https://doi.org/10.1021/acs.est.0c08443
63. Beriot N, Peek J, Zornoza R, Geissen V, Huerta Lwanga E (2020) Low density-microplastics detected in sheep faeces and soil: a case study from the intensive vegetable farming in Southeast Spain. Sci Total Environ:142653. https://doi.org/10.1016/j.scitotenv.2020.142653
64. Möller JN, Löder MGJ, Laforsch C (2020) Finding microplastics in soils: a review of analytical methods. Environ Sci Technol 54(4):2078–2090. https://doi.org/10.1021/acs.est.9b04618
65. Sait STL, Sørensen L, Kubowicz S, Vike-Jonas K, Gonzalez SV, Asimakopoulos AG, Booth AM (2021) Microplastic fibres from synthetic textiles: environmental degradation and additive chemical content. Environ Pollut 268:115745. https://doi.org/10.1016/j.envpol.2020.115745
66. Environment and Climate Change Canada (2020) Science assessment of plastic pollution. https://www.canada.ca/en/environment-climate-change/services/evaluating-existing-substances/science-assessment-plastic-pollution.html. Accessed 1 July 2021
67. Voukali I, Loizia P, Navarro-Pedreno J, Zorpas AA (2021) Urban strategies evaluation for waste management in coastal areas in the framework of area metabolism. Waste Manag Res 39(3):448–465. https://doi.org/10.1177/0734242X20972773
68. Souza Machado AA, Lau CW, Kloas W, Bergmann J, Bachelier JB, Faltin E, Becker R, Görlich AS, Rillig MC (2019) Microplastics can change soil properties and affect plant performance. Environ Sci Technol 53(10):6044–6052. https://doi.org/10.1021/acs.est.9b01339
69. Nizzetto L, Futter M, Langaas S (2016) Are agricultural soils dumps for microplastics of urban origin? https://doi.org/10.1021/acs.est.6b04140
70. Li X, Chen L, Ji Y, Li M, Dong B, Qian G, Zhou J, Dai X (2020) Effects of chemical pretreatments on microplastic extraction in sewage sludge and their physicochemical characteristics. Water Res 171:115379. https://doi.org/10.1016/j.watres.2019.115379
71. Kim SW, An YJ (2019) Soil microplastics inhibit the movement of springtail species. Environ Int 126:699–706. https://doi.org/10.1016/j.envint.2019.02.067
72. Kim SK, Kim JS, Lee H, Lee HJ (2021) Abundance and characteristics of microplastics in soils with different agricultural practices: importance of sources with internal origin and environmental fate. J Hazard Mater 403:123997. https://doi.org/10.1016/j.jhazmat.2020.123997
73. Lusher AL, Hollman PCH, Mendoza-Hill JJ (2017) Microplastics in fisheries and aquaculture: status of knowledge on their occurrence and implications for aquatic organisms and food safety. FAO Fisheries and Aquaculture Technical Paper. No. 615. Rome, Italy
74. WHO (2019) Microplastics in drinking-water. World Health Organization; 2019. Licence: CC BY-NC-SA 3.0 IGO. Geneva
75. Souza Machado AA, Kloas W, Zarfl C, Hempel S, Rillig MC (2018) Microplastics as an emerging threat to terrestrial ecosystems. Glob Chang Biol 24(4):1405–1416. https://doi.org/10.1111/gcb.14020

76. IAEA (2004) Soil sampling for environmental contaminants. IAEA-TECDOC-1415. Vienna. https://www.iaea.org/publications/7113/soil-sampling-for-environmental-contaminants. Accessed 1 July 2021
77. Yang L, Zhang Y, Kang S, Wang Z, Wu C (2021) Microplastics in soil: a review on methods, occurrence, sources, and potential risk. Sci Total Environ 780:146546. https://doi.org/10.1016/j.scitotenv.2021.146546
78. Scheurer M, Bigalke M (2018) Microplastics in Swiss floodplain soils. Environ Sci Technol 52(6):3591–3598. https://doi.org/10.1021/acs.est.7b06003
79. Zhang GS, Liu YF (2018) The distribution of microplastics in soil aggregate fractions in southwestern China. Sci Total Environ 642:12–20. https://doi.org/10.1016/j.scitotenv.2018.06.004

Environmental Transmission of Human Pathogens Associated with SARS-CoV-2 and the Effect on Soil and Aquatic Ecosystem: A Biological Way of Management

Mohana Devi Subramaniam, B. Abishek Kumar, Ruth Bright Chirayath, Aswathy P. Nair, Mahalaxmi Iyer, and Balachandar Vellingiri

Contents

M. D. Subramaniam (✉), B. Abishek Kumar, R. B. Chirayath, and A. P. Nair
SN ONGC, Department of Genetics and Molecular Biology, Vision Research Foundation, Sankara Nethralaya, Chennai, Tamil Nadu, India

M. Iyer
Livestock Farming and Bioresource Technology, Tamil Nadu, India

B. Vellingiri (✉)
Human Molecular Cytogenetics and Stem Cell Laboratory, Department of Human Genetics and Molecular Biology, Bharathiar University, Coimbatore, Tamil Nadu, India
e-mail: geneticbala@buc.edu.in

Avelino Núñez-Delgado and Manuel Arias-Estévez (eds.),
Emerging Pollutants in Sewage Sludge and Soils,
Hdb Env Chem (2023) 114: 61–88, DOI 10.1007/698_2022_849,

Published online: 16 April 2022

Abstract Severe acute respiratory syndrome coronavirus 2 (SARS-CoV-2) is a highly infectious virus which has caused the Coronavirus disease – 2019 (COVID-19) outbreak in 2019, and, in a very short time, it has infected people all over the world. In fact, WHO declared it as pandemic in March 2020. The adaptation of SARS-CoV-2 was found to be originated from interaction between several hosts, which makes it a potential zoonotic agent. Regarding the survival of virus on the environment, it is important to assess the period of infectivity, which differs among different surfaces. This chapter highlights the viability of virus on different contaminated surfaces, also focusing on metagenomic interactions. To investigate the potential directions for the evolution of this pandemic in the future, SARS-CoV-2 requires a reliable tool to access crucial details regarding the adaptation and evolution of the virus in the environment. Bioinformatic tools play an important role in identifying the similarity and variations in the Open Reading Frames (ORFs) of the viral genomes to elucidate the phytogenic relationships. In this regard, this chapter discusses about recent advances in bioinformatic tools to track the evolution of SARS-CoV-2 in the soil and aquatic ecosystem. New technologies are coming into existence with a new aspect of solving some social problems. Hence, in this work a reliable in silico tool for metagenomic analysis of SARS-CoV-2 variants in the environment has been discussed. Overall, this study tries to shed further light on aspect that could help to understand different types of robotics and techniques used for soil and aquatic system, with the main aim of accessing to relevant details on SARS-CoV-2 in the environment.

Keywords Aquatic ecosystem, Artificial intelligence, COVID-19, Environmental management, Metagenomics, SARS-CoV-2, Soil ecosystem

1 Introduction

Severe acute respiratory syndrome coronavirus 2 (SARS-CoV-2) is the virus that causes COVID-19 (Coronavirus disease 2019), the respiratory illness responsible for the COVID-19 pandemic. SARS-CoV-2 has a major impact on human health globally, infecting a large number of people, causing long-term health effects, and resulting in increased mortality and morbidity. As per the World Health Organization (WHO) guidelines, SARS-CoV-2 can be potent and active on numerous surfaces for several days. The survival of virions on the environment is important to assess the period of infectivity, elucidating for how long the droplets remain infectious [1]. The infectivity of virions varies among different surfaces in the environment, and this chapter highlights the viability of virus on different contaminated surfaces, as well as focus on metagenomic interactions of SARS-CoV-2. The use of disinfectants has prospectively rose, then increasing its presence in freshwater systems through runoff and wastewater discharge. Taking into account that this novel beta-coronavirus can be extremely fatal, causing severe pneumonia, leading to the current biggest ongoing pandemic, in the future, a reliable tool to access the adaptation and evolution of SARS-CoV-2 in the environment should be clearly useful. In addition, it must be noted that computational tools play a vital role in studying phylogenetic relationships, and this is the reason why this chapter discusses recent advances in bioinformatic tools to track the evolution of SARS-CoV-2 in the soil and aquatic ecosystems.

2 General Characteristics of SARS-CoV-2

SARS-CoV-2 is an enveloped virus with ~0.1 μm in diameter. It is a linear, positive-sense, single-stranded RNA virus. The structural proteins that build the viral architecture include outer Envelope protein (E), Spike protein (S), and Membrane proteins (M), which enclose the inner Nucleocapsid protein (N). The S protein is the viral epitope that binds to the host cell receptor, related to the angiotensin converting enzyme 2 (ACE2), for viral attachment. S1 subunit of the spike protein helps in binding to the ACE2 receptor and the S2 subunit facilitates the viral fusion for entry. The E and M proteins are derived from host cell during assembly and stabilize the inner capsid and viral RNA. On the other hand, it also makes the virus susceptible to lipid solvents like alcohol. The N protein is associated with the viral genome [2]. The genome length is ~29 kb, varying from virus to virus due to high genetic heterogenicity [3]. In the genomic region, 14 ORFs are distributed, which code for 27 viral proteins. Among them, 15 are non-structural proteins (NS), 4 are structural proteins, and 8 are accessory proteins [4, 5].The non-structural proteins NS1 to NS10 and NS12 to NS16 are encoded by orf1a and orf1b genes located at 5′ end. The structural proteins of the virus consist of N protein, M protein, E protein, and S protein, which are responsible for the structural integrity and assembly of the virus. The 8 accessory proteins are 3a, 3b, p6, 7a, 7b, 8b, 9b, & orf14, and the structural proteins are oriented near 3′ terminus [4, 5] (Fig. 1). SARS-CoV-2 is a respiratory virus that colonizes nasopharynx and is found in sputum, endotracheal aspirate, bronchoalveolar lavage, and saliva of humans. It is transmitted through respiratory droplets from infected patients during sneezing, coughing, and talking. These infectious droplets land on different surfaces in the surrounding environments, at a distance lower than 2 m, and infection spreads through indirect contact of the healthy individuals with infectious droplet contaminated surfaces [6, 7]. The landed droplets stay infectious as long as the virions are stable in it. The half-lives of the generated aerosols are around 3 h, but the risk of infection is highlighted only in closed areas, as in the open environment the viral loads in aerosols are low to cause potential infection [8]. In hospital environments, the prevalence of the virus in aerosols is common due to performing invasive medical procedures like intubation, tracheostomy, bronchoscopy, manual ventilation, nebulization, and few other aerosol generating procedures, which favor the spread of the infection. Hence, in healthcare environments the risk of airborne transmission of the infection is high,

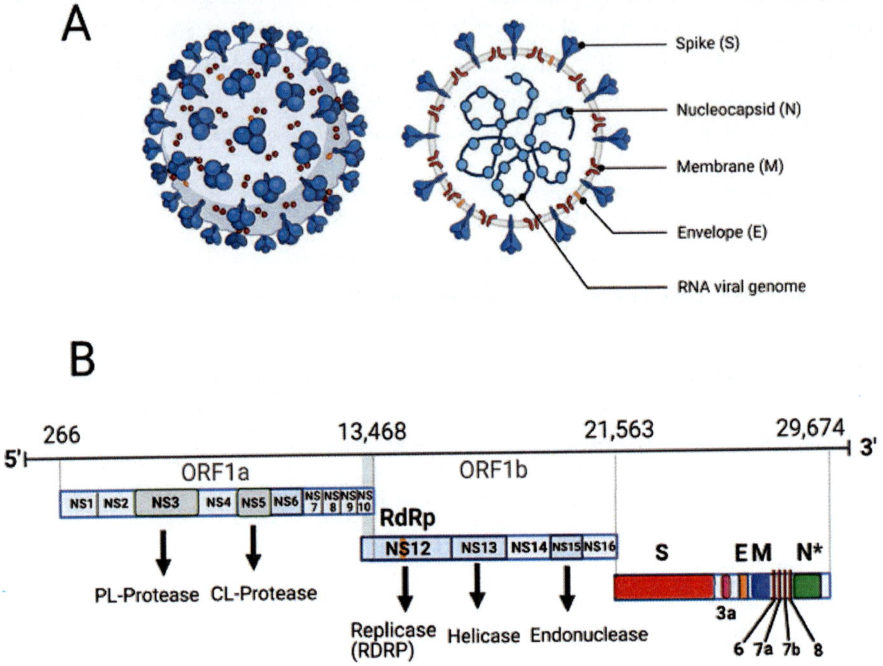

Fig. 1 SARS-CoV-2 structure and the arrangement of ORFs in its genome. (**a**) Structure of a complete SARS-CoV-2. The viral RNA is bound to nucleocapsid protein, and the complex is enclosed by the envelope. The spike and membrane proteins are found embedded in the lipid envelope. (**b**) The arrangement of SARS-CoV-2 genome. The viral genome consists of ORF1a, ORF1b followed by the genes coding for structural proteins and other accessory proteins. ORF1a located between 266 bp to 13,468 bp codes for NS1 to NS10, and the ORF1b codes for NS12 to NS16

and N95 masks should be used by healthcare workers to prevent the spread of the infection through aerosols [7, 9].

3 Chances of Survival of SARS-CoV-2 in Different Environmental Circumstances

The common surface materials we are exposed to, in day-to-day life, include plastics, stainless steel, cotton, copper, and cardboard, and the researchers have reported that this virus can easily survive on these materials when landed on them. The half-life of the virus in stainless steel is 5.6 h and on plastic is 6.8 h, which is found to be longer than that on copper and cardboard [8]. Other than the types of surfaces it lands on, the temperature and humidity also influence the survival of the virus. In stainless steel, plastics, and glass, the infective virions remain viable for

28 days at 20°C under 50% relative humidity. Under the same environmental conditions, the virions were infective for 7 days at 20°C, especially in porous material like cotton. The survivability of virus on different surfaces also changes with the change in temperature and relative humidity. Under elevated temperatures, of 30 and 40°C, the virus survived on steel, glass, and plastics only for 7 and 2 days. In cotton the virus remains viable only for 3 days at 30°C and less than 24 h at 40°C [10–13]. The overall periods of infectivity of virions on different surfaces are 48 h on steel, 72 h on plastic, 24 h on cardboard, and 4 h on copper [14]. Other external factors like sunlight, alcohol-based disinfectants, sodium hypochlorite, and hydrogen peroxide also affect the viability and infectivity of the virus. Hence, the survival rate of the virus on the surfaces reduces under external application of alcohol or 0.5% hydrogen peroxide or 0.1% sodium hypochlorite. Sunlight is found to inactivate SARS-CoV-2, as it disintegrates the structure of the virus [15].

In outdoor environments, the atmospheric pressure and pollution also affect the transmission of the virus. These factors, especially atmospheric pressure, favor the generation of aerosols. At higher atmospheric pressure, studies have shown generation of aerosols and possible transmission of virus in airborne mode. The formation of aerosols is less pronounced in areas with low atmospheric pressure [16]. Air pollution accelerates the spread of SARS-CoV-2 and increases the severity of the infection by affecting the pulmonary immunity. Particulate matter (PM) in the air serves as a carrier and polymerizes the virus particles to spread as airborne in polluted air [17]. Studies have shown that there is a positive correlation between increased content of pollutants like carbon monoxide, PM2.5, and nitrogen dioxide in the air, and the enhanced spreading of SARS-CoV-2 through aerosols [14, 18, 19]. Low humidity and temperature also favor the viability of virus particles in aerosols, and the transmission of virus through airborne route may promote its spread [20]. In summary, the survivability of the virus in the environment is affected by the environmental conditions like the kind of surfaces, temperature, relative humidity, atmospheric pressure, and air pollution. The viability of SARS-CoV-2 in co-relation with the environmental factors is necessary to be determined and considered in order to understand the potential transmission of the virus in different geographical conditions and to estimate demographically the risk of COVID-19.

4 Impacts of SARS-CoV-2 Virus, with Main Focus on Soil and Aquatic Ecosystems

As a preventive measure to control the community transmission of COVID-19, most of the countries had undergone nationwide lockdown. This lockdown measure results in certain positive, as well as in negative, impacts on the atmospheric environment and sustainable ecosystem. Rapid reduction in the atmospheric pollutants, clean beaches, and reduced environmental noise levels are some positive changes that occurred, at least in the short term, as part of COVID-19 effects

[21, 22]. On the other side, increased plastic wastes generation and reduction in waste recycling have damaged the environment during this period [22]. The potential risk to the environment takes special importance in the current situation of the worldwide socio-environmental crisis that is related to the COVID-19 pandemic. The dramatic global changes caused by COVID-19 are having prevalent effects on terrestrial as well as aquatic ecosystems.

Chlorine-based disinfectants have been the best solution for SARS-CoV-2 and its use has been on the rise, leading to adverse effects in aquatic systems. Chlorine toxicity affects water quality and vitality of aquatic ecosystem causing hypoxia, affecting organs and respiratory system and inflammation and steatosis in various living beings. Dissolved organic matter is very high in surface water, which could generate disinfection by-products, such as trihalomethanes or halo-acetic acids, which could exert toxic effects on aquatic organisms. Disinfectants, in combination with nitrogen, can form carcinogens like chloramine or N-nitrosodimenthylamine, and these chemical contaminants may move with water through the water cycle [23]. Sand mining is an emerging threat to freshwater ecosystems, and it has been increased during this pandemic period due to reduced enforcement mechanism [24]. Unemployment and arrival of migrant workers to rural areas have been increasing fishing effort and its negative effect on endangered fishes such as hump-backed mahseer [25]. Reduced monitoring, control and scrutiny of freshwater ecosystems could further increase pollution risks from human activities.

Throughout the pandemic period, massive amounts of medical waste such as face masks, gloves, empty bottles of hand wash and sanitizers, etc. have been produced and dumped unethically, which has led to soil pollution. Research scientists and environmentalists are concerned about sodium hypochlorite usage on polythene sheets and COVID-19 affected dead bodies, before funeral, as these events might have adverse effect on soil as well as its microbial community. Polythene would take very long time to degrade and along with the chemical sodium hypochlorite it might increase the soil pH level. To know more about these adverse impacts on the environment and ecosystem, regular soil sampling for research and analysis must be done. Scientists are intrigued about the future impact on soil and underground water contamination. During monsoon time, the groundwater level is high, and body takes longer time for decomposition. They are demanding scientific analysis and documentation of the grave and nearby soil for future reference to analyze the impact ([26] www.midday.com). The immediate impact of pandemic on soil environment is linked with increased risk of food insecurity and disruption of food supply chain. COVID-19 pandemic has intensified food insecurity in urban areas due to disrupted food supply chain, physical and economic barriers that limit the access to food and terrible increase in food waste as a result of labor shortage. Major impacts of disruption of food supply chain caused by this pandemic have heightened the already existing problems of hunger and malnutrition, especially in developing countries [27]. We need greater understanding on this relation, and we must take positive actions to protect it. Details explaining the impact of SARS-CoV-2 in aquatic and soil systems are presented in Table 1.

Table 1 Studies on impact of COVID-19 in soil and aquatic ecosystems

Sl. No.	Soil/ aquatic ecosystem	Content of study	Country	Concluding remarks	Reference
1	Aquatic	COVID-19 pandemic's impact on freshwater fish biodiversity	Canada	Evaluating the potential of pandemic to amplify or alleviate the threats to fish biodiversity in freshwater	Cooke et al. [28]
2	Aquatic	Improvement in surface water quality in terms of suspended particulate matter (SPM) in the Vembanad Lake	India	It was seen that the pollutant level had decreased tremendously due to shut industries, vehicles and no boating providing good quality of atmosphere and hydrosphere	Yunus et al. [29]
3	Aquatic	Effect of disinfectants in aquatic ecosystems	China	Due to pandemic the use of disinfectant has increased leading to worldwide problems affecting the aquatic ecosystem	Hong Zhang et al. [23]
4	Aquatic	Provides a data-driven global snapshot of expert-perceived impacts of COVID-19 on inland fisheries	USA	The pandemic caused due to COVID-19 has deeply affected on inland fisheries and reached the complexity of crisis mitigation causing small impacts	Stokes et al. [30]
5	Aquatic	Importance of protecting oceans during this pandemic and for future	Ocean	There is a discussion taking place here to understand how urgent it is to take measured for preventing ocean to be affected by COVID-19	Laffoley et al. [31]
6	Aquatic	Signs of rejuvenation and significant improvement in many parameters in Ganga River, following nationwide lockdown due to coronavirus pandemic	India	The improvement seen in water bodies during the lockdown is temporary and quality of water could deteriorate post lockdown due to which discussion is carried out to come up with ideas to rejuvenate the same wastewater related issues	Dutta et al. [32]
7	Aquatic	This improvement of water quality at Haridwar and Rishikesh was ascribed to the	India	The article is focused about river Ganga and other rivers in India. Mainly how they are	Singhal and Matto [33]

(continued)

Table 1 (continued)

Sl. No.	Soil/ aquatic ecosystem	Content of study	Country	Concluding remarks	Reference
		sudden drop of the number of visitors and 500% reduction of sewage and industrial effluents		benefiting due to pandemic lockdown leading to focus on the causes and reinvestigate the solution for these issues	
8	Aquatic	Explains about possible solutions for the sustainability of the aquatic environment and wastewater management	Nigeria	Making of central reservoirs and improving the plumbing system in order to avoid aerosols and leakages could reduce the negative impact on aquatic ecosystem	Iyiola et al. [34]
9	Aquatic	The study shows the positive impact of the COVID-19 pandemic consequences on the coastal environment quality in the study area and indicates the importance of decreasing the industrial discharge on the west coast of Tangier	Morocco	The study shows how SST measurements in collaboration with field sampling could demonstrate the reliability of bathing water and how industrial activity has affected the water quality in Boukhalef river estuary and their neighboring areas	Khalil Cherif et al. [35]
10	Aquatic	Analyzed the abrupt reduction in fishing pressure of the mixed small-scale and industrial fisheries in the Catalan Sea, Spanish Mediterranean, and resulting ecological and economic impacts during the first half of 2020	Spain	Overall, few evidence supports the statement that COVID-19 and its crisis has produced revenue losses in the sector of Mediterranean fisheries	Coll et al. [36]
11	Aquatic	To identify the main aspects of microplastic pollution sources in lakes and rivers, with a focus on freshwater sediments as a site of accumulation	Italy	There are ecotoxicological studies done in this article focusing on the benthic fauna and sediments leading to analysis of microplastics vaguely responsible for the pollution of soils	Bellasi et al. [37]
12	Aquatic	New strategies to detect SARS-CoV-2 in wastewater treatment	India	Authors have stated that there is a huge possibility that the SARS-CoV-2 could easily survive in	Venugopal et al. [38]

(continued)

Table 1 (continued)

Sl. No.	Soil/ aquatic ecosystem	Content of study	Country	Concluding remarks	Reference
				water. Thus, more novel strategies and techniques are required to clear them off the water	
13	Soil	Explaining about the mutual impact of COVID-19 on pollution of soil and air	Egypt	Article deals with many questions that correlate with the pandemic being directly proportional to soil and air	El-Ramady et al. [39]
14	Soil	Food insecurity as the result of COVID-19 and connection between soil	Global	The study ensures that soil is recognized as a medium that links human being with their needs such as food, security, and health	Aday and Aday [40]
15	Soil	The COVID-19 pandemic has disrupted the global food supply chain and exacerbated the problem of food and nutritional insecurity	Global	Soil quality is affected heavily due to COVID-19 outbreak causing huge drift in food consumer market and other ecosystem services	Lal [27]

5 Transmission Route of SARS-CoV-2 into Aquatic Systems

Globally, when studies were going on all possibilities related to the spreading of SARS-CoV-2, one study indicated that the RNA of SARS-CoV-2 had been detected in sewage and wastewater worldwide [41]. Therefore, it can be stated that hospital and household wastes can harbor the virus, but the ecotoxicology of coronavirus plus their effects are to be fully elucidated. Scientists understood the emergency of this situation and studied the effects of SARS-CoV-2 on aquatic systems. On March 17, 2020, researchers found that SARS-CoV-2 was unstable in the atmosphere, while, for wastewater treatment, the use of chlorine could easily oxidize the virus followed by disinfection and filtration method. However, by April 12, 2020, a group of researchers were able to track and detect the SARS-CoV-2 in wastewaters of Netherlands, Sweden, and the USA [42]. As mentioned earlier, the virus has potential of surviving in fecal samples of an individual infected by it and later this theory was hypothesized worldwide as a potential possibility to allow spreading, as it could migrate by surface run off from present host site to the aquatic environment and biota. There are severe repercussions that the aquatic organisms are currently facing and will be facing in higher amount due to infection of SARS-CoV-2 [43]. It has been stated that virus can be equally active in water as on surface, especially for a longer time in water instigating from different hospital facilities.

Some studies demonstrate that 2% to 10% of COVID-19 patients have been associated with diarrhea [44–46], which would automatically lead to contamination of wastewater. These studies are potential evidences encouraging to monitor and track the amount of virus circulation in different communities [47]. SARS-CoV-2 could be capable of infecting drinking water, which still remains a major concern for water departments where chlorination or other effective disinfection procedures are not implemented. Countries that have increased amount of water from supply systems would face difficulties to overcome the current phase of disinfection and filtrations [48]. Recent articles have discussed about the survival and persistence of coronavirus in water bodies, which mainly depends on the type of water (filtered water, tap water, reagent grade water, domestic wastewater, etc.). However, minimal information has been documented regarding this [47]. Following are the potential routes (as a summary) of contamination for aquatic systems:

1. Surface water or groundwater being one of the major sources of drinking water, they may contain the coronavirus, as these sources are getting contaminated. Therefore, strict monitoring is required to avoid discharge of not disinfected wastewater in the surface water bodies.
2. Municipal corporations in general scenario treat water with different chemical and physical processes to remove contamination; however, if these treatments are reduced for some reason, then the potential risk of water to stay contaminated is high.
3. Another possibility could be leaks in pipelines underground and water distribution system having low dynamic pressure.
4. Viruses can easily be passed on from one infected person to many healthy people via feces, vomit, and urine of both symptomatic and asymptomatic patients; therefore, personal hygiene and sanitation practices should be given proper importance.
5. Reuse of gray water can be one huge source of transmission and risk related to the domestic surfaces like plastic toilet seats, ceramic sinks, etc.
6. Reuse of rainwater systems, where the transmission risk could be focused because of the virus persistence in aerosols.
7. Release of virus could also take place due to overflow of sewage and usage of septic tanks in individual buildings, causing risks of purge service operators.

In one study it has been explained that the tadpoles exposed to synthetic SARS-CoV-2 spike peptide proteins showed to have high expression of neurotoxicity biomarkers and oxidative stress [49]. This clearly shows that tadpoles are strongly affected by SARS-CoV-2 protein fragments, but there are more areas to be explored in order to understand this aspect in detail. However, this kind of research can give a lot of insights in understanding the mechanism of SARS-CoV-2 transmission in our ecosystem and environment [50]. Natural as well as surface water systems can act as major pathways for transmission of SARS-CoV-2 in the environment, leading as a threat to not only human lives but also the overall aquatic system. There is high requirement to further explore and estimate the potential receptors in order to understand the interspecies transmission (human as well as biota in the aquatic system) [51].

6 Transmission Route of SARS-CoV-2 into the Soil System

According to current evidence, SARS-CoV-2 is transferred from one person to another person by the proposed two routes: direct and/or indirect contact [20]. Contaminated human body fluids such as saliva, mucus, feces, vomits, urine, and blood come under direct way of transmission. For SARS-CoV-2 and other different pathogenic microorganisms, the fecal-oral transmission route is a possibility that has to be considered [52, 53]. In addition, for microorganism suffering frequent mutations, new characteristics in transmission, and the potential to be infective and cause diseases through new routes must also be considered [54–56]. If this can be relevant for wastewater, it could be also for sewage sludge, and then for soils receiving the spreading of both materials, as well as for plants growing on these soils, and even for surface and groundwater in the area, which could be contaminated by means of runoff or leaching. Perceptibly, the risks of biotic pollution would be higher in areas where wastewater and sewage sludge treatments (including disinfection) are scarce or, simply, do not take place. Further, the generation of aerosols containing SARS-CoV-2 (or other pathogenic microbes) in any of the locations where wastewater and sludge are spread is another concern [47]. Similarly, the virus can also spread over the public areas via inefficient disposal management of the wastes generated from the hospitals, houses, municipalities, and quarantine centers, and that are contaminated with SARS-CoV-2 [57]. The transmission of the coronavirus may be increased by solid waste management through poor handling conditions with poor solid waste management, and in addition when solid wastes are dumped in poorly managed dumpsites, posing serious risk of infection by SARS-CoV-2 into the soil system. In view of that, studies on soil biodiversity, in relation to SARS-CoV-2 (and/or other pathogenic microorganisms) are necessary and may serve as a future direction of research.

7 Interaction Between Soil-SARS-CoV-2 and Water-SARS-CoV-2: A Metagenomic Overview

The viral load in the environment increases in response to increased cases in the community and may pose risks of infection to more susceptible individuals. Hence, the environmental surveillance for SARS-CoV-2 by means of a metagenomic approach is important to understand crucial details affecting the presence of virus in the environment [58]. Here we focus on the metagenomic interactions and methods to detect SARS-CoV-2 in soil and water. The viral load in these environments may not reach the required infectious dose of SARS-CoV-2 to establish infection. But it is important to study the metagenomic interaction of SARS-CoV-2 in different environments to understand the dynamic of genomic versatility and mutations that favors the emergence of potential variants in face of infection. The interaction of viral elements with the abiotic factors in the environment is necessary to be elucidated to understand the distribution of SARS-CoV-2 in aquatic ecosystems and its metagenomic interactions. As the viral concentrations in wastewater and aquatic systems are low, significant pre-processing of the sample with appropriate viral concentration technique is required for successful metagenomic detection [59, 60]. Electronegative membranes and ultra-filtrations are the viral concentration methods used for SARS-CoV-2 [61]. After viral concentration, the extraction of viral nucleic acid should be performed using suitable methods such as RNA extraction, further subjected to sequencing using Illumina MiSeq platform [62].

It is also believed that SARS-CoV-2 can be present in soil and the possibility has been discussed previously [55, 56]. The source for soil existence of SARS-CoV-2 is thought to come from unprocessed sewage sludge, as happens for other viruses [63], open sanitary use, direct encounter from infected individual in open places and from improper disposal of infectious healthcare wastes. These are the main possible ways of SARS-CoV-2 for distribution in soil [64]. Although the virus stability is much lesser in the soil, its presence in soil should be studied through metagenomic sequencing. The metagenomic interaction of SARS-CoV-2 with soil in different geographical area is necessary to be determined to predict the potential zoonotic adaptation between different species that the virus poses in the environment. The survivability of the SARS-CoV-2 in soil is quite challenging and it is affected by primary factors like temperature, exposure to sunlight, and pH of the soil [55, 56]. Other counter factors like ionic and organic content of the soil, moisture, and microbiome may also play a role in the viral survivability. The distribution of SARS-CoV-2, potentially affected by soil type and the species diversity of the surrounding area is important in order to understand the adaptation of SARS-CoV-2 and its outbreak prediction.

Determination of interaction of SARS-CoV-2 in soil requires a standardized procedure of metagenomic approach. We discuss on a suitable protocol for future metagenomic studies in SARS-CoV-2, from pre-existing isolation methods and metagenomic sequencing of soil viruses. The first step for virome isolation from soil involves assortment of solid particles. Centrifugation can help to settle the sand particles in the soil. Centrifugation at 500 $\times$ *g* for 5 min can pellet out large soil particles [65]. Then rinsing the suspension using mild detergent followed by vertexing will separate the viral particles. The further filtration of the suspension using 5 μm membrane filter will remove the solid particles [66]. The filtrate is then prepared for Fluorescent Activated Cell Sorting by staining with SYBER Green [67]. After viral sorting by FACS, the pools should undergo RNA extraction from the virus and whole genome amplification. Viral RNA can be extracted using MoBioPower Soil RNA kit. Multiple displacement amplification can be followed by whole genome amplification. The prepared metagenomic library can be

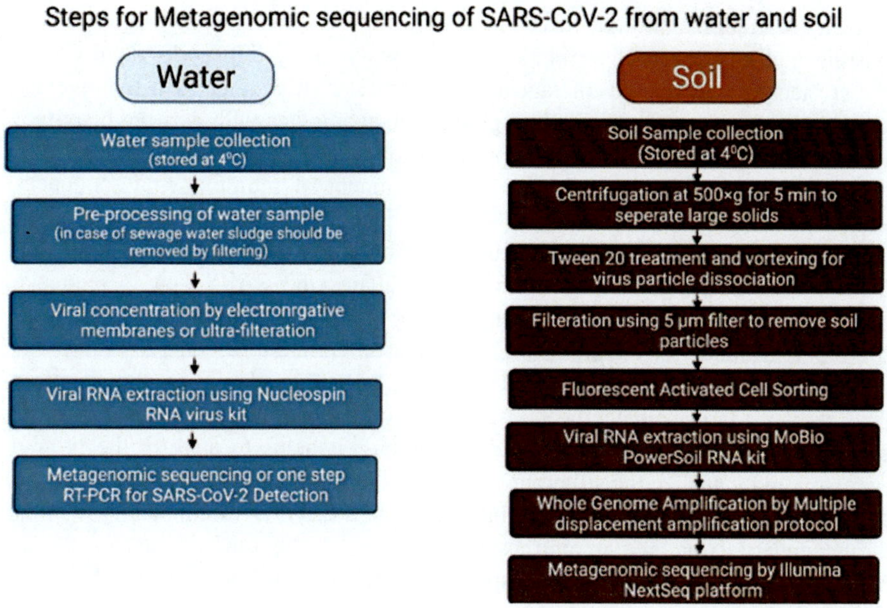

Fig. 2 Steps involved in the detection of SARS-CoV-2 by metagenomic sequencing from soil and water. Protocol for metagenomic detection of SARS-CoV-2 from soil and water. The sample storage, pre-processing, and viral concentration techniques to be adapted for successful metagenomic sequencing from soil and aquatic system are signified

sequenced using Illumina NextSeq platform and the sequences are assembled using suitable metagenomic assembler (Fig. 2).

7.1 Bioinformatic Tools in Recovery of SARS-CoV-2 from Soil and Aquatic Systems

To recover SARS-CoV-2 from soil and water, other than the viral concentration method in silico tools for designing metabarcoding primers and metagenomics analysis of Next Generation Sequencing (NSG) is necessary.

The following tools are involved in the detection and annotations of variants. PriSeT is a tool for metabarcoding SARS-CoV-2 which helps in designing RT-PCR primers for recovery of SARS-CoV-2 and its subspecies [68]. CoVPipe generates consensus sequences for NGS of SARS-CoV-2 for reconstruction of sequence variants. This pipeline provides sequence lineage of SARS-CoV-2 in the environment to recover the native and the variants that exist in soil and water [69]. PoreCov enables rapid processing of SARS-CoV-2 raw sequence data for nanopore sequencing with reduced bottleneck trafficking. It combines all the tools necessary for routine SARS-CoV-2 sequencing workflow and enables long read data for precision of reconstructing variants [70]. Haploflow tool assembles the SARS-CoV-2 sequence variations through novel flow algorithms, which is pre-feed from the existing multiple sequence variants in evolutionary lineage. It works rapidly with higher accuracy than the metagenomic assembler with increased complexity [71]. V-Pipe interprets the high-throughput sequencing data with existing heterogeneous virus populations and helps to probe for metagenomic sequencing [72]. VIRULIGN enables codon specific alignment of multiple viral sequences and annotations of SARS-CoV-2 genome. The functions of protein coding sequences of the closely related viruses can be accessed by exploring this tool [73]. VADR (Viral Annotation DefineR) validates the SARS-CoV-2 sequence homology and annotation of the sequence variants. It enables both NCBI RefSeq and input sequence for the annotation of sequence variation. It helps to predict the proteins encoded by the variants SARS-CoV-2 clades [74]. Pfam is another database that enables to investigate the consensus secondary structures of SARS-CoV-2 proteome in relation to the nucleotide variations. It helps to analyze the protein sequence and visualize the structure variants that originate from its native form and classify protein sequences into families [75]. Rfam COVID-19 enables to identify the secondary structures in untranslated regions of the mRNA and functional annotations of non-coding RNAs, especially miRNA [76]. For recovery of SARS-CoV-2 from soil and water through metagenomic sequencing, the computational tools shown in Fig. 3 can augment in the experiment designing or post analysis of viral sequences for complete recovery of variants. Numerous tools are available with a minor variation in the objectives, hence the in silico tools should be carefully selected to design a suitable experiment for recovery of SARS-CoV-2.

7.2 Robotics in Recovery of Soil and Aquatic Systems

Soil system condition can be tried to improve via various techniques. There are some ongoing systems using different forms of machine vision, artificial intelligence, and machine learning leading towards applied agriculture and robotic farming. There is a lot of research going on in the field of robotic management of plants changing landscaping of agriculture and soil system. Artificial intelligence (AI) is well studied to this point where one can be provided with opportunity to witness revolution in agriculture. Agricultural aspects of weeding, irrigation, spraying, and many more related programs are well embedded in drones and robots in order to carry out the work [77]. This AI and technology could help us save the unwanted and excess use of pesticides, herbicides, and water making it easy to maintain the fertility of the soil. Development of potential application of AI and robots carrying the programmed ability to detect infection and identify the disease was studied previously [78]. This development could give a different face to the disease management routine. Online tools like cloud-based solutions and internet of things are responsible for providing exceptional opportunity for the development of such systems that can benefit not only in disease management aspects but also in forestry and urban farming

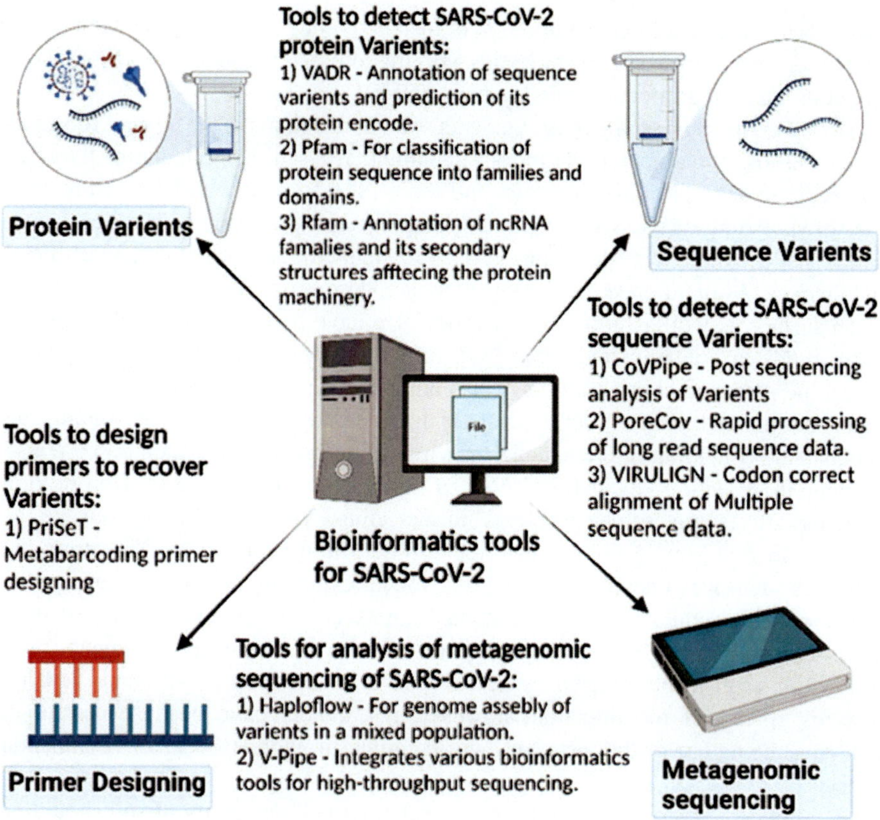

Fig. 3 Bioinformatic tools for pre- or post-analysis for recovery of SARS-CoV-2 from soil and water. Bioinformatic tools to identify the protein or sequence variants of SARS-CoV-2 and the online databases for designing metabarcoding or metagenomic experimentations to aid in the recovery of variants from soil and water

[79]. Researchers from various countries have already worked on this issue and are continuing to do so.

There has been development of multi robot systems, which are also termed as the autonomous aquatic vehicles. These robots are made to clean the aquatic system and water bodies in order to reduce the amount of contamination, without compromising the public health [80]. The initial sanctioned robot was for the removal and collection purpose for waste; but it was further developed on demand to clean the groundwater as well as surface water. There were different types of navigation algorithms which were considered in order to stimulate the de-contamination. Later, as the demand for the improved version of these robots increased, the precision for the making increased [81]. Different models and updates were produced in the market where different types of sensors, like trinocular sensors, acoustic sensors, were installed along with camera and environment specific modeling, where 3D model of specific object can be studied [82]. Various imaging techniques and vision-based localization techniques were installed along with mapping techniques. Moreover, many researchers are still working on making it better in terms of detecting virus and infection causing bacteria [60].

However, there are some improvements still needed to make the aquatic robots as efficient as the dynamic outdoor robots, including GPS settings, calculating the earth magnetic field, small-scale mapping, and more [83]. Research has shown possibilities of spreading of the SARS-CoV-2 via aquatic and soil systems. To avoid such situations, the aquatic system can be decontaminated using robotics and artificial intelligence. As mentioned previously, there are various developments of robots that clean surface as well as groundwater. They also have been programmed to release clean fluids in contaminated areas [83]. Similarly, this technique can be used to detect the presence of coronavirus or its components and release the disinfectant in that contaminated area in order to decrease the spread of contamination. Not only are robots and AI helping in eradicating the virus form the aquatic system but are also helping in detection of SARS-CoV-2 in sewage water [4, 5, 84]. Non-robotic systems are facing some problems in detecting SARS-CoV-2 in sewage as the viral load is low, leading to unsuccessful detection of the virus. Therefore, the robotic system was encouraged to be developed, expecting success because of the quick and different approach. Infectious agents can easily exist in our natural environment, especially cellular organisms/microorganisms. Soil and aquatic systems both seem to be affected by SARS-CoV-2 in one way or the other, leading towards spread of the infection; however, researchers have been successful in studying the nature of the virus and will continue to understand its environmental aspects.

In addition, it is important to bear in mind that the biodiversity of soil should be maintained, as that could aid in diminishing the propagation of bacteria that are antibiotic resistant, and similarly water bodies should be looked upon for its maintenance as they are the primary source of water.

7.3 Disinfection Methods and Techniques Used for Wastewater Treatment

It has been hypothesized that the current outbreak of coronavirus could be initiated in propagation due to saliva or feces of bats; so, infection through contaminated surface water or wastewater, though improbable, could be treated as a potential health hazard and must never be ruled out. As per the Centers for Disease Control and Prevention (CDC), the wastewater and standard water treatment along with its process of disinfectant are thought to be effective against SARS-CoV-2 [85]. Reusing water has been practiced for decades, where particularly gray water reduces the stress of water as it is a relatively clean form of wastewater from various sources like households (except toilets). The reusing of such wastewater could potentially tackle water scarcity and could provide (after appropriate treatment) clean and drinking water for the domestic use.

Below, we will further discuss appropriate techniques and procedures for disinfection.

7.3.1 Chlorine Gas

Chlorine gas is popularly used in water as a primary as well as secondary disinfectant. It is a greenish yellow gas which turns into liquid on high pressure. This technique is well known for the removal of most of the pathogenic microorganisms [86].

7.3.2 Chlorination

It is a technique used for the removal of enteric indicator organisms and pathogens. For about 70 years, chlorination has been the major method used for disinfection of drinking water. There are two main alternatives for using this technique, which are liquid chlorination using sodium hypochlorite solution, and solid chlorination using calcium hypochlorite. In liquid chlorination the bleaching agent is a sodium hypochlorite solution, which contains chlorine around 10–15%, but drops the force. On the other hand, the solid calcium hypochlorite can be feasibly used to replace the other chlorination methods. Surprisingly, this method is much safer in terms of handling and about 70% of chlorine is commercially available, which is used for both drinking and wastewater treatment [87].

7.3.3 Chloramines

These are designed by reacting free chlorine with ammonia. To quite an extent chloramine plays crucial role in protecting the system of distribution residually [88].

7.3.4 Ozone

In terms of oxidation potential, ozone is clearly higher than the hypochlorite ion or chlorine. It is only fluoride and hydroxyl radical that can surpass ozone in its oxidation capacity. Ozone is very often used as primary disinfectant in Canada and Europe. Ozone is a biocide that is highly efficient and has target to break the double bonds of fatty acids present in lipid envelop in viruses and in bacterial cell membrane [89].

7.3.5 Ultraviolet Light

Commonly known as UV light, which here is used as the treatment for drinking water, aquaculture, and wastewater. One of the major characteristics of UV light is that it can very easily change the biological components of any exposed microorganism by disrupting their RNA, protein, and DNA. The major advantage for using UV is that it is very easy to be maintained and extremely cost effective [90].

7.3.6 Copper Ionization

It is an uncommon method to reduce the amount of bacterial contamination in wastewater or drinking water. Apart from disinfecting, this technique also reduces scaling in holding tanks and pipes. There should be more attention drawn to this technique in combination with the ozone treatment, filtration, and UV. Agricultural and cooling towers wash could be a very good way to use this technique, as it is a very viable disinfection alternative. All that is required is low levels of copper to hold on to fungus, bacteria, and algae. Overall, for disinfecting bacteria and inactivating viruses from wastewater, it is best to go through chlorine, chlorine dioxide, ozone, and UV radiation. Whereas, for conventional disinfection against general microbial groups either of these methods would be used: chorine, chlorine dioxide, ozone, peracetic acid, UV radiation. Human handling as well as certain practices on the environment can cause repercussions leading to contamination of water, which may be tangible or not. In conclusion, it will be advantageous to use these techniques for problem areas and deal with wastewater management (Fig. 4).

8 Recommendations

COVID-19 is hitting hard, therefore, every individual is accountable for their own well-being. Researchers and clinicians have come up with certain recommendations keeping the current pandemic in mind, which will also be discussed further. Since the very beginning of the pandemic, several people experienced recurrent infection of COVID-19. The global war against the virus is still going on leading to understand the situation in an efficient manner. Considering the ongoing situation, many reinfections have been reported, making it recommendable to keep social distancing and safe spacing gear on. It is highly important to keep maintaining the culture of physical distancing and wearing masks, washing hands frequently, avoid crowding in closed setting and provide good ventilation, especially in closed spaces, while researchers continue to fight against the latest mutant and variant. Governments should be taking appropriate measures in order to provide vaccines as there has been a widespread of SARS-CoV-2, being exposed with increasing risk of new variant [91]. In view of the current situation, all the future interventions are going to be

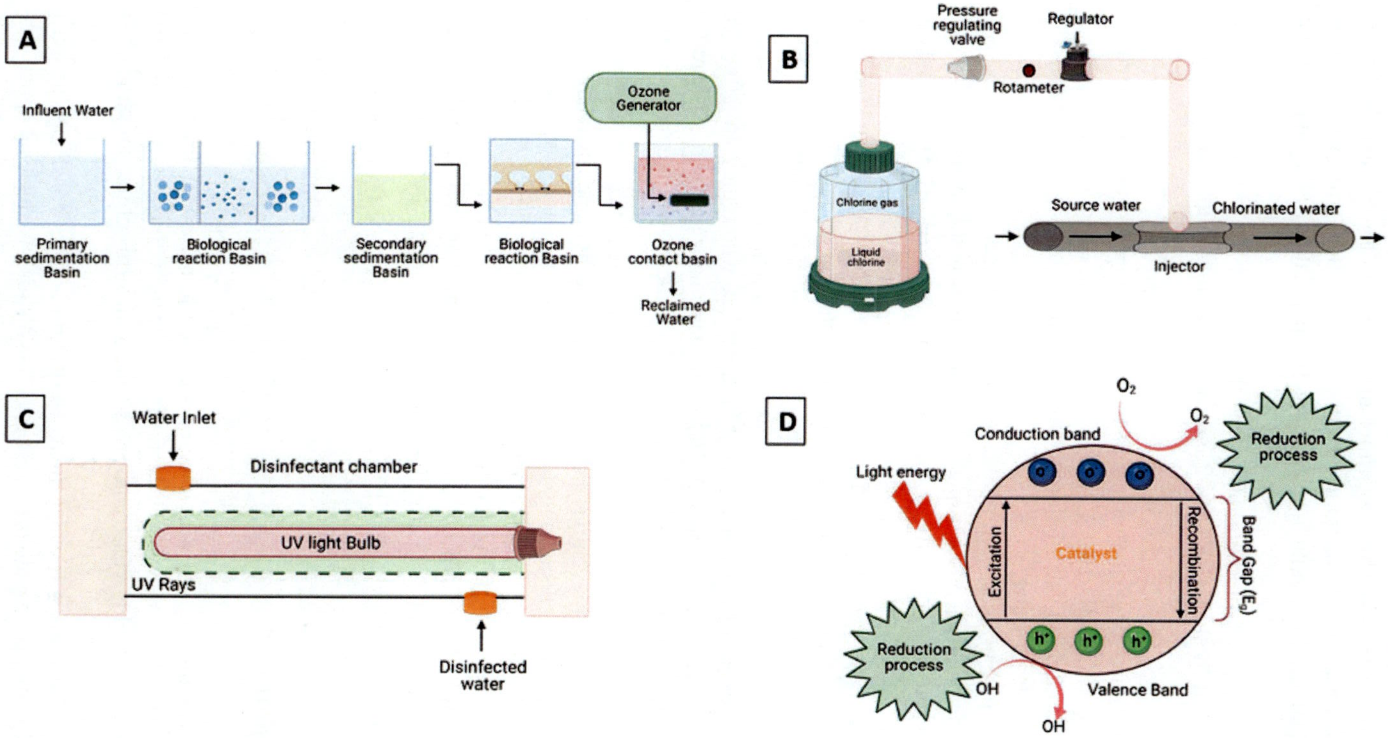

Fig. 4 Disinfection methods and techniques used for wastewater treatment. In the given figure, (**a**) represents disinfection using the ozone technique. Figure (**b**) explains how chlorination can help in disinfecting water. Figure (**c**) represents how UV rays can be very efficient in disinfecting water. Figure (**d**) shows the photocatalytic disinfection technique, where various catalytic reactions take place to clean infected water

directed towards maintaining safe social distance, and there will be a culture of improved hygienic practices. These measures upon being followed have the potential to delay the onset of wide community transmission. It is critical to consider acceptance for all the upcoming measures by the population, allowing that are followed.

Next are indicated the precautionary steps that could be taken up in order to reduce crisis caused by theSARS-CoV-2 infection spreading in the aquatic ecosystem [34]:

1. Using good practice for clean sanitation and clean water supply in order to reduce the occurrence of COVID-19.
2. Treatment of household water before the usage, as household treatments can also remove viruses from water.
3. Cleaning of toilet surfaces, both public and personal toilets.
4. Use of biosurfactants in soil and aquatic systems will be highly beneficial, as well as cost efficient in the recovery of effects due to SARS-CoV-2 in these systems [92].
5. Regularize the hand washing practice using detergent or soap of any kind, as it will reduce the risk of infection and its spread to quite an extent.
6. Designing of technologies that will help in ozonation, UV radiation, and membrane filtration in plants as treatment for water to help inactivate coronavirus.
7. Proper management of tap water should be taken up, where they are protected against the spread of coronavirus. WHO published major guidelines where the systems for elimination of virus from water and aquatic system were provided.

It is suggested to mobilize all the communities and various sectors, so that governments can impose preventive measures to increase hygiene, maintain respiratory etiquettes and social distancing. It is asked to reduce the mortality by adapting healthy culture of providing extensive care for patients affected by COVID-19 and by providing appropriate care for the frontline workers and vulnerable population [93]. Government and public healthcare workers are striving hard to provide and develop safe environment and effective therapeutics along with production of vaccination with the minimal side effects [94].

9 Conclusion

Every country is implementing wide-ranging measures to reduce the transmission of COVID-19 in every possible way. The detection of the infection must be considered, as that area would be highly vulnerable and sensitive in terms of transmission of the infection. Every country has strict quarantine rules and travel restriction for asymptomatic as well as positive tested patients [95]. Around the world, population is going through an unprecedented menace; but in this distress, we have an opportunity to emerge with better and improvised health system connecting various countries, so that we all are ready for the coming health related threat. One must learn all the lessons this pandemic has taught us, and continuing to do so, making the future a safer place. Furthermore, environmental pollution can be considered the most challenging issue facing the world today, along with COVID-19 pandemic. These pollution effects have closely linked to human activities and health. According to the world water council organization, our generation has generated the highest amount of wastewater than ever in the history of the planet. The ratio to access safe drinking water is more than 1:6 people. COVID-19 pandemic had increased pollution impacts in freshwater systems as well as in the soil system. Increased use of disinfectants has prospectively increased its presence in freshwater through runoff and wastewater discharge. COVID-19 impacted all aspects of human lives, including public health, socioeconomic issues, and environment. The tragedy of COVID-19 has once again reminded the need for protecting, healing, restoring, and support our planet.

Acknowledgements The author Dr. MDS would like to thank Vision Research Foundation, for providing the necessary infrastructure facility for carrying out this review process of the manuscript. The author Dr. VB would like to thank Bharathiar University for providing the necessary infrastructure facility for carrying out this review process of the manuscript.

References

1. Iyer M, Tiwari S, Renu K, Pasha MY, Pandit S, Singh B, Raj N, Krothapalli S, Kwak HJ, Balasubramanian V, Jang SB, Dileep Kumar G, Uttpal A, Narayanasamy A, Kinoshita M, Subramaniam MD, Nachimuthu SK, Roy A, Valsala Gopalakrishnan A, Ramakrishnan P, Vellingiri B (2021) Environmental survival of SARS-CoV-2 – a solid waste perspective. Environ Res 197:111015. https://doi.org/10.1016/j.envres.2021.111015
2. Vellingiri B, Jayaramayya K, Iyer M, Narayanasamy A, Govindasamy V, Giridharan B, Ganesan S, Venugopal A, Venkatesan D, Ganesan H, Rajagopalan K, Rahman PKSM, Cho SG, Kumar NS, Subramaniam MD (2020) COVID-19: a promising cure for the global panic. Sci Total Environ 725:138277. https://doi.org/10.1016/j.scitotenv.2020.138277
3. Rastogi M, Pandey N, Shukla A, Singh SK (2020) SARS coronavirus 2: from genome to infectome. Respir Res 21:318. https://doi.org/10.1186/s12931-020-01581-z
4. Wu A, Peng Y, Huang B, Ding X, Wang X, Niu P, Meng J, Zhu Z, Zhang Z, Wang J, Sheng J, Quan L, Xia Z, Ten W, Cheng G, Jiang T (2020) Genome composition and divergence of the novel coronavirus (2019-nCoV) originating in China. Cell Host Microbe 27:325–328. https://doi.org/10.1016/j.chom.2020.02.001
5. Wu F, Xiao A, Zhang J, Moniz K, Endo N, Armas F, Bonneau R, Brown MA, Bushman M, Chai PR, Duvallet C, Erickson TB, Foppe K, Ghaeli N, Gu X, Hanage WP, Huang KH, Lee WL, Matus M, McElroy KA, Nagler J, Rhode SF, Santillana M, Tucker JA, Wuertz S, Zhao S, Thompson J, Alm EJ (2020) SARS-CoV-2 titers in wastewater are higher than expected from clinically confirmed cases. mSystems 5:e00614–e00620. https://doi.org/10.1128/mSystems.00614-20
6. Liu J, Liao X, Qian S, Yuan J, Wang F, Liu Y, Wang Z, Wang FS, Liu L, Zhang Z (2020) Community transmission of severe acute respiratory syndrome coronavirus 2, Shenzhen, China. Emerg Infect Dis 26(6):1320–1323. https://doi.org/10.3201/eid2606.200239
7. World Health Organization (2020) Modes of transmission of virus causing COVID-19: implications for IPC precaution recommendations. https://www.who.int/news-room/commentaries/detail/modes-of-transmission-of-virus-causing-COVID-19-implications-for-ipc-precaution-recommendations

8. Van- Doremalen N, Bushmaker T, Morris DH, Holbrook MG, Gamble A, Williamson BN, Tamin A, Harcourt JL, Thornburg NJ, Gerber SI, Lloyd-Smith JO, de Wit E, Munster VJ (2020) Aerosol and surface stability of SARS-CoV-2 as compared with SARS-CoV-1. N Engl J Med 382(16):1564–1567. https://doi.org/10.1056/NEJMc2004973
9. Alhazzani W, Møller MH, Arabi YM, Loeb M, Gong MN, Fan E, Oczkowski S, Levy MM, Derde L, Dzierba A, Du B, Aboodi M, Wunsch H, Cecconi M, Koh Y, Chertow DS, Maitland K, Alshamsi F, Belley-Cote E, Greco M, Laundy M, Morgan JS, Kesecioglu J, McGeer A, Mermel L, Mammen MJ, Alexander PE, Arrington A, Centofanti JE, Citerio G, Baw B, Memish ZA, Hammond N, Hayden FG, Evans L, Rhodes A (2020) Surviving sepsis campaign: guidelines on the management of critically ill adults with coronavirus disease 2019 (COVID-19). Intensive Care Med 46(5):854–887. https://doi.org/10.1007/s00134-020-06022-5
10. Carraturo F, Del Giudice C, Morelli M, Cerullo V, Libralato G, Galdiero E, Guida M (2020) Persistence of SARS-CoV-2 in the environment and COVID-19 transmission risk from environmental matrices and surfaces. Environ Pollut 265:115010. https://doi.org/10.1016/j.envpol.2020.115010
11. Kumar S, Singh R, Kumari N, Karmakar S, Behera M, Siddiqui AJ, Rajput VD, Minkina T, Bauddh K, Kumar N (2021) Current understanding of the influence of environmental factors on SARS-CoV-2 transmission, persistence, and infectivity. Environ Sci Pollut Res 28(6): 6267–6288. https://doi.org/10.1007/s11356-020-12165-1
12. Ren SY, Wang WB, Hao YG, Zhang HR, Wang ZC, Chen YL, Gao RD (2020) Stability and infectivity of coronaviruses in inanimate environments. World J Clin Cases 8(8):1391–1399. https://doi.org/10.12998/wjcc.v8.i8.1391
13. Riddell S, Goldie S, Hill A, Eagles D, Drew TW (2020) The effect of temperature on persistence of SARS-CoV-2 on common surfaces. J Virol 17:145. https://doi.org/10.1186/s12985-020-01418-7
14. Marquès M, Domingo JL (2021) Contamination of inert surfaces by SARS-CoV-2: persistence, stability and infectivity. A review. Environ Res 193:110559.10.1016/j.envres.2020.110559
15. Ratnesar-Shumate S, Williams G, Green B, Krause M, Holland B, Wood S, Bohannon J, Boydston J, Freeburger D, Hooper I, Beck K, Yeager J, Altamura AL, Biryukov J, Yolitz J, Schuit M, Wahl V, Hevey M, Dabisch P (2020) Simulated sunlight rapidly inactivates SARS-CoV-2 on surfaces. J Infect Dis 222:214–222. https://doi.org/10.1093/infdis/jiaa274
16. Lolli S, Chen YC, Wang SH, Vivone G (2020) Impact of meteorological conditions and air pollution on COVID-19 pandemic transmission in Italy. Sci Rep 10:16213. https://doi.org/10.1038/s41598-020-73197-8
17. Comunian S, Dongo D, Milani C, Palestini P (2020) Air pollution and COVID-19: the role of particulate matter in the spread and increase of COVID-19's morbidity and mortality. Int J Environ Res Public Health 17(12):4487. https://doi.org/10.3390/ijerph17124487
18. Copat C, Cristaldi A, Fiore M, Grasso A, Zuccarello P, Signorelli SS, Conti GO, Ferrante M (2020) The role of air pollution (PM and NO_2) in COVID-19 spread and lethality: a systematic review. Environ Res 191:110129. https://doi.org/10.1016/j.envres.2020.110129
19. Yao Y, Pan J, Wang W, Liu Z, Kan H, Qiu Y, Meng X, Wang W (2020) Association of particulate matter pollution and case fatality rate of COVID-19 in 49 Chinese cities. Sci Total Environ 741:140396. https://doi.org/10.1016/j.scitotenv.2020.140396
20. Wiktorczyk-Kapischke N, Grudlewska-Buda K, Wałecka-Zacharska E, Kwiecińska-Piróg J, Radtke L, Gospodarek-Komkowska E, Skowron K (2021) SARS-CoV-2 in the environment-non-droplet spreading routes. Sci Total Environ 770:145260. https://doi.org/10.1016/j.scitotenv.2021.145260
21. Chen K, Wang M, Huang C, Kinney PL, Anastas PT (2020) Air pollution reduction and mortality benefit during the COVID-19 outbreak in China. Lancet Planet Health 4(6):e210–e212. https://doi.org/10.1016/s2542-5196(20)30107-8
22. Zambrano-Monserrate MA, Ruano MA, Sanchez-Alcade L (2020) Indirect effects of COVID-19 on the environment. Sci Total Environ 728:138813. https://doi.org/10.1016/j.scitotenv.2020

23. Zhang H, Tang W, Chen Y, Yin W (2020) Disinfection threatens aquatic ecosystems. Science 368(6487):146–147. https://doi.org/10.1126/science.abb8905
24. Kannan S (2020) Tamil Nadu: sand smugglers have a field day as police battle. https://timesofindia.indiatimes.com/city/chennai/sand-smugglers-have-a-field-day-as-police-battle-COVID/articleshow/76479627.cms
25. Pinder AC, Raghavan R, Britton JR, Cooke SJ (2020) COVID-19 and biodiversity: the paradox of cleaner rivers and elevated extinction risk to iconic fish species. Aquat Conserv Mar Freshw Ecosyst 30:1061–1062. https://doi.org/10.1002/aqc.3416
26. Mumbai News (2020). https://www.mid-day.com/mumbai/mumbai-news/article/Experts-raise-alarm-about-soil-and-water-contamination%2D%2DCOVID-19-22991245
27. Lal R (2020) Home gardening and urban agriculture for advancing food and nutritional security in response to the COVID-19 pandemic. Food Secur:1–6. https://doi.org/10.1007/s12571-020-01058-3
28. Cooke SJ, Twardek WM, Lynch AJ, Cowx IG, Olden JD, Funge-Smith S, Lorenzen K, Arlinghaus R, Chen Y, Weyl OLF, Nyboer EA, Pompeu PS, Carlson SM, Koehn JD, Pinder AC, Raghavan R, Phang S, Koning AA, Taylor WW, Bartley D, Robert Britton J (2021) A global perspective on the influence of the COVID-19 pandemic on freshwater fish biodiversity. Biol Conserv 253:108932. https://doi.org/10.1016/j.biocon.2020.108932
29. Yunus AP, Masago Y, Hijioka Y (2020) COVID-19 and surface water quality: improved lake water quality during the lockdown. Sci Total Environ 731:139012. https://doi.org/10.1016/j.scitotenv.2020.139012
30. Stokes GL, Lynch AJ, Lowe BS, Funge-Smith S, Valbo-Jørgensen J, Smidt SJ (2020) COVID-19 pandemic impacts on global inland fisheries. Proc Natl Acad Sci 117(47):29419–29421. https://doi.org/10.1073/pnas.2014016117
31. Laffoley D, Baxter J, Amon D, Currie D, Downs C, Hall-Spencer J et al (2019) Eight urgent, fundamental and simultaneous steps needed to restore ocean health, and the consequences for humanity and the planet of inaction or delay. Aquat Conserv Mar Freshwat Ecosyst 30(1):194–208. https://doi.org/10.1002/aqc.3182
32. Dutta V, Dubey D, Kumar S (2020) Cleaning the River Ganga: impact of lockdown on water quality and future implications on river rejuvenation strategies. Sci Total Environ 743:140756. https://doi.org/10.1016/j.scitotenv.2020.140756
33. Singhal S, Matto M (2020) COVID-19 lockdown: a ventilator for rivers. Downtoearth.org.in. from https://www.downtoearth.org.in/blog/covid-19-lockdown-aventilator-for-rivers-70771
34. Iyiola AO, Asiedu B, Fawole FJ (2020) Possible effects of COVID-19 on sustainability of aquatic ecosystems: an overview. Aquat Res 3(4):177–187. https://doi.org/10.3153/AR20016
35. Khalil Cherif E, Vodopivec M, Mejjad N, Esteves da Silva JCG, Simonovič S, Boulaassal H (2020) COVID-19 pandemic consequences on coastal water quality using WST Sentinel-3 data: case of Tangier, Morocco. Water 12(9):2638. https://doi.org/10.3390/w12092638
36. Coll M, Ortega-Cerdà M, Mascarell-Rocher Y (2021) Ecological and economic effects of COVID-19 in marine fisheries from the Northwestern Mediterranean Sea. Biol Conserv 255:108997. https://doi.org/10.1016/j.biocon.2021.108997
37. Bellasi A, Binda G, Pozzi A, Galafassi S, Volta P, Bettinetti R (2020) Microplastic contamination in freshwater environments: a review, focusing on interactions with sediments and benthic organisms. Environments 7(4):30. https://doi.org/10.3390/environments7040030
38. Venugopal A, Ganesan H, Sudalaimuthu Raja SS, Govindasamy V, Arunachalam M, Narayanasamy A, Sivaprakash P, Rahman P, Gopalakrishnan AV, Siama Z, Vellingiri B (2020) Novel wastewater surveillance strategy for early detection of coronavirus disease 2019 hotspots. Curr Opin Environ Sci Health 17:8–13. https://doi.org/10.1016/j.coesh.2020.05.003
39. El-Ramady H, Brevik E, Amer M, Elsakhawy T, Omara Ahmed A, Elbasiouny H, Elbehiry F, Mosa A, El-Ghamry A, Bayoumi Y, Shalaby T (2020) Soil and air pollution in the era of COVID-19: a global issue. Egypt J Soil Sci 60(4):437–448. https://doi.org/10.21608/EJSS.2020.49996.1411

40. Aday S, Aday MS (2020) Impact of COVID-19 on the food supply chain. Food Qual Saf 4 (4):167–180. https://doi.org/10.1093/fqsafe/fyaa024
41. Al Huraimel K, Alhosani M, Kunhabdulla S, Stietiya MH (2020) SARS-CoV-2 in the environment: modes of transmission, early detection and potential role of pollutions. Sci Total Environ 744:140946. https://doi.org/10.1016/j.scitotenv.2020.140946
42. Naddeo V, Liu H (2020) Correction: editorial perspectives: 2019 novel coronavirus (SARS CoV-2): what is its fate in urban water cycle and how can the water research community respond? Environ Sci Water Res Technol 6:1939. https://doi.org/10.1039/D0EW90030C
43. Cahill N, Morris D (2020) Recreational waters – a potential transmission route for SARS CoV-2 to humans? Sci Total Environ 740:140122. https://doi.org/10.1016/j.scitotenv.2020.140122
44. Guan W, Ni Z, Hu Y, Liang W, Ou C, He J, Liu L, Shan H, Lei C, Hui DSC, Du B, Li L, Zeng G, Yuen KY, Chen R, Tang C, Wang T, Chen P, Xiang J, Li S, Jin-lin W, Liang Z, Peng Y, Wei L, Liu L, Ya-hua H, Peng P, Jian-ming W, Liu J, Chen Z, Li G, Zheng Z, Qiu S, Luo J, Ye C, Zhu S, Zhong N (2020) Clinical characteristics of coronavirus disease 2019 in China. N Engl J Med 382:1708–1720. https://doi.org/10.1056/NEJMoa2002032
45. Huang C, Wang Y, Li X, Ren L, Zhao J, Hu Y, Zhang L, Fan G, Xu J, Gu X, Cheng Z, Yu T, Xia J, Wei Y, Wu W, Xie X, Yin W, Li H, Liu M, Xiao Y, Gao H, Guo L, Xie J, Wang G, Jiang R, Gao Z, Jin Q, Wang J, Cao B (2020) Clinical features of patients infected with 2019 novel coronavirus in Wuhan, China. Lancet 395(10223):497–506. https://doi.org/10.1016/S0140-6736(20)30183-5
46. Wang W, Xu Y, Gao R, Lu R, Han K, Wu G, Tan W (2020) Detection of SARS-CoV-2 in different types of clinical specimens. JAMA 323(18):1843–1844. https://doi.org/10.1001/jama.2020.3786
47. Kitajima M, Ahmed W, Bibby K, Carducci A, Gerba CP, Hamilton KA, Haramoto E, Rose JB (2020) SARS-CoV-2 in wastewater: state of the knowledge and research needs. Sci Total Environ 739:139076. https://doi.org/10.1016/j.scitotenv.2020.139076
48. Shutler JD, Zaraska K, Holding T, Machnik M, Uppuluri K, Ashton I, Migdał Ł, Dahiya RS (2021) Rapid assessment of SARS-CoV-2 transmission risk for fecally contaminated river water. ACS ES&T Water 1(4):949–957. https://doi.org/10.1021/acsestwater.0c00246
49. Charlie-Silva I, Araújo APC, Guimarães ATB, Veras FP, Braz HLB, de Pontes LG, Jorge RJB, Belo MAA, Fernandes BHV, Nóbrega RH, Galdino G, Condino-Neto A, Galindo-Villegas J, Machado-Santelli GM, Sanches PRS, Rezende RM, Cilli EM, Malafaia G (2021) An insight into neurotoxic and toxicity of spike fragments SARS-CoV-2 by exposure environment: a threat to aquatic health? bioRxiv:425914. https://doi.org/10.1101/2021.01.11.425914
50. Bivins A, Greaves J, Fischer R, Yinda KC, Ahmed W, Kitajima M, Munster VJ, Bibby K (2020) Persistence of SARS-CoV-2 in water and wastewater. Environ Sci Technol Lett 7 (12):937–942. https://doi.org/10.1021/acs.estlett.0c00730
51. Wartecki A, Rzymski P (2020) On the coronaviruses and their associations with the aquatic environment and wastewater. Water 12(6):1598. https://doi.org/10.3390/w12061598
52. Heller L, Mota CR, Greco DB (2020) COVID-19 faecal-oral transmission: are we asking the right questions? Sci Total Environ 729:138919. https://doi.org/10.1016/j.scitotenv.2020.138919
53. Hindson J (2020) COVID-19: faecal–oral transmission? Nat Rev Gastroenterol Hepatol 17:259. https://doi.org/10.1038/s41575-020-0295-7
54. Lyon GJ, Wang K (2012) Identifying disease mutations in genomic medicine settings: current challenges and how to accelerate progress. Genome Med 4(7):58. https://doi.org/10.1186/gm359
55. Núñez-Delgado A (2020) SARS-CoV-2 in soils. Environ Res 190:110045. https://doi.org/10.1016/j.envres.2020.110045
56. Núñez-Delgado A (2020) What do we know about the SARS-CoV-2 coronavirus in the environment? Sci Total Environ 727:138647. https://doi.org/10.1016/j.scitotenv.2020.138647

57. Behera BC (2021) Challenges in handling COVID-19 waste and its management mechanism: a review. Environ Nanotechnol Monit Manag 15:100432. https://doi.org/10.1016/j.enmm.2021.100432
58. Mohan SV, Hemalatha M, Kopperi H, Ranjith I, Kumar AK (2021) SARS-CoV-2 in environmental perspective: occurrence, persistence, surveillance, inactivation and challenges. Chem Eng J 405:126893. https://doi.org/10.1016/j.cej.2020.126893
59. Michael-Kordatou I, Karaolia P, Fatta-Kassinos D (2020) Sewage analysis as a tool for the COVID-19 pandemic response and management: the urgent need for optimised protocols for SARS-CoV-2 detection and quantification. J Environ Chem Eng 8:104306. https://doi.org/10.1016/j.jece.2020.104306
60. Patel M, Chaubey AK, Pittman Jr CU, Mlsna T, Mohan D (2021) Coronavirus (SARS-CoV-2) in the environment: occurrence, persistence, analysis in aquatic systems and possible management. Sci Total Environ 765:142698. https://doi.org/10.1016/j.scitotenv.2020.142698
61. Ahmed W, Angel N, Edson J, Bibby K, Bivins A, O'Brien JW, Choi PM, Kitajima M, Simpson SL, Li J, Tscharke B, Verhagen R, Smith W, Zaugg J, Dierens L, Hugenholtz P, Thomas KV, Mueller JF (2020) First confirmed detection of SARS-CoV-2 in untreated wastewater in Australia: a proof of concept for the wastewater surveillance of COVID-19 in the community. Sci Total Environ 728:138764. https://doi.org/10.1016/j.scitotenv.2020.138764
62. Hjelmsø MH, Hellmér M, Fernandez-Cassi X, Timoneda N, Lukjancenko O, Seidel M, Elsässer D, Aarestrup FM, Löfström C, Bofill-Mas S, Abril JF, Girones R, Schultz AC (2017) Evaluation of methods for the concentration and extraction of viruses from sewage in the context of metagenomic sequencing. PLoS One 12(1):e0170199. https://doi.org/10.1371/journal.pone.0170199
63. Bibby K, Peccia J (2013) Identification of viral pathogen diversity in sewage sludge by metagenome analysis. Environ Sci Technol 47:1945–1951. https://doi.org/10.1021/es305181x
64. Conde-Cid M, Arias-Estévez M, Núñez-Delgado A (2021) How to study SARS-CoV-2 in soils? Environ Res 198:110464. https://doi.org/10.1016/j.envres.2020.110464
65. Alteio LV, Schulz F, Seshadri R, Varghese N, Rodriguez-Reillo W, Ryan E, Goudeau D, Eichorst SA, Malmstrom RR, Bowers RM, Katz LA, Blanchard JL, Woyke T (2020) Complementary metagenomic approaches improve reconstruction of microbial diversity in a forest soil. mSystems 5(2):e00768–e00719. https://doi.org/10.1128/mSystems.00768-19
66. Schulz F, Alteio L, Goudeau D, Ryan EM, Yu FB, Malmstrom RR, Blanchard J, Woyke T (2018) Hidden diversity of soil giant viruses. Nat Commun 9(1):4881. https://doi.org/10.1038/s41467-018-07335-2
67. Khalil JY, Robert S, Reteno DG, Andreani J, Raoult D, La Scola B (2016) High-throughput isolation of giant viruses in liquid medium using automated flow cytometry and fluorescence staining. Front Microbiol 7:26. https://doi.org/10.3389/fmicb.2016.00026
68. Hoffmann M, Monaghan MT, Reinert K (2020) PriSeT: efficient de novo primer discovery. bioRxiv:027961. https://doi.org/10.1101/2020.04.06.027961, https://doi.org/10.3390/su9061010
69. Hufsky F, Lamkiewicz K, Almeida A, Aouacheria A, Arighi C, Bateman A, Baumbach J, Beerenwinkel N, Brandt C, Cacciabue M, Chuguransky S, Drechsel O, Finn RD, Fritz A, Fuchs S, Hattab G, Hauschild AC, Heider D, Hoffmann M, Hölzer M, Hoops S, Kaderali L, Kalvari I, von Kleist M, Kmiecinski R, Kühnert D, Lasso G, Libin P, List M, Löchel HF, Martin MJ, Martin R, Matschinske J, McHardy AC, Mendes P, Mistry J, Navratil V, Nawrocki EP, O'Toole ÁN, Ontiveros-Palacios N, Petrov AI, Rangel-Pineros G, Redaschi N, Reimering S, Reinert K, Reyes A, Richardson L, Robertson DL, Sadegh S, Singer JB, Theys K, Upton C, Welzel M, Williams L, Marz M (2021) Computational strategies to combat COVID-19: useful tools to accelerate SARS-CoV-2 and coronavirus research. Brief Bioinform 22(2):642–663. https://doi.org/10.1093/bib/bbaa232f
70. Brandt C, Krautwurst S, Spott R, Lohde M, Jundzill M, Marquet M, Hölzer M (2021) poreCov – an easy to use, fast, and robust workflow for SARS-CoV-2 genome reconstruction via nanopore sequencing. bioRxiv:443089. https://doi.org/10.1101/2021.05.07.443089

71. Fritz A, Bremges A, Deng ZL, Lesker TR, Götting J, Ganzenmüller T, Sczyrba A, Dilthey A, Klawonn F, McHardy AC (2021) Haploflow: strain-resolved *de novo* assembly of viral genomes. bioRxiv. https://doi.org/10.1101/2021.01.25.428049
72. Posada-Céspedes S, Seifert D, Topolsky I, Jablonski KP, Metzner KJ, Beerenwinkel N (2021) V-pipe: a computational pipeline for assessing viral genetic diversity from high-throughput data. Bioinformatics:btab015. https://doi.org/10.1093/bioinformatics/btab015
73. Libin PJK, Deforche K, Abecasis AB, Theys K (2019) VIRULIGN: fast codon-correct alignment and annotation of viral genomes. Bioinformatics 35:1763–1765. https://doi.org/10.1093/bioinformatics/bty851
74. Schäfer A, Muecksch F, Lorenzi J, Leist SR, Cipolla M, Bournazos S, Schmidt F, Maison RM, Gazumyan A, Martinez DR, Baric RS, Robbiani DF, Hatziioannou T, Ravetch JV, Bieniasz PD, Bowen RA, Nussenzweig MC, Sheahan TP (2021) Antibody potency, effector function, and combinations in protection and therapy for SARS-CoV-2 infection in vivo. J Exp Med 218(3): e20201993. https://doi.org/10.1084/jem.20201993
75. Mistry J, Chuguransky S, Williams L, Qureshi M, Salazar GA, Sonnhammer E, Tosatto S, Paladin L, Raj S, Richardson LJ, Finn RD, Bateman A (2021) Pfam: the protein families database in 2021. Nucleic Acids Res 49(D1):D412–D419. https://doi.org/10.1093/nar/gkaa913
76. Kalvari I, Nawrocki EP, Ontiveros-Palacios N, Argasinska J, Lamkiewicz K, Marz M, Griffiths-Jones S, Toffano-Nioche C, Gautheret D, Weinberg Z, Rivas E, Eddy SR, Finn RD, Bateman A, Petrov AI (2021) Rfam 14: expanded coverage of metagenomic, viral and microRNA families. Nucleic Acids Res 49(D1):D192–D200. https://doi.org/10.1093/nar/gkaa1047
77. Cresswell K, Sheikh A (2020) Can disinfection robots reduce the risk of transmission of SARS-CoV-2 in health care and educational settings? J Med Internet Res 22:e20896–e20896. https://doi.org/10.2196/20896
78. Ashraf MA, Maah MJ, Yusoff I (2014) Soil contamination, risk assessment and remediation https://doi.org/10.5772/57287
79. Ampatzidis Y, De Bellis L, Luvisi A (2017) iPathology: robotic applications and management of plants and plant diseases. Sustainability 9(6):1010
80. Legner CM, Tylka GL, Pandey S (2021) Robotic agricultural instrument for automated extraction of nematode cysts and eggs from soil to improve integrated pest management. Sci Rep 11:3212. https://doi.org/10.1038/s41598-021-82261-w
81. Garcia R, Gracias N, Nicosevici T, Prados R, Hurtos N, Campos R, Escartin J, Elibol A, Hegedus R (2017) Exploring the seafloor with underwater robots: land, sea & air. Comput Vision Vehicle Technol:75–99. https://doi.org/10.1002/9781118868065.ch4
82. Talaviya T, Shah D, Patel N, Yagnik H, Shah M (2020) Implementation of artificial intelligence in agriculture for optimisation of irrigation and application of pesticides and herbicides. Artif Intell Agric 4:58–73. https://doi.org/10.1016/j.aiia.2020.04.002
83. Agrawal P, Bhattacharya B (2013) Aquatic multi-robot system for lake cleaning. Comput Sci. https://doi.org/10.1142/9789814525534_0024
84. Hart OE, Halden RU (2020) Computational analysis of SARS-CoV-2/COVID-19 surveillance by wastewater-based epidemiology locally and globally: feasibility, economy, opportunities and challenges. Sci Total Environ 730:138875. https://doi.org/10.1016/j.scitotenv.2020.138875
85. Balamurugan M, Kasiviswanathan KS, Ilampooranan I, Soundharajan BS (2021) COVID-19 lockdown disruptions on water resources, wastewater, and agriculture in India. Front Water 3:24. https://doi.org/10.3389/frwa.2021.603531
86. Ishaq M (2018) Disinfection methods. https://doi.org/10.5772/intechopen.80999
87. Collivignarelli MC, Abbà A, Alloisio G, Gozio E, Benigna I (2017) Disinfection in wastewater treatment plants: evaluation of effectiveness and acute toxicity effects. Sustainability 9:1704. https://doi.org/10.3390/su9101704
88. Gamage J, Zhang Z (2010) Applications of photocatalytic disinfection. Int J Photoenergy 764870. https://doi.org/10.1155/2010/764870
89. Glaze WH (1987) Drinking-water treatment with ozone. Environ Sci Technol 21:224–230. https://doi.org/10.1021/es00157a001

90. Nguyen TMH, Suwan P, Koottatep T, Beck SE (2019) Application of a novel, continuous-feeding ultraviolet light emitting diode (UV-LED) system to disinfect domestic wastewater for discharge or agricultural reuse. Water Res 153:53–62. https://doi.org/10.1016/j.watres.2019.01.006
91. Dubey AK, Singh A, Prakash S, Kumar M, Singh AK (2020) Race to arsenal COVID-19 therapeutics: current alarming status and future directions. Chem Biol Interact 332:109298. https://doi.org/10.1016/j.cbi.2020.109298
92. Subramaniam MD, Venkatesan D, Iyer M, Subbarayan S, Govindasami V, Roy A, Narayanasamy A, Kamalakannan S, Gopalakrishnan AV, Thangarasu R, Kumar NS, Vellingiri B (2020) Biosurfactants and anti-inflammatory activity: a potential new approach towards COVID-19. Curr Opin Environ Sci Health 17:72–81. https://doi.org/10.1016/j.coesh.2020.09.002
93. Khanna RC, Cicinelli MV, Gilbert SS, Honavar SG, Murthy G (2020) COVID-19 pandemic: lessons learned and future directions. Indian J Ophthalmol 68(5):703–710. https://doi.org/10.4103/ijo.IJO_843_20
94. Dhanani R, Wasif M, Pasha HA, Hussain M, Ghaloo SK, Shah Vardag AB, Mahmood K, Hussain R (2020) COVID-19: implications, reactions and future directions. Turk Arch Otorhinolaryngol 58(2):122–126. https://doi.org/10.5152/tao.2020.5452
95. Balachandar V, Mahalaxmi I, Subramaniam M, Kaavya J, Senthil Kumar N, Laldinmawii G, Narayanasamy A, Janardhana Kumar Reddy P, Sivaprakash P, Kanchana S, Vivekanandhan G, Cho SG (2020) Follow-up studies in COVID-19 recovered patients – is it mandatory? Sci Total Environ 729:139021. https://doi.org/10.1016/j.scitotenv.2020.139021

Innovative Treatment Processes for Emerging Contaminants Removal from Sewage Sludge

A. Ferraro, A. Panico, H. Pirasteh-Anosheh, M. Race, D. Spasiano, G. Trancone, and F. Pirozzi

Contents

A. Ferraro and D. Spasiano
Department of Civil, Environmental, Land, Building Engineering and Chemistry, Polytechnic University of Bari, Bari, Italy

A. Panico
Department of Engineering, University of Campania "L. Vanvitelli", Aversa, Italy

H. Pirasteh-Anosheh
National Salinity Research Center, Agricultural Research, Education and Extension Organization, Yazd, Iran

M. Race (✉)
Department of Civil and Mechanical Engineering, University of Cassino and Southern Lazio, Cassino, Italy
e-mail: marco.race@unicas.it

G. Trancone and F. Pirozzi
Department of Civil, Architectural and Environmental Engineering, University of Naples "Federico II", Naples, Italy

Avelino Núñez-Delgado and Manuel Arias-Estévez (eds.),
Emerging Pollutants in Sewage Sludge and Soils,
Hdb Env Chem (2023) 114: 89–110, DOI 10.1007/698_2022_850,

Published online: 16 April 2022

Abstract This chapter deals with an environmental aspect not sufficiently taken into account by the scientific community as well as the economy productive sectors: the removal of emerging contaminants (ECs) from sewage sludge. The most common ECs that can be found in the sludge, their adsorption mechanisms, and the most suitable removal treatments, in such order, are analyzed and discussed in the following pages. An accurate scientific literature survey has highlighted, primarily, that the anaerobic digestion (AD) process is the most common treatment used to stabilize sewage sludge and, secondly, that such process is not appropriate for properly treating ECs. Therefore, further treatments are required to be coupled with AD for removing ECs. Among all possible treatments, those innovative, chemical (ozonation, hydrogen peroxide, Fenton) as well as thermal (hydrothermal liquefaction) are discussed and their performance is compared.

Keywords Fenton, Hydrogen peroxide, Hydrothermal liquefaction, Ozonation, Sewage sludge contamination

1 Introduction

Emerging contaminants (ECs) are chemicals that are produced either from natural phenomena (e.g., natural estrogenic hormones) or from human activities (e.g., pesticides and industrial compounds). In the past few years, such contaminants have been detected in the environment and represent a serious threat to the survival of living beings because of their toxicity and persistence [1–3]. Many ECs, such as antibiotics, pharmaceuticals, illegal drugs, and personal care products (PPCPs) are discharged into municipal sewers through domestic wastewaters and reach the wastewater treatment plants (WWTPs) [4, 5]. Most of these contaminants will not be completely removed by WWTPs processing and, depending on their specific physicochemical properties (e.g., hydrophobicity, water solubility, pK_a), they are discharged into the environment through treated wastewater or sewage sludge [6, 7]. Moreover, the presence of these contaminants in sewage sludge represents a concern for the biological treatment of the WWTPs. For instance, antibiotics at concentrations exceeding the inhibiting thresholds may hinder the regular development of both acidogenesis and methanogenesis phases during the anaerobic digestion (AD) process. When such an event occurs, the volatile solids (VS) conversion to methane is less efficient (methane production is lower than expected), even when the biomass has been acclimatized at low antibiotic concentrations [8–11]. Furthermore, the presence of ECs in sewage sludge can generate a diffuse environmental pollution when the sludge is reused as fertilizer in agriculture. Since ECs are generally scarcely biodegraded by microorganisms present in agricultural soils, they can contaminate surface water and groundwater due to water runoff and water infiltration, respectively [12, 13]. Moreover, ECs can accumulate into plant tissues, thus threatening the health of all species involved in the food chain [14, 15].

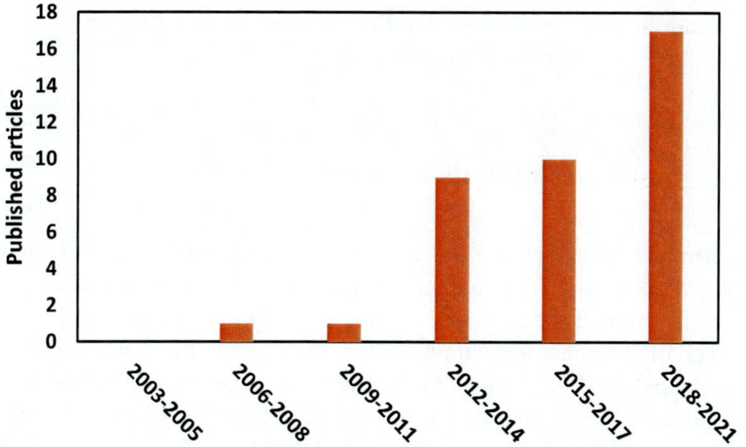

Fig. 1 Increasing trend of papers on emerging contaminants treatment from sewage sludge (Scopus). (String keywords: "emerging contaminants" and "sewage sludge" and "removal or treatment")

Therefore, prior to use sewage sludge as a fertilizer, controls on its characteristics have to be carefully conducted, and not only limited to the measurement of the original pollutant removal rate through chemical analysis. It is therefore essential to carry out bioassays that evaluate the potential adverse effects to ecosystems and human health due not only to the residual original contaminants but also to AD by-products and other compounds unregulated by law [16]. Once released into the environment, some ECs can directly interfere with the endocrine system or DNA of living organisms (mutagenic-carcinogenic risks), thus compromising reproductive and vital functions in humans and wildlife [17]. Even if some ECs may be partially removed by AD process [18, 19], some studies have demonstrated that pretreatments prior to AD, such as sonication or alkaline and thermal hydrolysis, can further increase the removal efficiency of ECs [20–22]. Nevertheless, to ensure a satisfactory removal of ECs from sewage sludge, further treatments such as advanced oxidation processes (AOPs) are required. For instance, ozonation is capable of removing nonylphenol ethoxylates (NPE) from AD sludge by 73% [23]. A successful oxidation of ECs contained in sewage sludge may be also achieved by Fenton process carried out with the addition of both Hydrogen Peroxide (H_2O_2) and Fe (II) salts or, alternatively, H_2O_2 solely if a sufficient amount of Fe(II) is already present in the sludge [24, 25]. Such methods are relatively new as the first studies concerning the treatment of ECs in sewage sludge date back only to 2003. Figure 1 shows an increasing trend of published papers (data from Scopus) from a total number of 1 between 2006 and 2008 to 17 between 2018 and 2021. These data show the growing interest in this topic. Hence, this chapter has been aimed at investigating the current state of the art and knowledge dealing with the fate of ECs in sewage sludge and treatments aimed at their removal. In particular, the following aspects have been examined and discussed in depth: (1) ECs that can be usually found in sewage sludge; (2) ECs adsorption mechanisms; (3) current removal methods; (4) AOP used to remove ECs.

2 Emerging Contaminants in Sewage Sludge

ECs are chemical compounds dissolved in wastewater that can have adverse effects on the environment as well as human health. They are usually found in very low concentrations, around units of ng/L. The extremely low concentrations coupled with the lack of a specific legislation are the main causes that hamper the use of a correct analytical procedure aimed at their detection and the implementation of an adequate monitoring system in WWTPs. The ECs present in the influents of WWTPs are usually treated by conventional processes and, if they are not adequately removed, their presence can be detected both in sewage sludge and in the WWTPs effluents, thus proving the inefficiency of the conventional processes for their removal. Therefore, ECs can bioaccumulate in sewage sludge and reach very high concentrations (such us > 1 mg/kg). ECs concentration in sewage sludge depends on chemical-physical characteristics of the specific compounds (i.e., octanol-water partition coefficient known as Log K_{ow}, water solubility, concentration, pK_a).

In general, the Log K_{ow} partition coefficient is often used as a predictive parameter of the tendency of an organic compound to be adsorbed on solids. In fact, when Log K_{ow} shows values below 2.5 adsorption capacity is negligible, when values range between 2.5 and 4 the adsorption capacity is moderate, whereas when Log K_{ow} shows values greater than 4, the adsorption capacity is really high. However, in the case of sewage sludge, Log K_{ow} is not the sole parameter contributing to the fate of ECs. For instance, also the ionization state of molecules plays a key role in the adsorption of ECs on sewage sludge accordingly to molecule pK_a and wastewater pH. For this reason, more specific research has been carried out on the adsorption of ECs in sewage sludge [26–28]. The adsorption capacity has been therefore evaluated by determining the Log K_d coefficient (sewage sludge-water partition coefficient) [29–31].

Table 1 reports the concentrations in sewage sludge of the most common ECs present in wastewater. ECs are divided into two categories: antibiotics/drugs and PPCPs. From Table 1, it can be clearly noticed that ECs concentrations vary from ng/kg to mg/kg. Regarding antibiotics, sulfamethoxazole is the compound that is most adsorbed on sewage sludge, thus reaching a concentration higher than 68,000 μg/kg although the Log K_{ow} is extremely low (0.89) [52, 65]. Such occurrence could be caused by several factors (i.e., concentration of contaminant in wastewater, WWTPs configurations, and chemical-physical characteristics of the sewage sludge). Along with sulfamethoxazole, other antibiotics are found adsorbed on sewage sludge and their presence, in addition to general concern about environmental pollution, can promote the specific serious issue of antibiotic resistance.

In recent years, such aspects have increasingly gained relevance. Moreover, the amount of pharmaceutical and PPCPs in the influent of WWTPs has been

Table 1 List of: (1) the most common drugs and PPCPs; (2) their range of concentrations detected in the sewage sludge; (3) their Log K_d and Log K_{ow} coefficients

Contaminant		Formula	Concentration range (μg/kg)	Log K_d	Log K_{ow}	Reference
Antibiotics/drugs	Ciprofloxacin	$C_{17}H_{18}FN_3O_3$	200–2030	4.3	0.28	[32–38]
	Carbamazepine	$C_{15}H_{12}N_2O$	0.036–11,060	1.31–1.83	2.45	[29, 32, 39–41]
	Diclofenac	$C_{14}H_{11}Cl_2NO_2$	1–11,060	2.7	4.51	[4, 32, 40, 42]
	Ibuprofen	$C_{13}H_{18}O_2$	0.11–208	1.00–1.78	3.97	[29, 32, 43–45]
	Naproxen	$C_{14}H_{14}O_3$	11.1–930	2.00–3.00	3.18	[29, 44–47]
	Norfloxacin	$C_{16}H_{18}FN_3O_3$	2–5,800	4.2	−1.03	[32, 47–49]
	Sulfamethoxazole	$C_{10}H_{11}N_3O_3S$	4,2–68,000	1.04	0.89	[46, 50–52]
	Trimethoprim	$C_{14}H_{18}N_4O_3$	26–41,000	2.2–2.57	0.91	[52–54]
Personal care products	Triclosan	$C_{12}H_7Cl_3O_2$	0.09–17,500	2.39–2.55	4.76	[55–61]
	Tonalide	$C_{18}H_{26}O$	2.3–1,600	3.4–3.7	5.7	[31, 55, 62, 63]
	Galaxolide	$C_{18}H_{26}O$	5.4–11,800	3.3–3.8	5.9	[31, 55, 62–65]

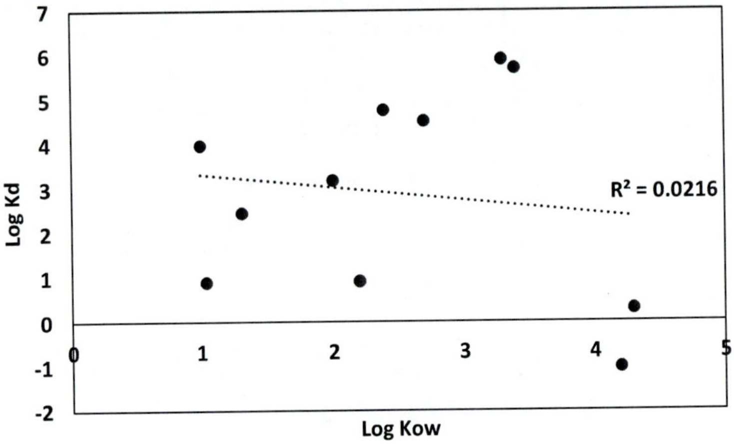

Fig. 2 Sludge sorption (Log K_d) plotted against hydrophobicity (Log K_{ow}), both referred to ECs listed in Table 1

progressively growing as much as their concentration in sewage sludge, since such compounds are easily adsorbed on sewage sludge according to their Log K_{ow} values higher than 5 (see Table 1). For instance, galaxolide has been found in concentration equal to 36,000 μg/kg [66]. The weak correlation (Fig. 2) between values of Log K_d and Log K_{ow} may be due to factors related to the nature of ECs. Indeed, many pharmaceutical products are weak acids or bases and, depending on the pK_a, they can be detected in the environment as ionized forms [67].

It has been observed that the adsorption of dissolved compounds on sewage sludge may depend on their ionic state. For instance, the adsorption capacity is higher when a compound is present in neutral form rather than in ionic form [67]. Nevertheless, a strong adsorption on substances such as soil organic matter has been reported for chemical species that are negatively charged, thus suggesting that adsorption may also occur primarily via molecular interactions that are independent of hydrophobic effects [67]. Highly hydrophilic compounds have higher Log K_d values than those expected from their Log K_{ow} values, while highly hydrophobic compounds have lower absorption coefficients than those expected (Table 1). Such evidence is likely due to specific features of these coefficients that properly describe the hydrophobic interactions, but fail to predict the absorption of polar and ionic compounds. In particular, Rybacka and Andersson [67] observed that the adsorption of neutral and cationic compounds on sewage sludge was mainly governed by intermolecular forces of similar intensities, hydrophobic interactions as well as pi–pi and dipole–dipole interactions. Conversely, the adsorption of anionic chemical species is mainly governed by covalent, ion–ion, ion–dipole, and dipole–dipole interactions.

2.1 Influence of WWTPs Characteristics on ECs Adsorption

The adsorption process is the main mechanism that promotes the removal of some contaminants present in the influent of WWTPs. Indeed, scarcely biodegradable drugs, antibiotics, and PPCPs are removed in WWTPs from wastewaters by adsorption on sludge. Li and Zhang [68] observed a removal efficiency through adsorption higher than 80% for the whole group of 11 drugs they tested. Similar results were found in other studies aimed at removing ECs from wastewater [69–71]. The adsorption process depends on chemical and physical properties of both adsorbate (sludge) and adsorbent (ECs) as well as on WWTPs configuration and operating conditions (HRT, SRT, etc.). Accounting for the sewage sludge, the most relevant chemical-physical properties that play a crucial role in adsorption mechanisms are the presence of organic substance, concentration of cations, and pH. Whereas, as regards ECs, the main parameters are the biodegradability, Log K_{ow} value, solubility, and hydrophobicity.

Specifically, hydrophobic ECs, characterized by Log $K_{ow} > 3.0$, such as azithromycin, miconazole, triclocarban, ibuprofen, and gemfibrozil, are ubiquitously found in sewage sludge (Table 2). On the other hand, it can also happen that some ECs such as chlortetracycline, ciprofloxacin, and tetracycline are adsorbed on sewage sludge even if their Log K_{ow} is negative (Table 2). This event can occur thanks to the electrostatic forces interacting between the negatively charged sewage sludge biosolids and those ECs compounds that, under neutral pH conditions, are speciated as cations or zwitterion [72, 73]. Among all ECs that can be usually detected in wastewaters, it is interesting to highlight the presence of salicylic acid at high concentrations in sewage sludge, although this compound is characterized by a Log K_{ow} equal to 2.26 and, furthermore, it is mainly present in solution in its anionic form (Table 2). This phenomenon could be attributed to the formation of a chemical complex between Fe(III) ions in the sediments and the phenolic group of the salicylic acid, as suggested by Ou et al. [74].

Regarding the properties of sewage sludge, Verlicchi et al. [75] found that the presence and the amount of adsorbed contaminants depend on the geographical position as well as the season. Some drugs (i.e., antibiotics such as paracetamol and ibuprofen) were, in fact, mainly detected during winter rather than other seasons as a consequence of the seasonal flu's occurrence.

Another important factor is the type of sewage sludge (primary or secondary) since it has been observed that, according to the type of sewage sludge, the ECs concentrations are different [76]. In particular, it was found that antibiotics are mostly found in secondary sludge while some types of drugs (i.e., antiepileptics) are mainly in primary sludge. This phenomenon is due to the density, the presence of proteins and cations as well as the structure of the extracellular polymeric substances (EPS), which can affect the adsorption capacity [37, 77]. Primary sludge is mainly composed of settling solids (i.e., feces, toilet paper), while secondary sludge is mainly composed of bacterial biomass. In terms of chemical and physical characteristics, primary sludge differs significantly from secondary sludge: total suspended

Table 2 Concentrations, Log K_{ow}, and ionization states of main ECs found in sewage sludge. Readapted from various publications [7, 72, 73]. *MQL* (method quantification limit)

	μg/kg$_{DS}$	logK_{ow}	pK_a	Ionization states at pH = 6.0 (%)				Ionization states at pH = 8.0 (%)			
				Anion	Neutral	Cation	Zwitterion	Anion	Neutral	Cation	Zwitterion
Antibiotics											
Azithromycin	<MQL - 1,220	4.02	8.74	0	0.1	99.9	0	0	0.3	99.7	0
Chlortetracycline	<MQL - 28,260	−0.62	3.3/7.4/9.3	0	0	0	100	0	0	0	100
Ciprofloxacin	1,400–160,000	−2.82	3.32/5.59/8.85	0	0	36.3	63.7	17.3	0	0.5	82.2
Clarithromycin	<MQL - 580	3.16	8.82/13.1	0	0.4	99.6	0	0	29.4	70.6	0
Enrofloxacin	7.6–11,560	0.83	5.69	–	–	–	–	–	–	–	–
Lincomycin	<MQL - 4,967	0.2	7.6	0.3	0	0.3	99.4	28	0	0	72
Minocycline	121–2,630	0.05	2.8/5.0/7.8/9.3	0	0	0	100	0	0	0	100
Ofloxacin	150–8,140	−0.39	5.97/9.28	–	–	–	–	–	–	–	–
Oxytetracycline	<MQL - 3,790	−0.90	3.27/7.3/9.6	3.8	0	0.1	96.1	52.9	0	0	47.1
Tetracycline	<MQL - 2,790	−1.37	3.3/7.7/9.7	0	0	0	100	5.2	0	0	94.8
Antifungal/antimicrobials											
Miconazole	<MQL - 1,100	6.1	6.77	–	–	–	–	–	–	–	–
Triclocarban	362–48,100	4.34	12.7	0	0	0	100	0	0	0	100
Triclosan	354–19,700	4.76	7.9	2	94	2	2	67.6	32.4	0	0
NSAIDs											
Acetaminophen	<MQL - 586	0.46	9.7	0	100	0	0	3.3	96.7	0	0
Ibuprofen	<MQL - 3,988	3.97	4.91	93.4	6.6	0	0	99.9	1	0	0
Naproxen	<MQL - 1,022	3.18	4.2	98.5	1.5	0	0	100	0	0	0
Salicylic acid	<MQL - 13,743	2.26	2.97	99.9	0.1	0	0	100	0	0	0
Lipid regulators											
Gemfibrozil	< MQL - 1,192	4.77	4.5	97.5	2.5	0	0	100	0	0	0
UV filters											
Octocrylene	1,060–9,170	7.04	14	0	100	0	0	0	100	0	0

Oxybenzone	<MQL - 790	3.79	9.6	7.8	92.2	0	0	89.4	10.6	0	0
Stimulant											
Caffeine	<MQL - 805	−0.07	14	0	100	0	0	0	100	0	0
Plasticizer											
Bisphenol A	<MQL - 4,700	3.32	9.6	0	100	0	0	1.6	98.4	0	0

solids (TSS), total volatile solids (TVS), oils and fats, pH, alkalinity, are, respectively, 6%, 3.9%, 2.3%, 6.5, and 1,000 mg/L of $CaCO_3$ for primary sludge and 1%, 0.8%, 0.1%, 7.25, and 840 mg/L of $CaCO_3$ for secondary sludge [78]. For instance, Ternes et al. [31] found an adsorption capacity of diclofenac in primary sludge (between 5% and 15%) higher than in mixed (mixture of primary and secondary sludge) sludge (<5%). Conversely, Hörsing et al. [79] tested 75 drugs in wastewaters and observed no difference about the adsorption capacity between primary and secondary sludge as well as no difference in secondary sludge with different sludge retention times (SRT).

3 Treatments

3.1 *Anaerobic Digestion (AD)*

The AD process is currently the most widely used technology for the stabilization of sewage sludge. AD is a biological process where, through the biological activity of bacteria, the organic substance is degraded and mainly converted to biogas. The performance of the AD process strongly depends on multiple factors (i.e., pH, temperature, OLR, HRT). This process is positively accepted because it is energy self-sustaining and even can produce a surplus of energy that can be sold. Furthermore, for the same amount of processing sewage sludge, the amount of sludge to be disposed resulting from AD is lower than that produced from an aerobic stabilization process, since the biomass yield rate is lower (e.g., 0.2 for anaerobic process, 0.6 for aerobic). The end products of AD are a gas mixture (biogas, useful to produce energy) and highly concentrated liquid (digestate, potentially useful to produce organic fertilizer) [80].

Different studies focused their attention on the evaluation of ECs removal by AD process (Table 3). Experimental results have shown a very high removal efficiency (>70%) of naproxen, sulfamethoxazole, roxithromycin, estrogen, galaxolide, tonalide, and diclofenac. Conversely, diazepam and ibuprofen are characterized by a lower removal efficiency (40–60%), whereas carbamazepine showed no removal. Samaras et al. [44] obtained the best results in removing ibuprofen (i.e., removal efficiency of 80%). Such results highlight the dependence of removal efficiencies on the specific characteristics of characteristics of the EC: estrogen, diazepam, and diclofenac were removed exclusively by properly acclimatized sludge [81, 82]. Therefore, a careful evaluation of the effective removal efficiency for each single contaminant and the possible occurrence of biological process inhibition have to be taken into account for each specific plant.

Paterakis et al. [83] obtained different removal efficiencies for several ECs (i.e., steroidal estrogens, nonylphenol ethoxylates) by comparatively testing primary and secondary sludge. On the contrary, no removal efficiency variation was noticed by using the same sludge with different SRT values, WTTPs plant size, and AD operating temperature (mesophilic and thermophilic conditions) [44]. Instead, the

Table 3 Removal efficiencies for several ECs by varying the biological process and the operating conditions

Type of sludge	Operating conditions	Contaminants	Removal (%)	Reference
Urban sewage treatment plants	Variations of SRT and temperature (mesophilic and thermophilic digestion)	Naproxen Sulfamethoxazole Roxithromycin Estrone+17-β-oestradiol 17α-ethinylestradiol Galaxolide Tonalide Diclofenac Diazepam Ibuprofen Iopromide Carbamazepine	88 ± 4 99 ± 1 94 ± 9 88 ± 6 (SA) 86 ± 9 (SA) 69 ± 9 63 ± 14 69 ± 10 (SA) 50 ± 16 (SA) 41 ± 15 22 ± 11 0	[81]
Primary sludge	Variations of SRT and temperature (mesophilic and thermophilic digestion)	17α-ethinylestradiol 17β-estradiol Estrone Estriol Estrone-3-sulfate Nonylphenols Nonylphenol ethoxycarboxylates Nonylphenol monoethoxylate and diethoxylate Nonylphenol polyethoxylates	34 (Mes) 43 (Term) −324 (Mes) −367 (Term) 79 (Mes) 96 (Term) 45 (Mes) 17 (Term) 36 (Mes) 30 (Term) 0 (Mes) 50 (Term) −215 (Mes) >−1,000 (Term) 88 (Mes) 2.5 (Term) 66 (Mes) 73 (Term)	[83]
Mixed sludge	Variations of SRT and temperature (mesophilic and thermophilic digestion)	17α-ethinylestradiol 17β-estradiol Estrone Estriol Estrone-3-sulfate Nonylphenols Nonylphenol ethoxycarboxylates Nonylphenol monoethoxylate and diethoxylate Nonylphenol polyethoxylates	4 (Mes) 14 (Term) −325 (Mes) −621 (Term) 70 (Mes) 68 (Term) 43 (Mes) 4 (Term) 21 (Mes) 28 (Term) 100 (Mes) 100 (Term) 0.0035 (Mes) −5,800 (Term) −0.0274	[83]

(continued)

Table 3 (continued)

Type of sludge	Operating conditions	Contaminants	Removal (%)	Reference
			(Mes) 100 (Term) 67 (Mes) 83 (Term)	
Urban sewage treatment plants	Variations of SRT and temperature (mesophilic and thermophilic digestion)	Galaxolide Tonalide Carbamazepine Diazepam Naproxen Diclofenac Iopromide Sulfamethoxazole Estrone +17-β-estradiol 17α-ethinylestradiol	65 ± 15 (Mes) 67 ± 16 (Term) 60 ± 8 (Mes) 67 ± 15 (Term) 0 (Mes) 0 (Term) 60 ± 18 (Mes) 38 ± 21 (Term) (SA) 87 ± 5 (Mes) 91 ± 5 (Term) 60 ± 18 (Mes) 73 ± 9 (Term) (SA) 23 ± 15 (Mes) 23 ± 11 (Term) 99 ± 1 (Mes) 99 ± 1 (Term) 85 ± 10 (Mes) 85 ± 5 (Term) (SA) 85 ± 5 (Mes) 75 ± 15 (Term) (SA)	[86]
Urban sewage sludge samples	Temperature (mesophilic and thermophilic digestion)	Triclosan Bisphenol A Nonylphenol Nonylphenol monoethoxylate Nonylphenol diethoxylate Ibuprofen Diclofenac Naproxen Ketoprofen Non-steroidal anti-inflammatory drugs	22 37 −17 47 31 81 <LOD 84 <LOD <LOD	[44]

removal efficiency of nonylphenols (NPs) and their nonylphenol ethoxylates (NPE) as well as carbamazepine was affected by the previously mentioned process operating parameters. Nevertheless, the AD process could be inhibited by toxic substances (i.e., antibiotics, drugs) and therefore the removal efficiencies could be largely different from those expected [84]. Different studies have been, in fact, focused on evaluating the effects of different contaminants on the biological kinetics. Boševski and Gotvajn [85] observed a decrease in biogas production by 44–45% when sewage sludge was artificially contaminated with antibiotics (tiamulin, amoxicillin, and levofloxacin). The worsening of the AD process performance was furthermore increased by the lower percentage of methane fraction in biogas.

A similar inhibiting effect was also observed for PPCPs. Chen et al. [80] found that chlorophenols (dichlorophenols, trichlorophenols and pentachlorophenol), although biodegradable, inhibited methanogenic and acetogenic microorganisms. However, not all ECs are responsible for the biogas yield decrease. Sulfamethazine and ampicillin, in fact, show no detrimental effect, although ampicillin can lead to a lag of the AD process start. On the contrary, other ECs (i.e., tylosin florfenicol) can impact seriously the biogas production. To avoid biogas production drops, various pre-treatment methods (i.e., ozonation, addition of nitrites) have been studied, aimed at totally removing contaminants or reducing their concentration.

The freezing pre-treatment of sewage sludge using activated nitrites resulted in an improvement of the AD process performance in terms of both biogas production and several contaminants removal. Liu et al. [87] observed, for a freezing time of 4 h at $-5^{\circ}C$ and with 600 mg of NO_2-N, a significant increase, by 30.6%, in methane production. Furthermore, the anaerobically digested sludge showed a better dehydration capacity and worse conditions for pathogens survival. Unfortunately, no investigation was carried out on the ECs detection. As regards the removal of ECs, the most common pre-treatment method involves ozone (O_3). This pre-treatment allows the AD hydrolysis phase by-pass thus improving the process performance. Antibiotics adsorbed on sewage sludge can also be partially degraded, thus reducing or even erasing their toxicity to anaerobic microorganisms. The O_3 dosage has to be carefully controlled because an overdosage causes partial or even total mineralization of the cells material released into the liquid bulk, thus adversely affecting the methane production. From the international literature, the optimal O_3 dosage to enhance sewage sludge solubilization ranges between 0.05 and 0.5 g O_3 g^{-1} TS (total solids), depending on the sludge properties and the operating conditions [88].

3.2 *Ozonation*

Ozonation is the most common technique to treat sewage sludge at both full and lab-scale. The oxidizing action is conducted by radicals ($\bullet O_2^-$, $HO_2^\bullet$, and $\bullet OH$) that O_3 forms as it is an unstable gas (as gas, at 20°C its half-life lasts 3 days, in aqueous solution the half-life lasts around 20 min) [89]. At the liquid state O_3 is explosive. Therefore, it is difficult to store, and therefore O_3 is produced in situ, prior to its use.

The first applications of O_3 for AD process were aimed at increasing the methane yield (compared to AD without ozonation the methane production was increased by 200% with 0.063 mg O_3/g TSS) [90]. Moreover, such treatment is also capable of removing ECs (mainly PPCPs, drugs/antibiotics) [91]. Thanks to its strong oxidizing power, O_3 is also widely used to remove pathogens from water showing a disinfection rate higher than 90–95%. The disinfectant action of O_3 is mainly due to two reactions (direct and indirect). The direct reaction depends on the structure of the reactants, whereas the indirect reaction is carried out by the hydroxyl radicals (•OH) that react with organic substances (i.e., sewage sludge biosolids).

Furthermore, O_3 is a selective electrophile and therefore reacts rapidly with chemical compounds characterized by double bonds, with phenols and amines as well as the most common ECs being included in these compounds [92]. The efficiency of O_3 pre-treatment in removing ECs from sewage sludge is largely proved by studies from the literature, where, in all of the investigated cases, it was found that the ECs removal efficiency was higher when AD was performed with O_3 pre-treatment compared with AD performed with no pre-treatment. Furthermore, Ak et al. [92] found higher removal performance with O_3 compared to other pre-treatment methods for all ECs (e.g., progesterone, acetaminophen) investigated, even those that are notoriously difficult to be biodegraded (carbamazepine, diltiazem). Moreover, Carballa et al. [91] observed a medium removal efficiency for ibuprofen (20–50%) and diazepam (50%), while removal efficiency was high (>80%) for sulfamethoxazole, 17α-ethinylestradiol, and natural estrogens (estrone and 17β-estradiol).

3.3 *Hydrogen Peroxide*

H_2O_2 is also commonly used to treat sewage sludge since the high contaminants removal efficiencies displayed. The •OH generation by AOP coupled to H_2O_2 promotes, in fact, the degradation of organic pollutants. The products of the oxidation reaction are CO_2 and specific by-products. H_2O_2 is also capable of efficiently removing odors [93]. Generally, the use of H_2O_2 is coupled with other processes such as ultraviolet rays (UV), microwaves (MWs) and Fenton [82, 93].

The MW process successfully removes chemical substances and pathogens from sewage sludge. Chang et al. [82] reported efficiencies higher than 90% for nonylphenol removal. Moreover, Yu et al. [94] with 0.08% H_2O_2 at 70°C observed a total degradation of fecal coliforms. The removal efficiency depends on MWs power, temperature, and H_2O_2 dosage. Higher efficiency results from temperature increase and higher H_2O_2 dosage. H_2O_2 is commonly combined with UVs. UVs are responsible for increasing the production of •OH radicals. Instead, if the H_2O_2 is coupled to MWs, the efficiencies and reaction times are increased. Zhang and Li [95] found, in fact, that using solely H_2O_2, the removal efficiency was 73%, 58%, 58%, 54% and 46% for estrone, 17α-ethinylestradiol, estriol, bisphenol A and NP, respectively. Instead, under the same operating conditions and using UVs, the removal efficiency increased up to values higher than 90%.

3.4 Fenton

H_2O_2 is most commonly used to perform the Fenton process. Reactions associated with the Fenton process which take place in solution are described as follows: (1) Fe (II) is oxidized by H_2O_2 to Fe (III) with the formation of •OH and a hydroxide ion (OH^-). On the other hand, reduction can take place in presence of UV–vis light (2) or in its absence (3). This latter reaction is slower than the first. Therefore, irradiance absorption is essential to the process performance. Nevertheless, the use of Fenton process at commercial scale presents various critical aspects mainly represented by: (1) high operative costs and (2) use of highly acid pH.

$$Fe^{2+} + H_2O_2 \rightarrow Fe^{3+} + \bullet OH + OH^- \quad (1)$$

$$Fe^{3+} + H_2O_2 \rightarrow Fe^{2+} + \bullet OOH + H^+ \quad (2)$$

$$Fe^{3+} + H_2O + h\nu \rightarrow Fe^{2+} + \bullet HO + H+ \quad (3)$$

The Fenton process can also occur with other metals capable of catalyzing the reactions (i.e., Cu (I), Co (II)). Process performance strongly depends on H_2O_2 concentration and pH conditions. Also, the amount of radicals produced is proportional to H_2O_2 dosage (the ideal Fe/H_2O_2 ratio is 1:25 (wt/wt)).

The use of Fenton process resulted in being efficient to remove some estrogens (i.e., estrone, estriol, 17 β-estradiol and 17 α-ethinylestradiol) from sewage sludge by oxidation (efficiencies are about 90%). Nevertheless, it has been observed that an excessive dosage of H_2O_2 could inhibit the degradation of estrogens, as the reaction between Fe (III) and •OH could be favored (efficiencies are about 70%) [96].

Good efficiencies were also obtained for polycyclic aromatic hydrocarbons (PAHs) and other contaminants removal from sewage sludge. Flotron et al. [24] reached removal efficiencies of 46%, 36%, and 48%, respectively, for fluoranthene, benzo [b] fluoranthene, benzo [a] pyrene. Pulicharla et al. [97] studied the degradation of chlortetracycline through the Fenton process and obtained a removal efficiency of 76%. Moreover, Fenton process combined with ultrasonification displayed higher removal efficiencies (up to 81%) than the ultrasonification treatment alone (i.e. 67%). Such result proves the remarkable efficiency of the integrated process (i.e., Fenton process coupled with ultrasonification). Studies conducted specifically on drugs are currently scarce, thus reflecting the need of increasing the research on this topic.

3.5 Hydrothermal Liquefaction

The valorization of sewage sludge by hydrothermal liquefaction (HTL) was studied in order to obtain high energy density fuels. This treatment, also called hydrous pyrolysis, is conducted by heating the sludge to a temperature of 200–400°C under pressures higher than the water vapor pressure (typically 10–25 MPa) with the aim to prevent water from evaporation. Through this process it has been possible to convert sewage sludge into bio-crude with weight ratio of 50–64% [98–100]. It has been observed that this process is also capable of removing ECs from sewage sludge thanks to the high temperatures and pressures. So far, a few studies have been conducted on this process to investigate its efficiency in removing ECs. Pham et al. [101] found excellent results in removing antibiotics (florfenicol, ceftiofur) from spirulina algae, already after 15 min of treatment at 250°C with removal efficiency higher than 95%. Furthermore, after 30 min of treatment the removal efficiency was almost 100%. Moreover, under a temperature of 300°C, the same removal efficiency (i.e., 100%) was reached in a shorter time (15 min), thus proving the positive effect of temperature on the process performance.

Thomsen et al. [102] tested the removal efficiencies of 37 drugs and biocides in primary sludge processed by HTL with different temperatures (between 300 and 350°C). In their study, they did not observe different removal efficiencies by varying temperatures, and removal efficiencies higher than 80% were found for almost all ECs. Only a few compounds showed to be resistant to the HTL process. An analytical study is required to better understand the recalcitrance of these compounds. Therefore, HTL can be a valid process to obtain a valorization of sewage sludge and concurrently reduce the ECs content.

4 Conclusions and Future Perspectives

The presence of ECs in wastewater is a current source of concern and legislation is beginning to consider and face this issue. Nevertheless, a secondary importance is given to their presence in sewage sludge. In fact, ECs can bioaccumulate in sewage sludge and need to be properly removed to avoid direct soil contamination and indirect water contamination. Conventional processes (i.e., anaerobic digestion) used for sludge treatments cannot ensure high removal efficiency values for ECs. Therefore, in this chapter current innovative processes that can be used to treat sewage sludge have been analyzed and discussed. It has been observed that some of them allow high removal efficiencies. Ozonation, the only common process used at industrial scale, shows an almost total ECs removal. Nevertheless, this process is difficult to manage when it is used in large treatment plants, so it is necessary to find treatments efficient as much as ozonation, but more easy to scale up. The use of H_2O_2 coupled with other processes could be a valid alternative (for instance, the Fenton process). HTL could also represent a valid technology to be used for the removal of ECs from sewage sludge. However, prior to their scaling-up, other scientific studies are required to evaluate the effective removal efficiencies and find the optimal operating conditions.

References

1. Bell CH, Gentile M, Kalve E, Ross I, Horst J, Suthersan S (2019) Emerging contaminants handbook. CRC Press
2. Mo L, Zheng J, Wang T, Shi Y-G, Chen B-J, Liu B, Ma Y-H, Li M, Zhuo L, Chen S-J (2019) Legacy and emerging contaminants in coastal surface sediments around Hainan Island in South China. Chemosphere 215:133–141
3. Sauvé S, Desrosiers M (2014) A review of what is an emerging contaminant. Chem Cent J 8: 1–7
4. Fijalkowski K, Rorat A, Grobelak A, Kacprzak MJ (2017) The presence of contaminations in sewage sludge–The current situation. J Environ Manage 203:1126–1136
5. Mailler R, Gasperi J, Chebbo G, Rocher V (2014) Priority and emerging pollutants in sewage sludge and fate during sludge treatment. Waste Manag 34:1217–1226
6. Stasinakis AS (2012) Review on the fate of emerging contaminants during sludge anaerobic digestion. Bioresour Technol 121:432–440
7. Tran NH, Reinhard M, Gin KY-H (2018) Occurrence and fate of emerging contaminants in municipal wastewater treatment plants from different geographical regions-a review. Water Res 133:182–207
8. Ni B-J, Zeng S, Wei W, Dai X, Sun J (2020) Impact of roxithromycin on waste activated sludge anaerobic digestion: methane production, carbon transformation and antibiotic resistance genes. Sci Total Environ 703:134899
9. Yin F, Dong H, Ji C, Tao X, Chen Y (2016) Effects of anaerobic digestion on chlortetracycline and oxytetracycline degradation efficiency for swine manure. Waste Manag 56:540–546
10. Yin F, Dong H, Zhang W, Zhu Z, Shang B, Wang Y (2019) Removal of combined antibiotic (florfenicol, tylosin and tilmicosin) during anaerobic digestion and their relative effect. Renew Energy 139:895–903
11. Zhao L, Ji Y, Sun P, Deng J, Wang H, Yang Y (2019) Effects of individual and combined zinc oxide nanoparticle, norfloxacin, and sulfamethazine contamination on sludge anaerobic digestion. Bioresour Technol 273:454–461
12. Lapen DR, Topp E, Metcalfe CD, Li H, Edwards M, Gottschall N, Bolton P, Curnoe W, Payne M, Beck A (2008) Pharmaceutical and personal care products in tile drainage following land application of municipal biosolids. Sci Total Environ 399:50–65
13. Sabourin L, Beck A, Duenk PW, Kleywegt S, Lapen DR, Li H, Metcalfe CD, Payne M, Topp E (2009) Runoff of pharmaceuticals and personal care products following application of dewatered municipal biosolids to an agricultural field. Sci Total Environ 407:4596–4604
14. Hospido A, Carballa M, Moreira M, Omil F, Lema JM, Feijoo G (2010) Environmental assessment of anaerobically digested sludge reuse in agriculture: potential impacts of emerging micropollutants. Water Res 44:3225–3233
15. Ivanová L, Mackuľak T, Grabic R, Golovko O, Koba O, Staňová AV, Szabová P, Grenčíková A, Bodík I (2018) Pharmaceuticals and illicit drugs–a new threat to the application of sewage sludge in agriculture. Sci Total Environ 634:606–615
16. Escher B, Leusch F (2011) Bioanalytical tools in water quality assessment. IWA Publishing
17. de Jesus Gaffney V, Almeida CMM, Rodrigues A, Ferreira E, Benoliel MJ, Cardoso VV (2015) Occurrence of pharmaceuticals in a water supply system and related human health risk assessment. Water Res 72:199–208

18. Barret M, Barcia GC, Guillon A, Carrère H, Patureau D (2010) Influence of feed characteristics on the removal of micropollutants during the anaerobic digestion of contaminated sludge. J Hazard Mater 181:241–247
19. Núñez-Delgado A, Pousada-Ferradás Y, Álvarez-Rodríguez E, Fernández-Sanjurjo MJ, Conde-Cid M, Nóvoa-Muñoz JC, Arias-Estévez M (2019) Effects of microbiological and non-microbiological treatments of sewage sludge on antibiotics as emerging pollutants present in wastewater: a review. Microb Wastewater Treat:1–17
20. Ibáñez GR, Esteban B, Ponce-Robles L, López JLC, Agüera A, Pérez JAS (2015) Fate of micropollutants during sewage sludge disintegration by low-frequency ultrasound. Chem Eng J 280:575–587
21. Reyes-Contreras C, Neumann P, Barriga F, Venegas M, Domínguez C, Bayona JM, Vidal G (2020) Organic micropollutants in sewage sludge: influence of thermal and ultrasound hydrolysis processes prior to anaerobic stabilization. Environ Technol 41:1358–1365
22. Zhang X, Li R (2018) Variation of antibiotics in sludge pretreatment and anaerobic digestion processes: Degradation and solid-liquid distribution. Bioresour Technol 255:266–272
23. Patureau D, Delgenes N, Delgenes J-P (2008) Impact of sewage sludge treatment processes on the removal of the endocrine disrupters nonylphenol ethoxylates. Chemosphere 72:586–591
24. Flotron V, Delteil C, Padellec Y, Camel V (2005) Removal of sorbed polycyclic aromatic hydrocarbons from soil, sludge and sediment samples using the Fenton's reagent process. Chemosphere 59:1427–1437
25. Zheng X-J, Blais J-F, Mercier G, Bergeron M, Drogui P (2007) PAH removal from spiked municipal wastewater sewage sludge using biological, chemical and electrochemical treatments. Chemosphere 68:1143–1152
26. Andersen HR, Hansen M, Kjølholt J, Stuer-Lauridsen F, Ternes T, Halling-Sørensen B (2005) Assessment of the importance of sorption for steroid estrogens removal during activated sludge treatment. Chemosphere 61:139–146
27. Goss K-U, Schwarzenbach RP (2001) Linear free energy relationships used to evaluate equilibrium partitioning of organic compounds. Environ Sci Technol 35:1–9
28. Schwarzenbach RP, Gschwend PM, Imboden DM (2016) Environmental organic chemistry. Wiley
29. Carballa M, Fink G, Omil F, Lema JM, Ternes T (2008) Determination of the solid–water distribution coefficient (Kd) for pharmaceuticals, estrogens and musk fragrances in digested sludge. Water Res 42:287–295
30. Simonich SL, Federle TW, Eckhoff WS, Rottiers A, Webb S, Sabaliunas D, De Wolf W (2002) Removal of fragrance materials during US and European wastewater treatment. Environ Sci Technol 36:2839–2847
31. Ternes TA, Herrmann N, Bonerz M, Knacker T, Siegrist H, Joss A (2004) A rapid method to measure the solid–water distribution coefficient (Kd) for pharmaceuticals and musk fragrances in sewage sludge. Water Res 38:4075–4084
32. Bourdat-Deschamps M, Ferhi S, Bernet N, Feder F, Crouzet O, Patureau D, Montenach D, Moussard GD, Mercier V, Benoit P (2017) Fate and impacts of pharmaceuticals and personal care products after repeated applications of organic waste products in long-term field experiments. Sci Total Environ 607:271–280
33. Gao P, Ding Y, Li H, Xagoraraki I (2012) Occurrence of pharmaceuticals in a municipal wastewater treatment plant: mass balance and removal processes. Chemosphere 88:17–24
34. Golet EM, Xifra I, Siegrist H, Alder AC, Giger W (2003) Environmental exposure assessment of fluoroquinolone antibacterial agents from sewage to soil. Environ Sci Technol 37:3243–3249
35. Golet EM, Strehler A, Alder AC, Giger W (2002) Determination of fluoroquinolone antibacterial agents in sewage sludge and sludge-treated soil using accelerated solvent extraction followed by solid-phase extraction. Anal Chem 74:5455–5462
36. Jia A, Wan Y, Xiao Y, Hu J (2012) Occurrence and fate of quinolone and fluoroquinolone antibiotics in a municipal sewage treatment plant. Water Res 46:387–394

37. Martín J, Santos JL, Aparicio I, Alonso E (2015) Pharmaceutically active compounds in sludge stabilization treatments: Anaerobic and aerobic digestion, wastewater stabilization ponds and composting. Sci Total Environ 503:97–104
38. Takács-Novák K, Józan M, Hermecz I, Szász G (1992) Lipophilicity of antibacterial fluoroquinolones. Int J Pharm 79:89–96
39. Dal Pozzo A, Donzelli G, Rodriquez L, Tajana A (1989) "In vitro" model for the evaluation of drug distribution and plasma protein-binding relationships. Int J Pharm 50:97–101
40. Dong R, Yu G, Guan Y, Wang B, Huang J, Deng S, Wang Y (2016) Occurrence and discharge of pharmaceuticals and personal care products in dewatered sludge from WWTPs in Beijing and Shenzhen. Emerg Contam 2:1–6
41. Subedi B, Balakrishna K, Joshua DI, Kannan K (2017) Mass loading and removal of pharmaceuticals and personal care products including psychoactives, antihypertensives, and antibiotics in two sewage treatment plants in southern India. Chemosphere 167:429–437
42. Avdeef A, Box KJ, Comer JEA, Hibbert C, Tam KY (1998) pH-Metric logP 10. Determination of liposomal membrane-water partition coefficients of lonizable drugs. Pharm Res 15: 209–215
43. Czyrski A (2019) Determination of the lipophilicity of ibuprofen, naproxen, ketoprofen, and flurbiprofen with thin-layer chromatography. J Chem 2019
44. Samaras VG, Stasinakis AS, Mamais D, Thomaidis NS, Lekkas TD (2013) Fate of selected pharmaceuticals and synthetic endocrine disrupting compounds during wastewater treatment and sludge anaerobic digestion. J Hazard Mater 244:259–267
45. Yu Y, Wu L (2012) Analysis of endocrine disrupting compounds, pharmaceuticals and personal care products in sewage sludge by gas chromatography–mass spectrometry. Talanta 89:258–263
46. Martín J, Camacho-Muñoz D, Santos JL, Aparicio I, Alonso E (2012) Occurrence of pharmaceutical compounds in wastewater and sludge from wastewater treatment plants: removal and ecotoxicological impact of wastewater discharges and sludge disposal. J Hazard Mater 239: 40–47
47. Prasad MNV, Vithanage M, Kapley A (2019) Pharmaceuticals and personal care products: waste management and treatment technology: emerging contaminants and micro pollutants. Butterworth-Heinemann
48. Kinney CA, Furlong ET, Zaugg SD, Burkhardt MR, Werner SL, Cahill JD, Jorgensen GR (2006) Survey of organic wastewater contaminants in biosolids destined for land application. Environ Sci Technol 40:7207–7215
49. Lindberg RH, Olofsson U, Rendahl P, Johansson MI, Tysklind M, Andersson BAV (2006) Behavior of fluoroquinolones and trimethoprim during mechanical, chemical, and active sludge treatment of sewage water and digestion of sludge. Environ Sci Technol 40:1042–1048
50. Ben W, Zhu B, Yuan X, Zhang Y, Yang M, Qiang Z (2018) Occurrence, removal and risk of organic micropollutants in wastewater treatment plants across China: Comparison of wastewater treatment processes. Water Res 130:38–46
51. Göbel A, Thomsen A, McArdell CS, Alder AC, Giger W, Theiß N, Löffler D, Ternes TA (2005) Extraction and determination of sulfonamides, macrolides, and trimethoprim in sewage sludge. J Chromatogr A 1085:179–189
52. Göbel A, Thomsen A, McArdell CS, Joss A, Giger W (2005) Occurrence and sorption behavior of sulfonamides, macrolides, and trimethoprim in activated sludge treatment. Environ Sci Technol 39:3981–3989
53. Hansch C, Leo A, Hoekman D (1995) Hydrophobic, electronic, and steric constants. American Chemical Society, Washington, pp 219–304
54. McClellan K, Halden RU (2010) Pharmaceuticals and personal care products in archived US biosolids from the 2001 EPA national sewage sludge survey. Water Res 44:658–668
55. Chen X, Bester K (2009) Determination of organic micro-pollutants such as personal care products, plasticizers and flame retardants in sludge. Anal Bioanal Chem 395:1877–1884

56. Chu S, Metcalfe CD (2007) Simultaneous determination of triclocarban and triclosan in municipal biosolids by liquid chromatography tandem mass spectrometry. J Chromatogr A 1164:212–218
57. Lyndall J, Fuchsman P, Bock M, Barber T, Lauren D, Leigh K, Perruchon E, Capdevielle M (2010) Probabilistic risk evaluation for triclosan in surface water, sediments, and aquatic biota tissues. Integr Environ Assess Manag An Int J 6(3):419–440
58. Peysson W, Vulliet E (2013) Determination of 136 pharmaceuticals and hormones in sewage sludge using quick, easy, cheap, effective, rugged and safe extraction followed by analysis with liquid chromatography–time-of-flight-mass spectrometry. J Chromatogr A 1290:46–61
59. Shi L, Zhou X, Zhang Y, Jiang T (2011) Comparison of sorption coefficients of triclosan in different activated sludge and application in sewage treatment system. In: 2011 Third international conference on measuring technology and mechatronics automation. IEEE, pp 749–752
60. Stasinakis AS, Gatidou G, Mamais D, Thomaidis NS, Lekkas TD (2008) Occurrence and fate of endocrine disrupters in Greek sewage treatment plants. Water Res 42:1796–1804
61. Ying G-G, Kookana RS (2007) Triclosan in wastewaters and biosolids from Australian wastewater treatment plants. Environ Int 33:199–205
62. Balk F, Ford RA (1999) Environmental risk assessment for the polycyclic musks AHTN and HHCB in the EU: I. Fate and exposure assessment. Toxicol Lett 111:57–79
63. Ternes TA, Bonerz M, Herrmann N, Löffler D, Keller E, Lacida BB, Alder AC (2005) Determination of pharmaceuticals, iodinated contrast media and musk fragrances in sludge by LC tandem MS and GC/MS. J Chromatogr A 1067:213–223
64. Stevens JL, Northcott GL, Stern GA, Tomy GT, Jones KC (2003) PAHs, PCBs, PCNs, organochlorine pesticides, synthetic musks, and polychlorinated n-alkanes in UK sewage sludge: survey results and implications. Environ Sci Technol 37:462–467
65. Venegas M, Leiva AM, Reyes-Contreras C, Neumann P, Piña B, Vidal G (2020) Presence and fate of micropollutants during anaerobic digestion of sewage and their implications for the circular economy: a short review. J Environ Chem Eng:104931
66. De Maria IC, Chiba MK, Costa A, Berton RS (2010) Sewage sludge application to agricultural land as soil physical conditioner. Rev Bras Ciência do Solo 34:967–974
67. Rybacka A, Andersson PL (2016) Considering ionic state in modeling sorption of pharmaceuticals to sewage sludge. Chemosphere 165:284–293
68. Li B, Zhang T (2010) Biodegradation and adsorption of antibiotics in the activated sludge process. Environ Sci Technol 44:3468–3473
69. Ben W, Qiang Z, Yin X, Qu J, Pan X (2014) Adsorption behavior of sulfamethazine in an activated sludge process treating swine wastewater. J Environ Sci 26:1623–1629
70. Galán MJG, Díaz-Cruz MS, Barceló D (2012) Removal of sulfonamide antibiotics upon conventional activated sludge and advanced membrane bioreactor treatment. Anal Bioanal Chem 404:1505–1515
71. Song X, Liu D, Zhang G, Frigon M, Meng X, Li K (2014) Adsorption mechanisms and the effect of oxytetracycline on activated sludge. Bioresour Technol 151:428–431
72. Tran NH, Chen H, Reinhard M, Mao F, Gin KY-H (2016) Occurrence and removal of multiple classes of antibiotics and antimicrobial agents in biological wastewater treatment processes. Water Res 104:461–472
73. Tran NH, Gin KY-H (2017) Occurrence and removal of pharmaceuticals, hormones, personal care products, and endocrine disrupters in a full-scale water reclamation plant. Sci Total Environ 599:1503–1516
74. Ou X, Wang C, Zhang F, Quan X, Ma Y, Liu H (2010) Complexation of iron by salicylic acid and its effect on atrazine photodegradation in aqueous solution. Front Environ Sci Eng China 4:157–163
75. Verlicchi P, Al Aukidy M, Zambello E (2012) Occurrence of pharmaceutical compounds in urban wastewater: removal, mass load and environmental risk after a secondary treatment—a review. Sci Total Environ 429:123–155

76. Dubey M, Mohapatra S, Tyagi VK, Suthar S, Kazmi AA (2021) Occurrence, fate, and persistence of emerging micropollutants in sewage sludge treatment. Environ Pollut 273: 116515
77. Oberoi AS, Jia Y, Zhang H, Khanal SK, Lu H (2019) Insights into the fate and removal of antibiotics in engineered biological treatment systems: a critical review. Environ Sci Technol 53:7234–7264
78. Eddy M, Abu-Orf M, Bowden G, Burton FL, Pfrang W, Stensel HD, Tchobanoglous G, Tsuchihashi R, (Firm) A (2014) Wastewater engineering: treatment and resource recovery. McGraw Hill Education
79. Hörsing M, Ledin A, Grabic R, Fick J, Tysklind M, la Cour Jansen J, Andersen HR (2011) Determination of sorption of seventy-five pharmaceuticals in sewage sludge. Water Res 45: 4470–4482
80. Chen Y, Cheng JJ, Creamer KS (2008) Inhibition of anaerobic digestion process: a review. Bioresour Technol 99:4044–4064
81. Carballa M, Omil F, Ternes T, Lema JM (2007) Fate of pharmaceutical and personal care products (PPCPs) during anaerobic digestion of sewage sludge. Water Res 41:2139–2150
82. Chang Y, Tsai K, Tseng C, Chen J, Kao C, Lin K (2015) Rapid nonylphenol degradation in wastewater sludge using microwave peroxide oxidation with nitric acid. Environ Prog Sustain Energy 34:520–525
83. Paterakis N, Chiu TY, Koh YKK, Lester JN, McAdam EJ, Scrimshaw MD, Soares A, Cartmell E (2012) The effectiveness of anaerobic digestion in removing estrogens and nonylphenol ethoxylates. J Hazard Mater 199:88–95
84. Mitchell SM, Ullman JL, Teel AL, Watts RJ, Frear C (2013) The effects of the antibiotics ampicillin, florfenicol, sulfamethazine, and tylosin on biogas production and their degradation efficiency during anaerobic digestion. Bioresour Technol 149:244–252
85. Boševski I, Gotvajn AŽ (2021) The impact of single step ozonation of antibiotics-contaminated waste sludge to biogas production. Chemosphere 271:129527
86. Carballa M, Omil F, Alder AC, Lema JM (2006) Comparison between the conventional anaerobic digestion of sewage sludge and its combination with a chemical or thermal pre-treatment concerning the removal of pharmaceuticals and personal care products. Water Sci Technol 53:109–117
87. Liu X, Huang X, Wu Y, Xu Q, Du M, Wang D, Yang Q, Liu Y, Ni B-J, Yang G (2020) Activation of nitrite by freezing process for anaerobic digestion enhancement of waste activated sludge: performance and mechanisms. Chem Eng J 387:124147
88. Zhen G, Lu X, Kato H, Zhao Y, Li YY (2017) Overview of pretreatment strategies for enhancing sewage sludge disintegration and subsequent anaerobic digestion: current advances, full-scale application and future perspectives. Renew Sustain Energy Rev 69:559–577. https://doi.org/10.1016/j.rser.2016.11.187
89. Khare P, Patel RK, Sharan S, Shankar R (2021) Recent trends in advanced oxidation process for treatment of recalcitrant industrial effluents. In: Advanced oxidation processes for effluent treatment plants. Elsevier, pp 137–160
90. Silvestre G, Ruiz B, Fiter M, Ferrer C, Berlanga JG, Alonso S, Canut A (2015) Ozonation as a pre-treatment for anaerobic digestion of waste-activated sludge: effect of the ozone doses. Ozone Sci Eng 37:316–322
91. Carballa M, Manterola G, Larrea L, Ternes T, Omil F, Lema JM (2007) Influence of ozone pre-treatment on sludge anaerobic digestion: removal of pharmaceutical and personal care products. Chemosphere 67:1444–1452
92. Ak MS, Muz M, Komesli OT, Gökçay CF (2013) Enhancement of bio-gas production and xenobiotics degradation during anaerobic sludge digestion by ozone treated feed sludge. Chem Eng J 230:499–505
93. Guan R, Yuan X, Wu Z, Jiang L, Li Y, Zeng G (2018) Principle and application of hydrogen peroxide based advanced oxidation processes in activated sludge treatment: A review. Chem Eng J 339:519–530

94. Yu Y, Chan WI, Liao PH, Lo KV (2010) Disinfection and solubilization of sewage sludge using the microwave enhanced advanced oxidation process. J Hazard Mater 181:1143–1147
95. Zhang A, Li Y (2014) Removal of phenolic endocrine disrupting compounds from waste activated sludge using UV, H2O2, and UV/H2O2 oxidation processes: effects of reaction conditions and sludge matrix. Sci Total Environ 493:307–323
96. Li Y, Zhang A (2014) Removal of steroid estrogens from waste activated sludge using Fenton oxidation: influencing factors and degradation intermediates. Chemosphere 105:24–30
97. Pulicharla R, Brar SK, Rouissi T, Auger S, Drogui P, Verma M, Surampalli RY (2017) Degradation of chlortetracycline in wastewater sludge by ultrasonication, Fenton oxidation, and ferro-sonication. Ultrason Sonochem 34:332–342
98. Biller P, Johannsen I, Dos Passos JS, Ottosen LDM (2018) Primary sewage sludge filtration using biomass filter aids and subsequent hydrothermal co-liquefaction. Water Res 130:58–68
99. Madsen RB, Glasius M (2019) How do hydrothermal liquefaction conditions and feedstock type influence product distribution and elemental composition? Ind Eng Chem Res 58:17583–17600
100. Marrone PA, Elliott DC, Billing JM, Hallen RT, Hart TR, Kadota P, Moeller JC, Randel MA, Schmidt AJ (2018) Bench-scale evaluation of hydrothermal processing technology for conversion of wastewater solids to fuels: Marrone et al. Water Environ Res 90:329–342
101. Pham M, Schideman L, Sharma BK, Zhang Y, Chen W-T (2013) Effects of hydrothermal liquefaction on the fate of bioactive contaminants in manure and algal feedstocks. Bioresour Technol 149:126–135
102. Thomsen LBS, Carvalho PN, Dos Passos JS, Anastasakis K, Bester K, Biller P (2020) Hydrothermal liquefaction of sewage sludge; energy considerations and fate of micropollutants during pilot scale processing. Water Res 183:116101

Emerging Pollutants That Can Be Transformed into PCDD/Fs

S. Schröder, I. Ortiz, and M.-F. San-Román

Contents

Abstract Water and soil pollution is one of the top problems of environmental concern that has been worsened in the last years due to the lack of available water and overall resources. Owing to the difficulty in eliminating pollutants from the environment, applying cost-effective treatments to the anthropogenic sources of origin (i.e., wastewaters and industrial effluents) before they are discharged to the environment becomes a very good alternative. The presence of emerging pollutants (EPs) deserves special attention and is being considered in new regulations due to their adverse effects and/or persistence. Among them, a group with increasing presence is that related to the use of Pharmaceutical and Personal Care Products (PPCPs). Triclosan (TCS; 5-Chloro-2-(2,4-dichlorophenoxy)phenol) is an emerging antimicrobial contaminant that belongs to PPCPs. The extensive use of TCS has caused its presence in different water sources, especially in WWTPs effluents, lakes, rivers, and also in sediments throughout the world, posing a risk to the environment and human health. Advanced Oxidation Processes (AOPs), defined as aqueous phase

S. Schröder, I. Ortiz, and M.-F. San-Román (✉)
Departamento de Ingenierías Química y Biomolecular, ETSIIyT, Universidad de Cantabria, Santander, Spain
e-mail: sanromm@unican.es

Avelino Núñez-Delgado and Manuel Arias-Estévez (eds.),
Emerging Pollutants in Sewage Sludge and Soils,
Hdb Env Chem (2023) 114: 111–136, DOI 10.1007/698_2022_851,

Published online: 21 July 2022

oxidation technologies based on the formation of highly reactive species, such as hydroxyl radicals ($^{\bullet}OH$), have been successfully applied to the remediation of wastewaters containing recalcitrant organic compounds. However, depending on the operating conditions, AOPs can lead to the generation of by-products during the degradation of the primary pollutants that may increase the final toxicity of the treated water. In this way, TCS, along with its degradation/oxidation routes, has been related to the formation of persistent organic pollutants (POPs) such as polychlorinated dibenzo-*p*-dioxins and dibenzofurans (PCDD/Fs). Specifically, this chapter focuses on the potential of electrochemical and photocatalytic oxidation, in terms of TCS degradation rate and by-products formation, paying special attention to the potential formation of PCDD/Fs. The role of two electrolytes widely applied in electrochemical oxidation, NaCl and Na_2SO_4, as well as the influence of the concentration of a catalyst (TiO_2), on the photocatalytic medium, has been analyzed. The toxicity in the treated waters was assessed through the toxic equivalents (TEQ) expressed as the 2,3,7,8-TCDD TEQ concentration, where 2,3,7,8-TCDD is the most toxic congener (NATO Committee on the Challenges of Modern Society (NATO/CCMS)). After complete TCS degradation, TEQ reached a maximum value of 4.78 pg L^{-1} in electrochemically remediated waters, using NaCl as electrolyte (19 times higher with respect to untreated water containing TCS), whereas the value of the photocatalytically treated waters with TiO_2 0.75 g L^{-1} was 6.25 pg L^{-1} (more than 25 times higher with respect to untreated water containing TCS). These results emphasize the importance of the quantitative traceability of the formation of by-products and the resulting toxicity (PCDD/Fs, TEQ,) when applying AOPs to polluted waters containing EPs, especially organochlorines, considered precursors of PCDD/Fs.

Keywords Advanced oxidation processes (AOPs), Electrochemical oxidation, Emerging pollutants (EPs), Pharmaceutical and personal care products (PPCPs), Photocatalysis, Polychlorinated dibenzo-*p*-dioxins and Polychlorinated dibenzofurans (PCDD/Fs), Triclosan (TCS)

1 Introduction

Water and soil pollution are some of the top problems of environmental concern that have been worsened in the last years due to the lack of available resources. This motivates the promotion of higher quality requirements of treated waters before they enter the environment [1]. As reported by the United Nations, currently, the world population is 7.8 billion people and is projected to reach 8.5 billion by 2030 and to raise to more than 9.7 billion by 2050 [2]. In the next decade, due to the increase of population and with this, of the world urbanization and industrial activity, an increase in total waste generation and in the water demand is expected [2]. Specifically, the world population will need high amounts of potable water.

According to The United Nations World Water Development Report 2021, approximately more than 80% of all world's wastewater is released into the environment, without any previous treatment, entering lakes, rivers, aquifers, and the sea and deteriorating water quality [3]. So far, there is no equilibrium between the quick industrialization and the appropriate water treatments. In addition, more than 2 billion people live in countries experiencing water stress and the continuous exploitation of the water resources will continue increasing these levels in coming years [4]. In this context, the United Nations declare, in the Goal number 6 of the 2030 Agenda and Sustainable Development Goals (SDGs), that by 2030 it is needed to "*Improve water quality by reducing pollution, eliminating dumping and minimizing the release of hazardous chemicals and materials*" and increase water recycling and safe reuse in a global way [5].

The soil quality is included in Goal 6 of the SDGs; it plays an important role in the movement, storage, and transformation of water and influences the quality and accessibility of water supply. When the pollution of surface water occurs, it does not only affect the aquatic organisms or the surrounding vegetation, but also the soils due to the transport of the contaminants through them. Although soils have an important role as natural filter, some pollutants travel to groundwater reservoirs, and then the problem is even more crucial, as there are no simple methods to solve this situation [6]. Owing to the difficulty in eliminating pollutants present in soils, cost-effective remediation methods should be applied to the sources of origin before the pollutants are discharged into the environment. Soil contamination can be produced by far too much fertilizers and pesticides utilization, discharges of industry, or the waters coming from WWTPs, that have not been able to eliminate EPs, among other sources. Nowadays, soil and land are increasingly considered scarce and limited resources worldwide. For this reason, Goal 6 can be related to other Goals of SDGs, Goal 15: Life on Land and Goal 2: Zero Hunger, where soil is essential to ensure the production of healthy food but without excess abuse of the land [7].

This chapter focuses on a problem closely related to water and soil contamination, as manifested in the SDGs, numbers 2, 6, and 15, specifically considering the impact of remediation technologies to decrease the concentration of EPs that are discharged to the environment. The emerging pollutant TCS, an antimicrobial and antifungal disinfectant agent belonging to PPCPs, has been selected as a case of study. Electrochemical and photocatalytic oxidation treatments have been applied to emphasize the importance of the quantitative traceability of by-products formation and determination of the resulting toxicity (TEQ, PCDD/Fs) depending on the selected treatment and operational conditions.

2 Emerging Pollutants (EPs)

As a result of uncontrolled urbanization, industrial development, health care and personal care activities, agriculture and transport, the production and use of a wide range of substances considered indispensable for the modern society and known as emerging pollutants has increased [8]. EPs constitute a group of wide spectrum chemicals that are present at trace concentration (ranges from μg L^{-1} to ng L^{-1}), and they are not habitually tracked but they have the capability to penetrate into the ecosystems because of their physical and chemical properties; they are responsible for negative environmental and harmful human health impacts, such as endocrine system disruption effects or the development of pathogen resistance [9]. These pollutants are candidates for future legislation that will try to control their adverse effects and/or persistency [10]. The growth of the population in the last few decades, the higher rate of goods consumption, and other anthropogenic activities have led to the increased release of EPs to the environment; industrial and household effluents, agriculture and farms water discharge, hospital effluents and landfill leachates are the main sources of these pollutants. EPs can be classified, according to the NORMAN database [11], into 20 groups: antibiotics and other pharmaceuticals, steroid hormones and other endocrine-disrupting compounds (EDCs), fire retardants, sunscreens, disinfection by-products, etc. Among them, Pharmaceutical and Personal Care Products (PPCPs) is one of the largest groups and includes some of the most frequently identified EPs in the different ecosystems [8]. Table 1 collects a short classification of EPs, together with the most common compounds of each group. Figure 1 shows the structural shape of some examples of PPCPs molecules.

The available data on EPs are relatively new and there is scarce regulation in general terms. However, priority pollutants (PPs), which are mostly part of EPs, are regulated at international and national levels due to their high risk toward the aquatic biota and human health, hence their "priority" status [13]. Within EPs, the Persistent Organic Pollutants (POPs) are also toxic chemicals (intentionally or unintentionally produced) that also affect negatively to human health and are very persistent remaining long time in the environment; they can be accumulated and transported long distances through air or water, to finally entering in the food chain [14]. In the 2001 Stockholm Convention, 12 key POPs, the so-called "Dirty Dozen" were highlighted and the countries determined and accepted the reduction and/or elimination of the manufacture, use and discharge of those 12 POPs. Currently, there are 36 POPs in the list [15], and Polychlorinated dibenzo-*p*-dioxins (PCDDs) and Polychlorinated dibenzofurans (PCDFs) compounds are among them (Fig. 2).

Table 1 Main categories of EPs [8, 12]

EPs group	Representative compounds
Pesticides	Atrazine, 2,4-D, fenoprop, tebuthiuron, pentachlorophenol, diuron
Pharmaceutical and personal care products (PPCPs)	*β-blockers:* atenolol, metoprolol, propranolol, timolol, sotalol *Analgesics, anti-inflammatory drugs*: ibuprofen, diclofenac, fenoprofen *Antibiotics*: azithromycin, clarithromycin, erythromycin, ofloxacin *Disinfectants*: triclosan *Lipid regulators*: fenofibric acid, gemfibrozil, bezafibrate *X-ray contrasting agents*: iopromide, iopamidol, diatrizoate
Flame retardants	Polybrominated diphenyl ethers (PBDEs): polybrominated biphenyls (PBBs), polybrominated dibenzo-*p*-dioxins (PBDDs), polybrominated dibenzofurans (PBDFs). Tetrabromo bisphenol A, C10-C13 chloroalkanes, Tris (2-chloroethyl) phosphate, Hexabromocyclododecanes (HBCDs)
Hormones and steroids	Estradiol, estrone, estriol, diethylstilbestrol (DES)
Surfactants	Alkylphenol ethoxylates, 4-nonylphenol, 4-octylphenol
Food additives	Acesulfame
Perfluorinated compounds	Perfluoro octane sulfonates (PFOs), perfluorooctanoic acid (PFOA)
Gasoline additives	Dialkyl ethers, methyl-t-butyl ether (MTBE)
Biocides	Parabens, neonicotinoids
UV filters	Benzophenone-3, homosalate, octocrylene, 4-MBC
Particles	Microplastics
Anthropogenic markers	Bromoform, nicotine, Benzotriazole methyl-1H

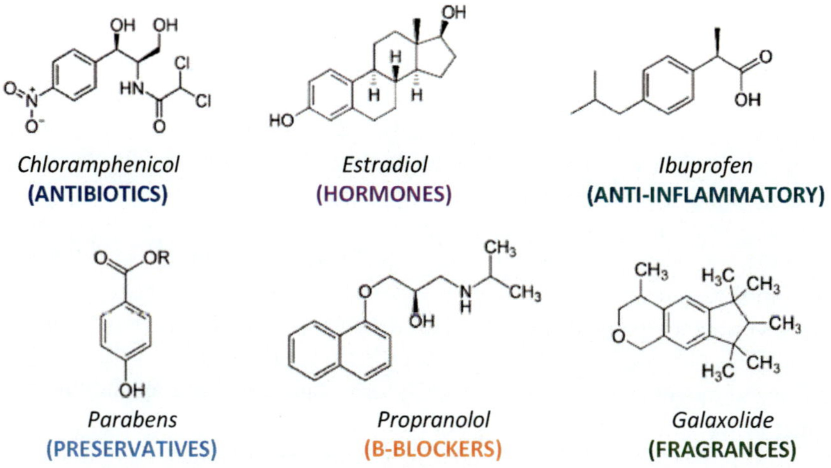

Fig. 1 Some examples of EPs belonging to PPCPs

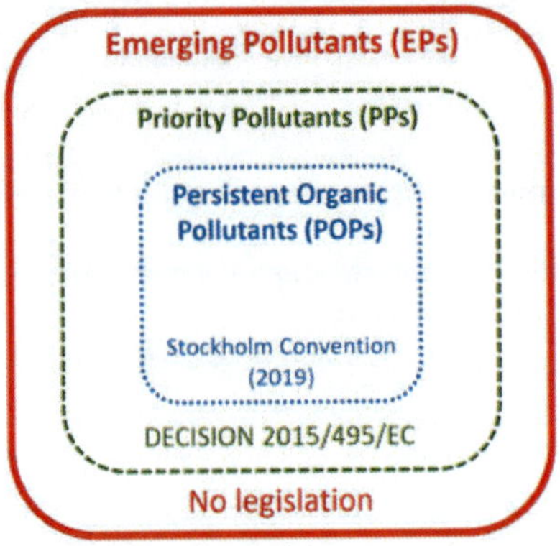

Fig. 2 Scheme of emerging, priority and persistent pollutants and their respective current regulations

2.1 Pharmaceutical and Personal Care Products (PPCPs): Triclosan (TCS) as a Case of Study

Among EPs, the group of PPCPs deserves special attention due to its wide-ranging presence in comparison with the other groups. PPCPs contain, on the one hand, several pharmaceutical compounds and drugs, and on the other hand, chemicals that are present in many cosmetic and personal care products, in all cases, to improve human or livestock health. All of them are broadly available commercial products, therefore, they are easily accessible by the population and every human being is an active contributor due to the presence of several PPCPs in many daily use products, like soaps or hand wash gels. Because of the extended use, and a very stable chemical structure, which is the reason why they are environmentally persistent and are not affected by biodegradation, there is a growing environmental concern [16]. PPCPs are present in many environmental places such as rivers, seas, treated waters, soils, etc., coming from several industrial sites, wastewater treatment plants (WWTPs), households, or agriculture [12]. However, their fate in the environment is not fully defined. A well-known procedure to solve this issue and in this way prevent the contamination of waters and soils is the treatment through specific technologies within the WWTPs, thereby preventing the PPCPs entry into the environmental cycle [17]. The major sources, distribution pathways, and prevention of PPCPs are shown in Fig. 3.

It is worth anticipating the necessity of deepening into the comprehension of the chemical behavior of PPCPs in soils, since their sorption makes a critical impact on the fate and movement of these chemicals; for instance, higher clay and lower sand content in the soil can lead to a reduction in the transport rate of these chemicals. Other soil properties like pH, minerals, or organic matter content have influence as well [18]. Kodešová et al. [19] studied the effects of soil properties on pharmaceutically active compounds sorption/desorption finding that the sorption of ionizable

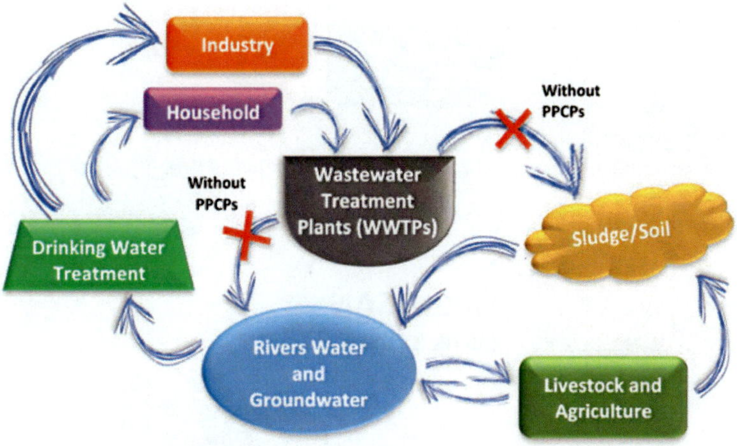

Fig. 3 Major sources and distribution pathways of PPCPs in the environment. Prevention of contamination by PPCPs in WWTPs

compounds was highly affected by soil pH. Positive correlations between sorption coefficients and clay content were found for clarithromycin, clindamycin, atenolol, and metoprolol; carbamazepine and sulfamethoxazole exhibited the lowest sorption and therefore higher mobility in soils, which may result in contamination of groundwater. Le Guet et al. [20] reported different effects of organic matter on pharmaceutically active compounds sorption capacity; when organic matter is present, the apparent solubility of the solute increases enhancing their mobility, but the mobility can be decreased due to cumulative sorption. The principal ways of the contaminants to enter into soils, apart from direct discharge, are, through the deposition of the suspended particles from the atmosphere, deposition of polluted sludge or via the circulation of treated wastewater from WWTPs [21] (Fig. 3).

Delving into the common paths of PPCPs in soil penetration, it is needed to highlight that irrigation with polluted water represents an important source of soil pollution. Nowadays, due to water scarcity, water reuse for irrigation purposes shows an increasing practice; however, PPCPs are not completely eliminated in WWTPs, so that they are present in waters to be used for irrigation purposes, agricultural or urban parks and gardens. Moreover, depending on the kind of chemical compounds and the streams of water, they can be transported through the soil profile, unintentionally polluting groundwater [22]. One of the negative consequences of the presence of PPCPs is the change of soil activity; for example, TCS has been reported to disturb the nitrogen cycle or the microbial activity in some soils [23] (Fig. 3).

Among all PPCPs (Table 1), TCS appears as one of the most employed additives in several consumer goods, such as cosmetics, hand wash soaps and gels, deodorant soaps, detergents, toothpaste, textiles, plastic consumer goods, as well as toys, etc., due to its action as an antimicrobial and antifungal disinfectant agent (Fig. 4).

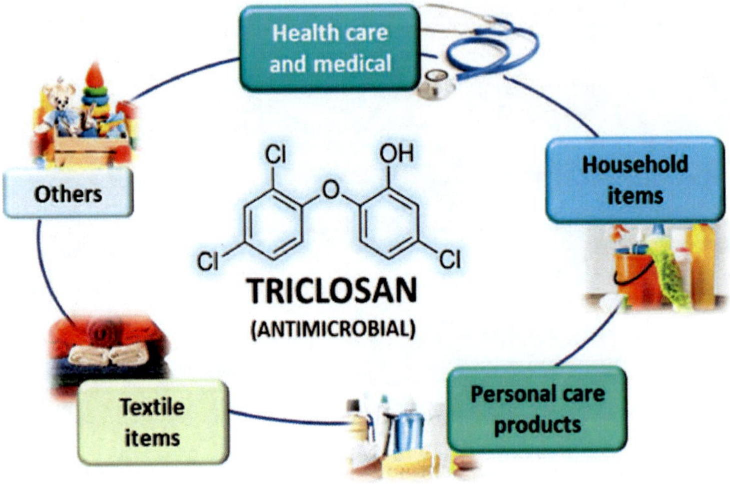

Fig. 4 Common uses of Triclosan

As a reference of its use, the production of TCS reached around 3,200 t per year in Asia, meanwhile in the European Union was 850 t in 2015. Nevertheless, in the last 5 years, a falling trend in global production and consumption has been observed [24]. TCS is highly toxic and presents relatively high lipophilicity; it is also very resistant to biodegradation and thus, environmentally persistent. It has been detected in surface water, seawater, lakes and rivers, sediments, and soils [25–28]. Furthermore, its presence has been identified in the influent and effluent of WWTPs, sewage sludge and urban sewers, soils and sediments [18, 29, 30] which evidences that sometimes the treatments applied in WWTPs are inefficient and exhibit certain drawbacks for its degradation (Table 2).

TCS reaches aquatic organisms where its presence has been demonstrated [42, 43], and consequently, because of the human consumption of those marine species, it enters eventually in the human body; it has been detected in several tissues (adipose tissue, liver, or brain), skin and even in human breast milk [44–46]. TCS has proven to be the cause of several human health problems, as it is an endocrine disruptor in multiple species including humans, and it has influence in cancer development, decrease of cardiovascular functions or reproductive disorders [47]. All of these attributes have resulted in its partial prohibition in several countries like Canada, its complete prohibition like over-the-counter (OTC) consumer antiseptic products, such as different forms of hand soap (liquid, foam or gel), body soaps, by the U.S. Food and Drug Administration (US FDA) in the Federal Rule 84 FR 14847 [48] or disapproval for the employment in human hygiene biocidal products by the European Community [49, 50].

In general, the basis for the destruction of EPs using green chemistry processes includes high efficiency processes operated preferentially at room temperature and atmospheric pressure using non-toxic and environmentally benign materials and reagents. Moreover, processes should release non-toxic and biodegradable

Table 2 Real matrices containing TCS in different sources: water, soil, dust, and sediment

Sources	Matrix	[TCS] (min.–max.)	Description/other	Reference
Water	Surface water/Sri Lanka	<MDL-5.85 ng L^{-1}	Analysis of 72 PPCPs to understand the distribution and environmental risk. Nineteen samples were collected. Forty-one PPCPs were detected, and risk quotients (RQ) were calculated (RQ TCS = 0.09–0.17, low-medium risk)	[25]
	Surface water/China	35–1,023 ng L^{-1}	9 PPCPs are studied in three urban streams and in Major Pearl River in Guangzhou, China. Samples were collected during six different months. No ecotoxicological study was carried out	[31]
	Influent/effluent WWTPs/ Korea	247–785 ng L^{-1}; 79–149 ng L^{-1}	Occurrence and removal of 20 PPCPs were investigated in 5 WWTPs in Ulsan, Korea, and each WWTPs possesses different treatments. Good degradation rates were achieved for 7 PPCPs	[32]
	River water/Mexico	0.31–15.32 mg L^{-1}	A review of several pollutants detected during the past 10 years in waters and bed sediments of the Atoyac system, and the implication of this contamination over the human health.	[27]
	Tap water/Taiwan	ND-49 ng L^{-1}	13 pharmaceuticals are monitored in several points of drinking water of Kaohsiung City, Taiwan. No significant differences were found between the concentrations of pharmaceuticals in both types of water samples	[33]
	Drinking fountain water/ Taiwan	ND-103 ng L^{-1}		
	Gray water	21 μg L^{-1}	22 micropollutants were detected in samples taken from a household with gray water system.	[34]
	Seawater/China	ND-5 ng L^{-1}	32 PPCPS were detected in samples taken around 24 sites in Xiamen island (China). Potential ecological risks were assessed through the RQ values of the 23 investigated PPCP categories	[35]
	Groundwater/USA	ND-45 ng L^{-1}	Assess the fate of PPCPs at field scale. They were measured systematically in WWTPs (liquid and sludge lines), and in soil and groundwater receiving treated effluent from the WWTPs. In general, various PPCPS are eliminated during the WWT, but certain compounds remain and were occasionally detected in the effluent in elevated concentrations than in the effluent	[36]
Soil	Soil/USA	ND-19.15 ng g^{-1}		
	Sludge/USA	3.52–18.6 μg g^{-1}		

(continued)

Table 2 (continued)

Sources	Matrix	[TCS] (min.–max.)	Description/other	Reference
	Sewage sludge/Norway	1.9–2.2 μg g^{-1}	With the purpose of investigating the dissipation, earthworm and plant accumulation of organic contaminants in soil amended with three types of sewage sludge were collected. Transfer of TCS to worms was significant	[37]
	Sludge/Greece	9,850 ng g^{-1}	50 pharmaceuticals and illicit drugs were studied in sludge samples from Athens sewage treatment plant. TCS possesses an environmental risk on the soil (RQ > 1)	[38]
Dust	Indoor dust/Vietnam	<5–1,090 ng g^{-1}	Contamination status, spatial variability, and exposure risk of TCS were studied in indoor dust in Vietnam. Human exposure through dust ingestion was investigated	[39]
	Indoor dust/China	<MDL-1,180 ng g^{-1}	129 indoor dust samples were measured collected from two cities in China, and the contribution of indoor dust ingestion to the total exposure was investigated	[40]
Sediment	Estuarine/USA	ND-800 μg kg^{-1}	Sediment samples were collected in eight locations from two sites in the U.S. east coast	[41]

<MDL: Below method detection limit, *N.D.* not detected

substances and avoid the formation of hazardous by-products that could end up in the environment [51]. However, sometimes the partial oxidation of EPs during the remediation of wastewaters may produce transformation products being more toxic than their parent compounds [52]. Accordingly, wastes released to aquatic systems should be subjected to more stringent standards and the assessment of toxicity due to potential formation of POPs and other toxic by-products, specifically, if highly toxic PCDD/Fs are formed, assessing under which operating conditions of the applied treatment their formation is relevant [53].

3 Polychlorinated Dibenzo-*p*-Dioxins and Dibenzofurans (PCDD/Fs)

PCDD/Fs (or better known as dioxins and furans) constitute a family of organic chemical compounds that are categorized as POPs. They possess similar structures, being a group of practically planar tricyclic aromatic compounds and they are characterized by low vapor pressure, high thermal stability, high solubility in organic/fatty matrices (including human fatty tissues), and inclination to bind to organic matter in soil and sediments [54]. PCDD/Fs are divided into PCDDs, which include 75 congeners, and PCDFs, with 135 congeners; in total, there are 210 congeners, which can have from none to eight chlorine atoms attached to the molecule. Congeners with the same number of chlorine atoms are called homologues; the groups of homologues are differentiated according to the number of chlorine atoms in the molecule. Table 3 collects the names and number of congeners in each homologue group. PCDD/Fs are related to a high number of harmful diseases in humans, including immune disorders, reproductive and development problems, or cancer when long-term exposure takes place. In accordance with several animal and human epidemiology data, TCDD was categorized by the International Agency for Research on Cancer (IARC) as a "known human carcinogen" [55].

Generally, PCDD/Fs can be produced in a natural manner because of combustion processes like forest fires or volcanic eruptions (natural sources) or in an unintentional manner as unwanted by-products in many thermal and industrial processes (primary sources); during incineration (municipal, hospital and hazardous wastes, . . .), combustion sources like cement kilns, diesel vehicles or wood burning, industrial processes (chemical and metal industry, pulp and paper mills) or reservoirs (biochemical and photocatalytic processes, accidental origin or landfills) [54, 56–60]. Another common route of formation PCDD/Fs is via transformation of their precursors (secondary sources), which comprise organochloride compounds such as chlorophenols (CPs), chlorobenzenes (CBs), chlorinated diphenyl ethers and polychlorinated biphenyls (PCBs) [61–63], pentachlorophenol (PCP) or polychlorinated phenoxy phenols such as TCS [64, 65].

PCDD/Fs are, as POPs, ubiquitous in the environment for long periods of time due to their physical, chemical, and biological stability [51]. Dioxins and furans have

Table 3 PCDDs and PCDFs homologue classification

	PCDDs		PCDFs	
	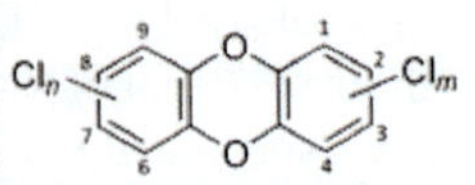		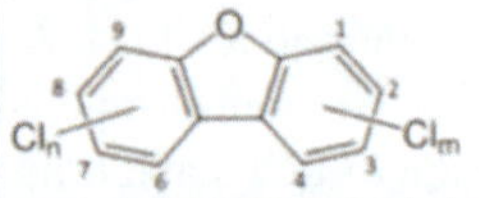	
No. of chlorines	Name	No. of congeners	Name	No. of congeners
1	Monochlorodibenzo-*p*-dioxins	2	Monochlorodibenzofurans	4
2	Dichlorodibenzo-*p*-dioxins	10	Dichlorodibenzofurans	16
3	Trichlorodibenzo-*p*-dioxins	14	Trichlorodibenzofurans	28
4	Tetrachlorodibenzo-*p*-dioxins	22	Tetrachlorodibenzofurans	38
5	Pentachlorodibenzo-*p*-dioxins	14	Pentachlorodibenzofurans	28
6	Hexachlorodibenzo-*p*-dioxins	10	Hexachlorodibenzofurans	16
7	Heptachlorodibenzo-*p*-dioxins	2	Heptachlorodibenzofurans	4
8	Octachlorodibenzo-*p*-dioxin	1	Octachlorodibenzofuran	1
	Total	75	Total	135

been detected in the aquatic environment, despite their low solubility in water. They enter into the environment from different sources; air from thermal processes, superficial water (rivers, lakes) and groundwater by employment of chemicals in agriculture and as direct discharge from wastewater treatment plants or industrial sources, soil/sediment from landfills, waste dumps, pesticides uses, and sewage sludge [66, 67]. Due to their lipophilic nature, their accumulation together with the inherent toxicity represents a matter of severe concern for animal and human life, as well as for the environment. Further, PCDD/Fs have been found worldwide also in secondary media like food (generally, in fatty foods such as fish and shellfish, meat, dairy products) and consumer goods [67, 68].

Commonly, PCDD/Fs appear as a mixture of all the possible congeners, but not all possess equal risk, those with four or more chlorine atoms in the molecule deserve special consideration, due to the high toxicity and special resistance to various degradation processes, such as chemical and biological processes. This type of PCDD/Fs possesses chlorine atoms at the positions 2, 3, 7 and 8, being a total of 17 congeners with these characteristics. In this way, TEQ defines the toxicity of a mixture of dioxins, where each of these 17 congeners weights its toxicity referred to the most toxic dioxin, 2,3,7,8-TCDD, which has a reference value equal to one [69]. Each extra chlorine atom to 2, 3, 7, and 8 position commonly lowers the

Table 4 Toxicity equivalency factors for the 17 2,3,7,8-PCDD/Fs congeners of the PCDD/Fs

Congener	I-TEF	WHO-TEF 2005	Congener	I-TEF	WHO-TEF 2005
PCDDs			PCDFs		
2,3,7,8-TCDD	1	1	2,3,7,8-TCDF	0.1	0.1
1,2,3,7,8-PeCDD	0.5	1	1,2,3,7,8-PeCDF	0.05	0.03
1,2,3,4,7,8-HxCDD	0.1	0.1	2,3,4,7,8-PeCDF	0.5	0.3
1,2,3,6,7,8-HxCDD	0.1	0.1	1,2,3,4,7,8-HxCDF	0.1	0.1
1,2,3,7,8,9-HxCDD	0.1	0.1	1,2,3,6,7,8-HxCDF	0.1	0.1
1,2,3,4,6,7,8-HpCDD	0.01	0.01	1,2,3,7,8,9-HxCDF	0.1	0.1
OCDD	0.001	0.0003	2,3,4,6,7,8-HxCDF	0.1	0.1
			1,2,3,4,6,7,8-HpCDF	0.01	0.01
			1,2,3,4,7,8,9-HpCDF	0.01	0.01
			OCDF	0.001	0.0003

toxicity by several times. The total TEQ is calculated by adding the product of the congener concentration times the individual Toxicity Equivalence Factor (TEF) of the congener, offering a global view of the toxicity of the sample, which is shown in Eq. 1:

$$\mathrm{TEQ} = \sum_{\mathrm{i}} (\mathrm{PCDD_i} \cdot \mathrm{TEF_i}) + \sum_{\mathrm{j}} (\mathrm{PCDF_j} \cdot \mathrm{TEF_j}) \tag{1}$$

Two different TEF have been defined: the International Toxicity Equivalency Factor (I-TEF), established by a scientific research committee called NATO Committee on the Challenges of Modern Society (NATO/CCMS); and the most recently reported, the World Health Organization Toxicity Equivalency Factor (WHO-TEF). Table 4 summarizes the values of the International Toxicity Equivalency Factor (I-TEF) and World Health Organization Toxicity Equivalency Factor (WHO-TEF).

Commonly, the analysis of PCDD/Fs in water samples is carried out by application of the Standard Method USEPA 1613 following the isotope dilution method and high-resolution gas chromatography/high-resolution mass spectrometry (HRGC/HRMS) [70]. Prior to the measurement in the gas chromatograph, the method requires a rigorous series of steps pre-treatment and concentration of the target sample, such as extraction, concentration, and purification, among others.

Lately, PCDD/Fs have been incorporated as new priority substances in the field of water policy in the Directive 2013/39/EU [71]. Referring to the directive of the U.S. EPA, the Safe Drinking Water Act (SDWA), in the framework of The National Primary Drinking Water Regulations (NPDWR), limits the maximum pollutant level for 2,3,7,8-TCDD in drinking water as zero; nevertheless, taking into account that the best existing treatment technologies are being used, and taking cost into consideration, a new maximum level of 30 pg L^{-1} has been established [72].

4 Advanced Oxidation Processes (AOPs)

Conventional WWTPs are considered the most significant emerging pollutant emitters [73]. EPs, due to the ineffectiveness of conventional physical and biological treatments, are partially removed and remain in WWTPs effluents, being discharged into the environment [74]. A well-known procedure to prevent the contamination of waters, and with this, the contamination of soils, is the application of effective remediation technologies such as AOPs. AOPs came out as new technologies that provide effective and powerful treatments to wastewater containing EPs, such as pesticides, PPCPs, industrial chemicals, among others [75–77]. The main challenge so far in the selection of the optimum treatment alternative is to ensure the successful implementation regarding their operational costs (energy consumption, chemicals required), sustainability (carbon footprint, resources uses), or common feasibility (by-products formation) [78]. AOPs, defined as aqueous phase oxidation technologies based on the formation of highly reactive species, especially hydroxyl radicals ($^{\bullet}OH$), have been successfully applied to wastewaters containing recalcitrant organic compounds. Among the numerous advantages of applying AOPs, it should be mentioned the fast reaction rates, the potential to reduce toxicity and the possibility to achieve a complete mineralization of organic compounds, or the non-selectivity oxidation, offering the possibility to treat more than one pollutant at the same time. However, together with the complete removal of the pollutant, the generation of by-products during the degradation of the primary pollutant is another issue of concern in the application of AOPs. In fact, the presence of these by-products could increase the final toxicity of the treated wastewater. In this way, chlorinated organic compounds have been related to the formation of polychlorinated dibenzo-*p*-dioxins and dibenzofurans (PCDD/Fs), since they are considered an important group of substances that act as precursors of these highly toxic compounds.

4.1 Electrochemical Oxidation of TCS

Also called anodic oxidation, electrochemical oxidation is an AOP based on redox chemistry, where the application of electric energy produces an electron flux that passes through an electrochemical cell, oxidizing a chemical compound by electrons losing and reducing other species, which gain electrons. Two main types of electrochemical oxidation can take place when oxidizing organic matter, direct and indirect electrolysis [64]: (1) direct electrolysis, where direct electron transfer is produced from the contaminant to the anode surface without the contribution of other compounds, (2) indirect electrolysis, where another species can exchange electrons with the anode to produce electroactive species carrying out the oxidation of the pollutants. This involves the production of hydroxyl radicals ($^{\bullet}OH$) from water on the anode surface that behave as intermediary species between the electrodes and the organic matter. The major strengths of this technology are the simple equipment required, robustness, its simple manipulation of the operating variables, its operation at ambient temperature and pressure, and it is considered a clean technology, as it does not produce wastes, normally [52].

This technology has been successfully applied to the remediation of several organo-chlorinated compounds, such as 2-chlorophenol [62, 63, 79, 80] or TCS [64]. In both cases, boron doped diamond (BDD) anodes have been employed due to the highest overpotential for the oxygen evolution reaction, which results in a higher efficiency to oxidize pollutants in comparison with other electrodes; BDD is corrosion resistant and stable and has excellent conductive properties [81, 82]. However, due to the powerful oxidizing conditions, the potential formation of new POPs and other toxic by-products, such as PCDDs and PCDFs, might constitute an adverse consequence of the application of electrochemical oxidation process [83]. In this way, TCS is structurally similar to other polychlorinated phenoxy phenols that have been shown to cyclicize to the toxic PCDD/Fs, so that it is considered a precursor of these compounds [64].

Solá-Gutiérrez et al. [65] demonstrated the PCDD/Fs formation employing two different electrolytes, sodium chloride (NaCl) and sodium sulfate (Na_2SO_4), with 10.0 mg L^{-1} of TCS, when electrochemical oxidation was applied. The calculation of the limiting current density (J_{lim}) gave a value of 4.8 A m^{-2}, whereas the working current density (J) was 6 A m^{-2}. Complete TCS degradation for 10 mg L^{-1} of initial concentration and for both electrolytes (NaCl: 56.3 mM; Na_2SO_4: 21.1 mM) was obtained after 45 min and 1 h, for NaCl and Na_2SO_4, respectively. Figure 5a, b depicts the dioxins and furans homologues groups when NaCl and Na_2SO_4 were employed as electrolyte, respectively.

As it can be seen, when using NaCl as electrolyte, the predominant group among all PCDD/Fs homologues is the tetrachlorodibenzodioxin group, TCDD, accounting for 59.3% of the total PCDD/Fs (218.46 pg L^{-1} of the total amount, 368.48 pg L^{-1}); it is followed by the PeCDD group, which accounts for 17.1% of the total, with 62.91 pg L^{-1}. In the case of Na_2SO_4 as electrolyte, once again the TCDD group is the largest group of all PCDD/Fs, accounting for 62.3% of the total amount (184.12 pg L^{-1} of 295.53 pg L^{-1}). Following this, PeCDD is the second largest group, with 58.65 pg L^{-1}, around 20% of the total.

Along with the quantitative determination of the change in Triclosan concentration, GC–MS analyses were performed to identify the intermediate products that could be formed during the electrochemical oxidation of TCS. The qualitative analysis confirmed the formation of chloro-aromatic intermediary species, such as 2,4-dichlorophenol, 4-chlorocatechol, 2-chlorohydroquinone, 2-chloro-4-methoxyphenol, and 1-chloro-2,5-dimethoxybenzene, for both electrolytes. Solá-Gutiérrez et al. [64] described the formation mechanism of PCDD/Fs from TCS, which is corroborated with a theoretical simulation modelling previously reported in the literature for gas and aqueous phase [84, 85].

Finally, Fig. 6 depicts TEQ values. As can be seen, when NaCl is employed, TEQ values increased almost 19 times with respect to the untreated sample, reaching a value of 4.78 pg L^{-1}. When using Na_2SO_4, a value of 4.36 pg L^{-1} of TEQ was obtained (around 18 times more than for untreated sample). It is remarkable that with

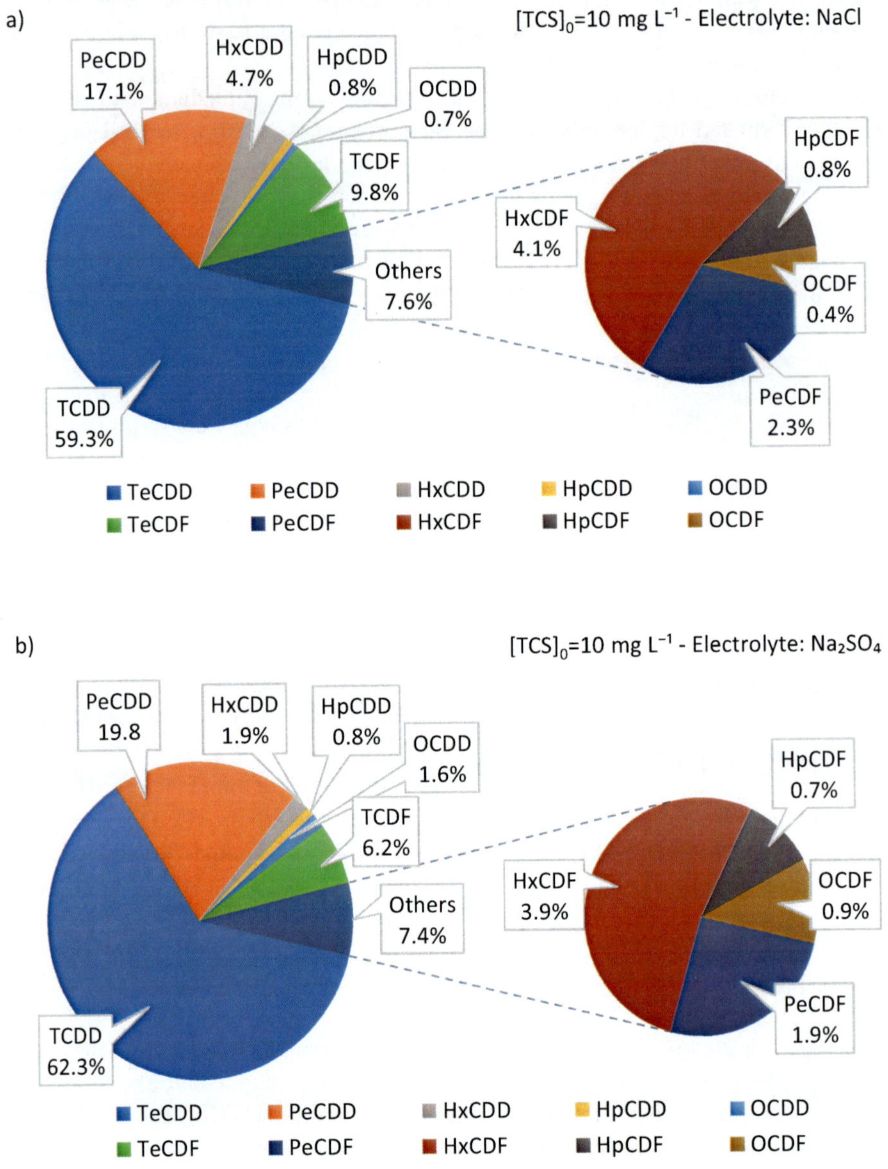

Fig. 5 % PCDD/Fs homologue groups formed at the end of the electrochemical experiments of TCS employing (**a**) NaCl and (**b**) Na_2SO_4 as electrolytes [65]

both electrolytes, values lower than the EPA recommended limit (30 pg L^{-1}) were reached. Despite the low formation of PCDD/Fs for the case of study, the application of the electrochemical oxidation favors the potential formation of these species, and thus of the toxicity (TEQ) in relation to the untreated sample.

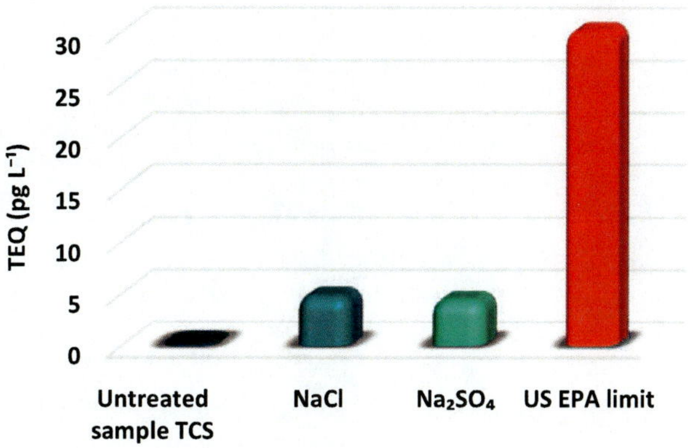

Fig. 6 TEQ values for both electrolytes when electrochemical oxidation is applied to degrade TCS [65]

4.2 Indirect Oxidation of TCS: Photocatalysis

Indirect photolysis using a catalyst, most-known as photocatalysis, takes place when a photocatalyst, a semiconductor, accelerates a photochemical reaction. The photocatalyst absorbs photons that possess equal or higher energy than the band gap, producing electron-hole pairs (e^- and h^+) when an electron is promoted from the conduction band to the valence band. The latter facilitate the formation of reactive oxygen species (ROS) when reacting with H_2O, such as hydroxyl radicals $^{\bullet}OH$, which are responsible for the breakage of the respective molecule of concern. Figure 7 depicts the schematic of the photocatalytic process [47, 52].

One of the most employed and extensively studied catalysts is TiO_2, because of its advantageous properties, like excellent oxidizing capacity together with its chemical stability, long durability, non-toxicity, and low price. But, this compound is only active under UV radiation; in particular, wavelengths between 300–400 nm were established for a correct operation and, because of this, artificial lamps are essential to its employment [50, 52].

This technology has been effectively employed in the remediation of different organic pollutants, such as Non-Steroidal Anti-Inflammatory Drugs (NSAIDs) naproxen, diclofenac, or ibuprofen, Analgesic drugs (ANs) like paracetamol, or studies conducted to assess the potential of TiO_2 photocatalytic degradation on important classes of antibiotics, namely b-lactam antibiotics, quinolones, tetracyclines, and sulfonamides [86]. In the case of TCS, Schröder et al. [50] studied several TiO_2 doses (0.75, 1.0 and 1.5 g L^{-1}) in order to eliminate TCS from aqueous solutions using commercial TiO_2, Aeroxide® P25. An initial concentration of 10 mg L^{-1} of TCS was employed, reaching its complete degradation after 4 h for 1.5 g L^{-1} of TiO_2, and after 5 h for the rest of the catalyst doses employed (0.75 and 1.0 g L^{-1}). Some differences could be observed in the degradation kinetics during the first 30 min, when almost 50% of TCS had been degraded; when employing 1.5 g L^{-1} of catalyst, a fastest degradation rate was achieved. The compound 2,4-dichlorophenol was the intermediate detected with the highest concentration of 0.79 mg L^{-1} at 1 h of experiment. Other intermediates were detected, such as TCS

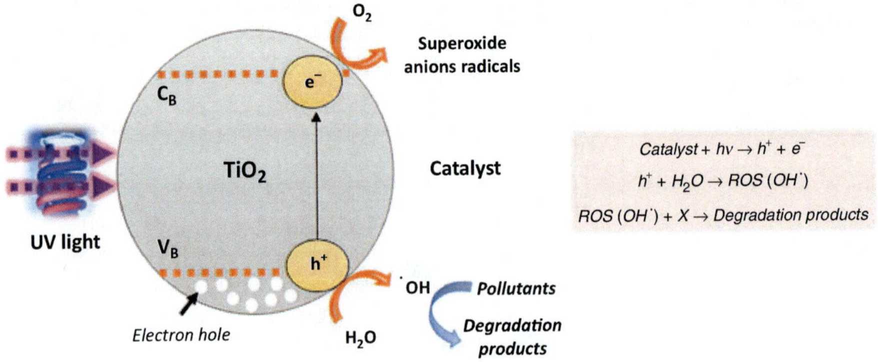

Fig. 7 Scheme and reactions of the photocatalytic oxidation

quinone, different hydroxylated TCS or small molecules like 2-chlorohydroquinone or 4-chlorocatechol. Finally, Schröder et al. [50] analyzed the PCDD/Fs formation at the end of the experiments for all the catalyst concentrations employed. Figures 8a, b represents the PCDD/Fs formed for the highest and the lowest amount of catalyst (0.75 and 1.5 g L^{-1} of TiO_2).

For 0.75 g L^{-1} of catalyst, it can be appreciated that the highest amount corresponds to the TCDF homologue group, that accounts for 98% of the total amount of PCDD/Fs, which is $1.02 \cdot 10^5$ pg L^{-1}; it is followed by the TCDD group, with 1.92% of the total. PCDD/Fs follow a similar trend when using 1.5 g L^{-1} of TiO_2, being the total amount 975.12 pg L^{-1}. Likewise, TCDF represents the group with the highest concentration, accounting for 90% of the total (881.5 pg L^{-1}), followed by TCDD with 8.86% (86.8 pg L^{-1}). Finally, in both cases, the rest of the groups have negligible concentrations. Thus, it is concluded that the total concentration of PCDD/Fs decreases when increasing the catalyst concentration.

Finally, 2,3,7,8-congeners were analyzed, being 2,3,7,8-TCDD the one with the highest concentration when using 0.75 g L^{-1} of catalyst, with a value of 5.31 pg L^{-1} which increases considerably the toxicity due to the I-TEF associated value (with a score of 1.0). The final toxicity for the three different catalyst doses employed was also calculated and is represented in Fig. 9. TEQ values increased with respect to the untreated sample (0.25 pg L^{-1}), being the highest value of TEQ 6.25 pg L^{-1} that was obtained working with 0.75 g L^{-1} of catalyst, that is TEQ values decreased when increasing the catalyst loading. For any of the situations studied, values lower than the EPA recommended limit (30 pg L^{-1}) were reached. As observed during the electrochemical oxidation, despite the low formation of PCDD/Fs, the application of the photocatalytic oxidation favors the potential formation of these species, and thus of the increase of toxicity (TEQ) in comparison with the untreated sample; therefore, it is very important the correct selection of the oxidation method, as well as of the operating conditions, especially when starting with chlorinated organic compounds that are potential precursors of PCDD/Fs.

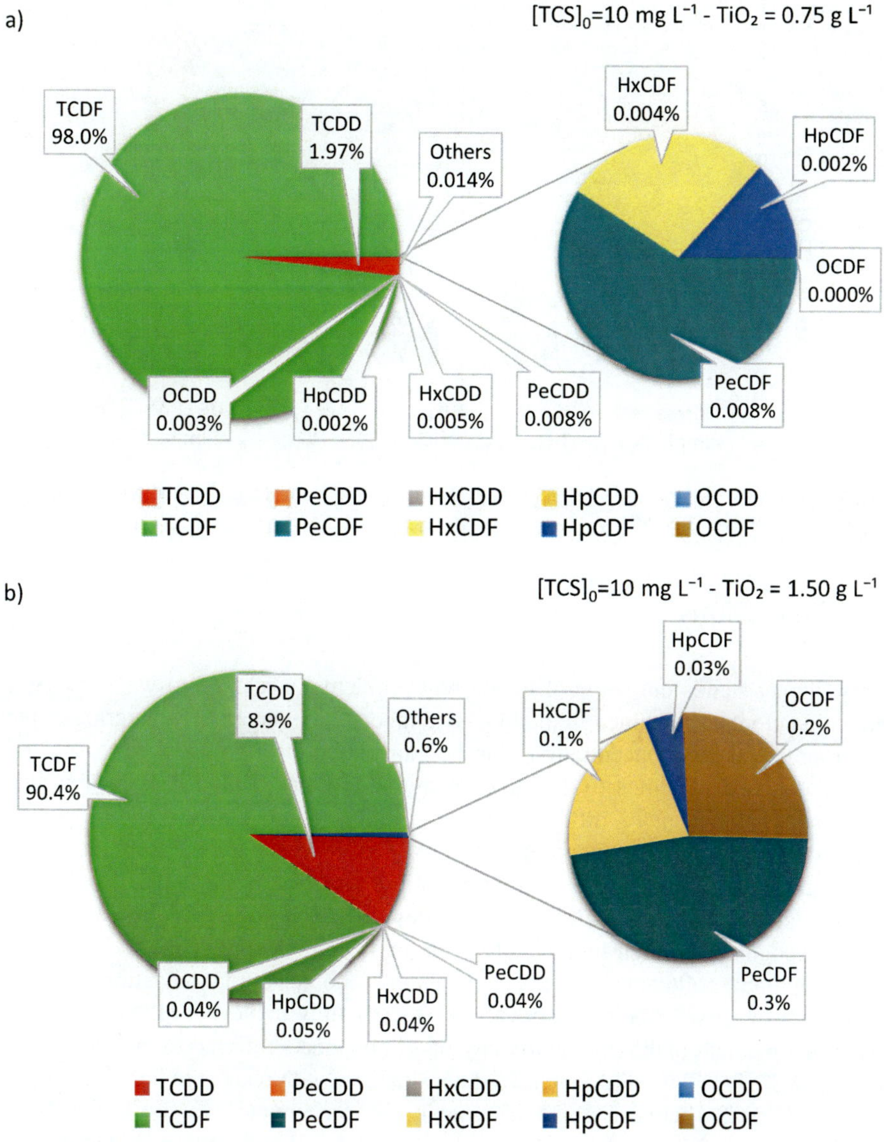

Fig. 8 PCDD/Fs profile at the end of the photocatalytic experiments using (**a**) 0.75 g L^{-1} and (**b**) 1.5 g L^{-1} of TiO_2 [50]

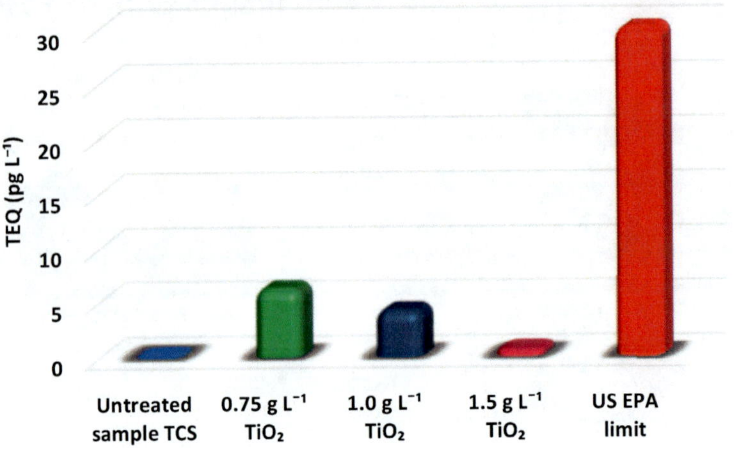

Fig. 9 TEQ values for the three different catalyst doses employed when photocatalytic oxidation is applied to eliminate TCS [50]

5 Conclusions

This chapter stands out some of the current problems in relation to water and soil contamination, as contained in the SDGs, numbers 2, 6, and 15. EPs entering in the environment at low concentrations can be transported through the water cycle and constitute a serious threat to the environment and human health. As their removal from soils is especially difficult, the application of cost-effective remediation methods to the sources containing EPs, WWTPs effluents among them, is highly recommendable. Among EPs, the TCS antimicrobial and antifungal disinfectant agent that belongs to the PPCPs group deserves attention. After applying two AOPs recognized by their high effectiveness in the mineralization of a wide number of recalcitrant pollutants, electrochemical and photocatalytic oxidation to TCS containing waters, complete removal was successfully achieved. However, under the experimental conditions the toxicity of 17 congeners referred to the most toxic dioxin, 2,3,7,8-TCDD, TEQ, increased considerably with respect to untreated water containing TCS (between 18 to 25 times); electrochemical oxidation with NaCl resulted in a maximum value of 4.78 pg L^{-1}, whereas photocatalytic oxidation with TiO_2 0.75 g L^{-1} gave a value of 6.25 pg L^{-1}. On the other hand, the analysis of by-products in both technologies confirmed the formation of organo-chlorinated intermediate species, such as 2,4-dichlorophenol; this compound has been reported to be precursor in the formation of PCDD/Fs. These results emphasize the importance of the quantitative traceability of by-products formation and the resulting toxicity (TEQ, PCDD/Fs) after application of advanced oxidation processes to waters containing EPs, with special attention to the compounds that can lead to the formation of PCDD/Fs precursors in the degradation pathways.

Acknowledgements This research was developed in the framework of the project CTM2017-87740-R financed by the Ministry of Science, Innovation and Universities (Spain). Sophie Schröder is also grateful for the predoctoral grant (FPI), PRE2018-083526.

References

1. NIH, National Institute of Environmental Health Sciences (2021) Safe water and your health. https://www.niehs.nih.gov/health/topics/agents/water-poll/index.cfm. Accessed 7 Aug 2021
2. UN, United Nations (2019) Global issue: population. https://www.un.org/en/global-issues/population. Accessed 7 Aug 2021
3. UNESCO (2021) The United Nations world water development report 2021: valuing water. UNESCO
4. UNESCO (2019) The United Nations world water development report 2019: leaving no one behind. UNESCO
5. UN, United Nations (2016) The 2030 agenda for sustainable development (A/RES/70/1), United Nations. https://doi.org/10.1201/b20466-7
6. Lal R, Bouma J, Brevik E, Dawson L, Field DJ, Glaser B, Hatano R, Hartemink AE, Kosaki T, Lascelles B, Monger C, Muggler C, Martial G, Norra S, Pan X, Paradelo R, Reyes-sánchez LB, Sandén T, Ram B, Spiegel H, Yanai J, Zhang J (2021) Soils and sustainable development goals of the United Nations: an international union of soil sciences perspective. Geoderma Reg 25. https://doi.org/10.1016/j.geodrs.2021.e00398
7. European Environment Agency (2019) Governance – acting together for sustainable land management. https://www.eea.europa.eu/signals/signals-2019-content-list/articles/governance-2014-acting-together-for. Accessed 7 Aug 2021
8. Vasilachi IC, Asiminicesei DM, Fertu DI, Gavrilescu M (2021) Occurrence and fate of emerging pollutants in water environment and options for their removal. Water 13:1–34. https://doi.org/10.3390/w13020181
9. Hube S, Wu B (2021) Mitigation of emerging pollutants and pathogens in decentralized wastewater treatment processes: a review. Sci Total Environ 779:146545. https://doi.org/10.1016/j.scitotenv.2021.146545
10. Teodosiu C, Gilca AF, Barjoveanu G, Fiore S (2018) Emerging pollutants removal through advanced drinking water treatment: a review on processes and environmental performances assessment. J Clean Prod 197:1210–1221. https://doi.org/10.1016/j.jclepro.2018.06.247
11. Norman Network (2021) Network of reference laboratories, research centres and related organisations for monitoring of emerging environmental substances. https://www.norman-network.net/. Accessed 7 Aug 2021
12. Shahid MK, Kashif A, Fuwad A, Choi Y (2021) Current advances in treatment technologies for removal of emerging contaminants from water – a critical review. Coord Chem Rev 442:213993. https://doi.org/10.1016/j.ccr.2021.213993
13. DECISION 2015/495/EU (2015). Commission implementing decision (EU) 2015/495 of 20 March 2015 establishing a watch list of substances for Union-wide monitoring in the field of water policy pursuant to Directive 2008/105/EC of the European Parliament and of the Council. Off J Eur Union L78
14. US EPA (2009) persistent organic pollutants: a global issue, a global response. https://www.epa.gov/international-cooperation/persistent-organic-pollutants-global-issue-global-response. Accessed 21 Jun 2021
15. Stockholm Convention (2019) Stockholm Convention on persistent organic pollutants (POPs) – texts and annexes, United Nations Environment Programme (UNEP). Châtelaine, Switzerland. http://chm.pops.int/Convention/ConventionText/tabid/2232/Default.aspx

16. WHO, World Health Organization (2020) Antibiotic resistance [WWW Document]. https://www.who.int/news-room/fact-sheets/detail/antibiotic-resistance. Accessed 7 Aug 2021
17. Xu Y, Yu X, Xu B, Peng D, Guo X (2021) Science of the total environment sorption of pharmaceuticals and personal care products on soil and soil components: influencing factors and mechanisms. Sci Total Environ 753:141891. https://doi.org/10.1016/j.scitotenv.2020.141891
18. Mejías C, Martín J, Santos JL, Aparicio I, Alonso E (2021) Occurrence of pharmaceuticals and their metabolites in sewage sludge and soil: a review on their distribution and environmental risk assessment. Trends Environ Anal Chem 30. https://doi.org/10.1016/j.teac.2021.e0015
19. Kodešová R, Grabic R, Kocárek M, Klement A, Golovko O, Fér M, Nikodem A, Jakšík O (2015) Pharmaceuticals' sorptions relative to properties of thirteen different soils. Sci Total Environ 511:435–443. https://doi.org/10.1016/j.scitotenv.2014.12.088
20. Le Guet T, Hsini I, Labanowski J, Mondamert L (2018) Sorption of selected pharmaceuticals by a river sediment: role and mechanisms of sediment or Aldrich humic substances. Environ Sci Pollut Res Int 25:14532–14543. https://doi.org/10.1007/s11356-018-1684-0
21. Zhang K, Zhao Y, Fent K (2020) Science of the total environment cardiovascular drugs and lipid regulating agents in surface waters at global scale: occurrence, ecotoxicity and risk assessment. Sci Total Environ 729:138770. https://doi.org/10.1016/j.scitotenv.2020.138770
22. Chen W, Xu J, Lu S, Jiao W, Wu L, Chang AC (2013) Fates and transport of PPCPs in soil receiving reclaimed water irrigation. Chemosphere 93:2621–2630. https://doi.org/10.1016/j.chemosphere.2013.09.088
23. Horswell J, Prosser JA, Siggins A, van Schaik A, Ying L, Ross C, McGill A, Northcott G (2014) Assessing the impacts of chemical cocktails on the soil ecosystem. Soil Biol Biochem 75:64–72. https://doi.org/10.1016/j.soilbio.2014.03.013
24. Warming M, Lassen C, Kjølholt J (2016) Survey of triclosan in cosmetic products. Danish EPA
25. Guruge KS, Goswami P, Tanoue R, Nomiyama K, Wijesekara RGS, Dharmaratne TS (2019) First nationwide investigation and environmental risk assessment of 72 pharmaceuticals and personal care products from Sri Lankan surface waterways. Sci Total Environ 690:683–695. https://doi.org/10.1016/j.scitotenv.2019.07.042
26. Kobusińska ME, Lewandowski KK, Panasiuk A, Łęczyński L, Urbaniak M, Ossowski T, Niemirycz E (2020) Precursors of polychlorinated dibenzo-p-dioxins and dibenzofurans in Arctic and Antarctic marine sediments: environmental concern in the face of climate change. Chemosphere 260. https://doi.org/10.1016/j.chemosphere.2020.127605
27. Mora A, García-Gamboa M, Sánchez-Luna MS, Gloria-García L, Cervantes-Avilés P, Mahlknecht J (2021) A review of the current environmental status and human health implications of one of the most polluted rivers of Mexico: the Atoyac River, Puebla. Sci Total Environ 782:146788. https://doi.org/10.1016/j.scitotenv.2021.146788
28. Xiang Y, Wu H, Li L, Ren M, Qie H, Lin A (2021) A review of distribution and risk of pharmaceuticals and personal care products in the aquatic environment in China. Ecotoxicol Environ Saf 213. https://doi.org/10.1016/j.ecoenv.2021.112044
29. Komolafe O, Mrozik W, Dolfing J, Acharya K, Vassalle L, Mota CR, Davenport R (2021) Occurrence and removal of micropollutants in full-scale aerobic, anaerobic and facultative wastewater treatment plants in Brazil. J Environ Manag 287:112286. https://doi.org/10.1016/j.jenvman.2021.112286
30. Ren B, Shi X, Jin X, Wang XC, Jin P (2021) Comprehensive evaluation of pharmaceuticals and personal care products (PPCPs) in urban sewers: degradation, intermediate products and environmental risk. Chem Eng J 404:127024. https://doi.org/10.1016/j.cej.2020.127024
31. Peng X, Yu Y, Tang C, Tan J, Huang Q, Wang Z (2008) Occurrence of steroid estrogens, endocrine-disrupting phenols, and acid pharmaceutical residues in urban riverine water of the Pearl River Delta, South China. Sci Total Environ 397:158–166. https://doi.org/10.1016/j.scitotenv.2008.02.059
32. Behera SK, Kim HW, Oh JE, Park HS (2011) Occurrence and removal of antibiotics, hormones and several other pharmaceuticals in wastewater treatment plants of the largest industrial city of Korea. Sci Total Environ 409:4351–4360. https://doi.org/10.1016/j.scitotenv.2011.07.015

33. Yang GCC, Yen CH, Wang CL (2014) Monitoring and removal of residual phthalate esters and pharmaceuticals in the drinking water of Kaohsiung City, Taiwan. J Hazard Mater 277:53–61. https://doi.org/10.1016/j.jhazmat.2014.03.005
34. Turner RDR, Warne MSJ, Dawes LA, Thompson K, Will GD (2019) Greywater irrigation as a source of organic micro-pollutants to shallow groundwater and nearby surface water. Sci Total Environ 669:570–578. https://doi.org/10.1016/j.scitotenv.2019.03.073
35. Chen H, Chen W, Guo H, Lin H, Zhang Y (2021) Pharmaceuticals and personal care products in the seawater around a typical subtropical tourist city of China and associated ecological risk. Environ Sci Pollut Res 28:22716–22728. https://doi.org/10.1007/s11356-020-12335-1
36. Karnjanapiboonwong A, Suski JG, Shah AA, Cai Q, Morse AN, Anderson TA (2011) Occurrence of PPCPs at a wastewater treatment plant and in soil and groundwater at a land application site. Water Air Soil Pollut 216:257–273. https://doi.org/10.1007/s11270-010-0532-8
37. Rivier PA, Havranek I, Coutris C, Norli HR, Joner EJ (2019) Transfer of organic pollutants from sewage sludge to earthworms and barley under field conditions. Chemosphere 222:954–960. https://doi.org/10.1016/j.chemosphere.2019.02.010
38. Thomaidi VS, Stasinakis AS, Borova VL, Thomaidis NS (2016) Assessing the risk associated with the presence of emerging organic contaminants in sludge-amended soil: a country-level analysis. Sci Total Environ 548–549:280–288. https://doi.org/10.1016/j.scitotenv.2016.01.043
39. Tran TM, Trinh HT, Anh HQ, Van Le T, Le SN, Minh TB (2020) Characterization of triclosan and triclocarban in indoor dust from home micro-environments in Vietnam and relevance of non-dietary exposure. Sci Total Environ 732:139326. https://doi.org/10.1016/j.scitotenv.2020.139326
40. Wang Y, Li G, Zhu Q, Liao C (2021) Occurrence of parabens, triclosan and triclocarban in paired human urine and indoor dust from two typical cities in China and its implications for human exposure. Sci Total Environ 786:147485. https://doi.org/10.1016/j.scitotenv.2021.147485
41. Miller TR, Heidler J, Chillrud SN, Delaquil A, Ritchie JC, Mihalic JN, Bopp R, Halden RU (2008) Fate of triclosan and evidence for reductive dechlorination of triclocarban in estuarine sediments. Environ Sci Technol 42:4570–4576. https://doi.org/10.1021/es702882g
42. Oliveira R, Domingues I, Grisolia CK, Soares AMVM (2009) Effects of triclosan on zebrafish early-life stages and adults. Environ Sci Pollut Res 16:679–688. https://doi.org/10.1007/s11356-009-0119-3
43. Ruszkiewicz JA, Li S, Rodriguez MB, Aschner M (2017) Is triclosan a neurotoxic agent?. J Toxicol Environ Heal Part B Crit Rev 20:104–117. https://doi.org/10.1080/10937404.2017.1281181
44. Geens T, Neels H, Covaci A (2012) Distribution of bisphenol-A, triclosan and n-nonylphenol in human adipose tissue, liver and brain. Chemosphere 87:796–802. https://doi.org/10.1016/j.chemosphere.2012.01.002
45. Hewitt NJ, Grégoire S, Cubberley R, Duplan H, Eilstein J, Ellison C, Lester C, Fabian E, Fernandez J, Géniès C, Jacques-Jamin C, Klaric M, Rothe H, Sorrell I, Lange D, Schepky A (2019) Measurement of the penetration of 56 cosmetic relevant chemicals into and through human skin using a standardized protocol. J Appl Toxicol 40:403–415. https://doi.org/10.1002/jat.3913
46. Toms LML, Allmyr M, Mueller JF, Adolfsson-Erici M, McLachlan M, Murby J, Harden FA (2011) Triclosan in individual human milk samples from Australia. Chemosphere 85:1682–1686. https://doi.org/10.1016/j.chemosphere.2011.08.009
47. Solá-Gutiérrez C, Schröder S, San-Román MF, Ortiz I (2020) Critical review on the mechanistic photolytic and photocatalytic degradation of triclosan. J Environ Manag 260:110101. https://doi.org/10.1016/j.jenvman.2020.110101
48. US FDA (2019) 84 FR 14847 – Safety and effectiveness of consumer antiseptic rubs; topical antimicrobial drug products for over-the-counter human use. Vol. 84, no. 71/Friday, April 12, 2019

49. DECISION 2016/110/EU (2016) Commission implementing decision (EU) 2016/110 of 27 January 2016 not approving triclosan as an existing active substance for use in biocidal products for product-type 1. Off J Eur Union L21/86
50. Schröder S, San-Román MF, Ortiz I (2020) Photocatalytic transformation of triclosan. Reaction products and kinetics. Catalysts 10:1–15. https://doi.org/10.3390/catal10121468
51. Vallejo M, San Román M-F, Ortiz I, Irabien A (2015) Overview of the PCDD/Fs degradation potential and formation risk in the application of advanced oxidation processes (AOPs) to wastewater treatment. Chemosphere 118:44–56. https://doi.org/10.1016/j.chemosphere.2014.05.077
52. Fernández-Castro F, Vallejo MS, Román MF, Ortiz I (2015) Insight on the fundamentals of advanced oxidation processes. Role and review of the determination methods of reactive oxygen species. J Chem Technol Biotechnol 90:796–820. https://doi.org/10.1002/jctb.4634
53. Karci A (2014) Degradation of chlorophenols and alkylphenol ethoxylates, two representative textile chemicals, in water by advanced oxidation processes: the state of the art on transformation products and toxicity. Chemosphere 99:1–18. https://doi.org/10.1016/j.chemosphere.2013.10.034
54. Fiedler H (2003) Dioxins and furans (PCDD/PCDF). In: Fiedler H (ed) Persistent organic pollutants. The handbook of environmental chemistry (vol. 3 series: anthropogenic compounds), vol vol 30. Springer, Berlin. https://doi.org/10.1007/10751132_6
55. WHO, World Health Organization (2016) Dioxins and their effects on human health. https://www.who.int/news-room/fact-sheets/detail/dioxins-and-their-effects-on-human-health. Accessed 21 Jun 2021
56. Baker JI, Hites RA (2000) Is combustion the major source of polychlorinated dibenzo-p-dioxins and dibenzofurans to the environment? A mass balance investigation. Environ Sci Technol 34:2879–2886. https://doi.org/10.1021/es9912325
57. Balmer ME, Goss KU, Schwarzenbach RP (2000) Photolytic transformation of organic pollutants on soil surfaces – an experimental approach. Environ Sci Technol 34:1240–1245. https://doi.org/10.1021/es990910k
58. Holt E, Weber R, Stevenson G, Gaus C (2012) Formation of dioxins during exposure of pesticide formulations to sunlight. Chemosphere 88:364–370. https://doi.org/10.1016/j.chemosphere.2012.03.058
59. Wang X, Ni Y, Zhang H, Zhang X, Chen J (2012) Formation and emission of PCDD/Fs in Chinese non-wood pulp and paper mills. Environ Sci Technol 46:12234–12240. https://doi.org/10.1021/es303373b
60. Weber R, Watson A, Forter M, Oliaei F (2011) Persistent organic pollutants and landfills – a review of past experiences and future challenges. Waste Manag Res 29:107–121. https://doi.org/10.1177/0734242X10390730
61. Fernández-Castro P, San Román MF, Ortiz I (2016) Theoretical and experimental formation of low chlorinated dibenzo-p-dioxins and dibenzofurans in the Fenton oxidation of chlorophenol solutions. Chemosphere 161:136–144. https://doi.org/10.1016/j.chemosphere.2016.07.011
62. Vallejo M, Fernández-Castro P, San Román MF, Ortiz I (2015) Assessment of PCDD/fs formation in the Fenton oxidation of 2-chlorophenol: influence of the iron dose applied. Chemosphere 137:135–141. https://doi.org/10.1016/j.chemosphere.2015.06.056
63. Vallejo M, San Román MF, Ortiz I, Irabien A (2014) The critical role of the operating conditions on the Fenton oxidation of 2-chlorophenol: assessment of PCDD/Fs formation. J Hazard Mater 279:579–585. https://doi.org/10.1016/j.jhazmat.2014.07.020
64. Solá-Gutiérrez C, San Román MF, Ortiz I (2018) Fate and hazard of the electrochemical oxidation of triclosan. Evaluation of polychlorodibenzo-p-dioxins and polychlorodibenzofurans (PCDD/Fs) formation. Sci Total Environ 626:126–133. https://doi.org/10.1016/j.scitotenv.2018.01.082
65. Solá-Gutiérrez C, Schröder S, San Román MF, Ortiz I (2019) PCDD/Fs traceability during triclosan electrochemical oxidation. J Hazard Mater 369:584–592. https://doi.org/10.1016/j.jhazmat.2019.02.066

66. Kim HK, Masaki H, Matsumura T, Kamei T, Magara Y (2002) Removal efficiency and homologue patterns of dioxins in drinking water treatment. Water Res 36:4861–4869. https://doi.org/10.1016/S0043-1354(02)00202-6
67. Weber R, Herold C, Hollert H, Kamphues J, Blepp M, Ballschmiter K (2018) Reviewing the relevance of dioxin and PCB sources for food from animal origin and the need for their inventory, control and management. Environ Sci Eur 30. https://doi.org/10.1186/s12302-018-0166-9
68. WHO, World Health Organization (2010) Exposure to dioxins and dioxin-like substances: a major public health concern. What are dioxins and dioxin-like substances? Derivation of toxic equivalency factors (TEFs). Public Health and Environment, World Health Organization, Public Health and Environment World Health Organization 20 Avenue Appia, 1211 Geneva 27, Switzerland
69. Tuomisto J (2011) The toxic equivalency principle and its application in dioxin risk assessment. AH Recept Biol Toxicol 317–330. https://doi.org/10.1002/9781118140574.ch23
70. US EPA (1994) Method 1613 Tetra- through octa-chlorinated dioxins and furans by isotope dilution HRGC/HRMS October 1994 U. S. Environmental Protection Agency Office of Water Engineering and Analysis Division (4303) 401 M Street S. W. Washington, D. C. 20460. Environ Prot
71. Directive 2013/39/EU of the European Parliament and of the Council of 12 August 2013 amending Directives 2000/60/EC and 2008/105/EC as regards priority substances in the field of water policy, 2013. Off J Eur Union L 226/1
72. US EPA (2009) National primary drinking water guidelines. EPA 816-F-09-004 1,7
73. Luo Y, Guo W, Ngo HH, Nghiem LD, Hai FI, Zhang J, Liang S, Wang XC (2014) A review on the occurrence of micropollutants in the aquatic environment and their fate and removal during wastewater treatment. Sci Total Environ 473–474:619–641. https://doi.org/10.1016/j.scitotenv.2013.12.065
74. Zhang Y, Geißen SU, Gal C (2008) Carbamazepine and diclofenac: removal in wastewater treatment plants and occurrence in water bodies. Chemosphere 73:1151–1161. https://doi.org/10.1016/j.chemosphere.2008.07.086
75. Comninellis C, Kapalka A, Malato S, Parsons SA, Poulis I, Mantzavinos D (2008) Advanced oxidation processes for water treatment: advances and trends for R&D. J Chem Technol Biotechnol 83:769–776. https://doi.org/10.1002/jctb.1873
76. Giannakis S, Gamarra Vives FA, Grandjean D, Magnet A, De Alencastro LF, Pulgarin C (2015) Effect of advanced oxidation processes on the micropollutants and the effluent organic matter contained in municipal wastewater previously treated by three different secondary methods. Water Res 84:295–306. https://doi.org/10.1016/j.watres.2015.07.030
77. Klavarioti M, Mantzavinos D, Kassinos D (2009) Removal of residual pharmaceuticals from aqueous systems by advanced oxidation processes. Environ Int 35:402–417. https://doi.org/10.1016/j.envint.2008.07.009
78. Miklos DB, Remy C, Jekel M, Linden KG, Drewes JE, Hübner U (2018) Evaluation of advanced oxidation processes for water and wastewater treatment – a critical review. Water Res 139:118–131. https://doi.org/10.1016/j.watres.2018.03.042
79. Vallejo M, San Román MF, Irabien A, Ortiz I (2013) Comparative study of the destruction of polychlorinated dibenzo-p-dioxins and dibenzofurans during Fenton and electrochemical oxidation of landfill leachates. Chemosphere 90:132–138. https://doi.org/10.1016/j.chemosphere.2012.08.018
80. Vallejo M, San Román MF, Ortiz I (2013) Quantitative assessment of the formation of polychlorinated derivatives, PCDD/Fs, in the electrochemical oxidation of 2-chlorophenol as function of the electrolyte type. Environ Sci Technol 47:12400–12408. https://doi.org/10.1021/es403246g
81. Haidar M, Dirany A, Sirés I, Oturan N, Oturan MA (2013) Electrochemical degradation of the antibiotic sulfachloropyridazine by hydroxyl radicals generated at a BDD anode. Chemosphere 91:1304–1309. https://doi.org/10.1016/j.chemosphere.2013.02.058

82. Labiadh L, Barbucci A, Carpanese MP, Gadri A, Ammar S, Panizza M (2016) Comparative depollution of methyl orange aqueous solutions by electrochemical incineration using $TiRuSnO_2$, BDD and PbO_2 as high oxidation power anodes. J Electroanal Chem 766:94–99. https://doi.org/10.1016/j.jelechem.2016.01.036
83. Weber R (2007) Relevance of PCDD/PCDF formation for the evaluation of POPs destruction technologies – review on current status and assessment gaps. Chemosphere 67:109–117. https://doi.org/10.1016/j.chemosphere.2006.05.094
84. Gao Y, Ji Y, Li G, An T (2014) Mechanism, kinetics and toxicity assessment of OH-initiated transformation of triclosan in aquatic environments. Water Res 49:360–370. https://doi.org/10.1016/j.watres.2013.10.027
85. Zhang X, Zhang C, Sun X, Kang L, Zhao Y (2015) Chemical conversion pathways and kinetic modeling for the OH-initiated reaction of triclosan in gas-phase. Int J Mol Sci 16:8128–8141. https://doi.org/10.3390/ijms16048128
86. Kanakaraju D, Glass BD, Oelgemöller M (2014) Titanium dioxide photocatalysis for pharmaceutical wastewater treatment. Environ Chem Lett 12:27–47. https://doi.org/10.1007/s10311-013-0428-0

An Innovative Technology to Minimize Biological Sludge Production and Improve Its Quality in a Circular Economy Perspective

Maria Cristina Collivignarelli, Francesca Maria Caccamo, and Marco Carnevale Miino

Contents

Abstract In the coming years, the production of biological sewage sludge is set to increase. According to the European legislation, the management of sludge, as well as other waste, must follow a hierarchical approach according to which the first place in order of priority is represented by the prevention/minimization of the production. Over the last few years, thermophilic aerobic processes proved to be effective in minimizing the production of sludge within wastewater treatment plants (WWTPs). Thermophilic aerobic/anoxic membrane reactor (TAMR) technology combines the advantages of thermophilic aerobic treatments with those of biological membrane processes. This work reviews the literature concerning the application of TAMR

M. C. Collivignarelli
Department of Civil Engineering and Architecture, University of Pavia, Pavia, Italy

Interdepartmental Centre for Water Research, University of Pavia, Pavia, Italy

F. M. Caccamo and M. Carnevale Miino (✉)
Department of Civil Engineering and Architecture, University of Pavia, Pavia, Italy
e-mail: marco.carnevalemiino01@universitadipavia.it

Avelino Núñez-Delgado and Manuel Arias-Estévez (eds.),
Emerging Pollutants in Sewage Sludge and Soils,
Hdb Env Chem (2023) 114: 137–146, DOI 10.1007/698_2022_852,

Published online: 26 February 2022

focusing on the prevention of the production of biological sludge and on the improvement of its quality for the purpose of a possible recovery in agriculture in a circular economy perspective. The results show that the process is mature and effective for full-scale application in conventional WWTPs.

Keywords Agricultural reuse, Aqueous waste, Circular economy, Sludge minimization, Thermophilic membrane reactor

1 Introduction

A growing production of biological sewage sludge (BSS) and a simultaneous worsening of the qualitative characteristics are the consequences of the imposition of more restrictive limits, as European Directive 91/271/EEC and subsequent amendments [1, 2], on the effluents of wastewater treatment plants (WWTPs) [3]. In 2015, European urban WWTPs produced 9.7 million tons of dry matter of BSS [4]. Therefore, a sustainable management of sludge is nowadays desirable and, above all, mandatory objective.

In fact, Directive 2018/851/EC [5] identified a hierarchy in waste management, therefore also applicable to BSS: (1) prevention and minimization of the production, (2) matter recovery and reuse, (3) energy recovery, and finally (4) safe disposal of waste. The prevention/minimization of the production of BSS at the source is an aspect of primary importance not only because the legislation requires it, but also because it can guarantee many non-negligible benefits including the reduction of costs incurred by WWTPs. As reported in literature [6–9], the management of sludge represents about 50% of the total operating costs of WWTPs. In addition to the economic aspect, the environmental impact linked to the treatments, transport and final disposal of the sludge must also be considered.

According to the Italian Higher Institute for Environmental Protection and Research, in 2017, Italian urban WWTPs produced about 3.2 million tons of sludge (about 0.8 million tons of dry matter) [10, 11], with 47.7% being sent for recovery and 50.6% for disposal, recording a 1.4% decrease in landfill disposal in favour of recovery compared to the previous year [10]. The European Directive 86/278/EEC [12] aimed at encouraging the use of good quality sludge in agriculture by banning the use of untreated sludge on agricultural land to avoid any harmful effects caused by the presence of pathogens and organic contaminants [13, 14]. The practice of reuse can be fully integrated into a circular economy vision [15, 16]. Concerning this aspect, in 2020 the European Commission adopted a new Action Plan for the Circular Economy to promote the sustainable use and reuse of resources [17]. In the urban water management system, one of the main actions needed to implement a circular economy approach is the transformation of WWTPs into water resource recovery plants (WRRFs) [18, 19]. To do this, the prevention and minimization of the production of BSS represents the first step that can be pursued in two distinct ways: (1) adopting processes capable of treating the water with a minimum production of residual sludge; (2) providing in situ treatments to minimize the quantities of sludge produced [4].

This chapter aims to provide an overview of the results obtained testing the thermophilic aerobic/anoxic membrane reactor (TAMR) technology which can guarantee both approaches described above.

2 The Technology

TAMR is an advanced biological process that simultaneously combines a pure oxygen membrane bioreactor (MBR) system and a thermophilic treatment in autothermal conditions. According to previous publications [20–24], the application of this combined process, in addition to having a low environmental impact as a biological technology, has the following advantages: (1) drastic reduction of the sludge produced, (2) high removal rates of slowly biodegradable compounds in mesophilic conditions, (3) excellent flexibility in case of organic overload, (4) high compactness of the system, (5) inhibition of pathogens, and (6) possibility of energy recovery.

The process can be applied both in the water line and in the sludge line of WWTPs. In case that aqueous waste is fed, TAMR operates only in aerobic conditions while BSS also require an anoxic phase to effectively minimize sludge production. Thermophilic conditions (47–53°C) are maintained thanks to the exothermic degradation processes of the thermophilic microorganisms. To ensure the self-heating of the process, the feed must be rich in organic matter and therefore, the water line application should be in WWTPs authorized for the treatment of aqueous waste, as an urban sewage would not guarantee the self-heating of the thermophilic process (Fig. 1). In the case of sludge line application, the TAMR can be used both to co-treat sewage sludge and aqueous waste and to treat only BSS from conventional active sludge (CAS) systems.

The TAMR produces (1) residual sludge (Sects. 3.1 and 3.2) and (2) aqueous permeate (Sect. 4). In Lombardy (Italy), there are currently two full-scale TAMR plants for the treatment of aqueous waste (sludge prevention through water line intervention).

3 Sludge Prevention/Minimization

3.1 Residual Sludge Production

Residual thermophilic sludge represents the excess sludge of the thermophilic biological system and can have a percentage of dry matter up to 19% [25–27]. Its production is lower in terms of mass and volume than that of the permeate. Table 1 shows the results of the specific production of thermophilic sludge obtained mainly

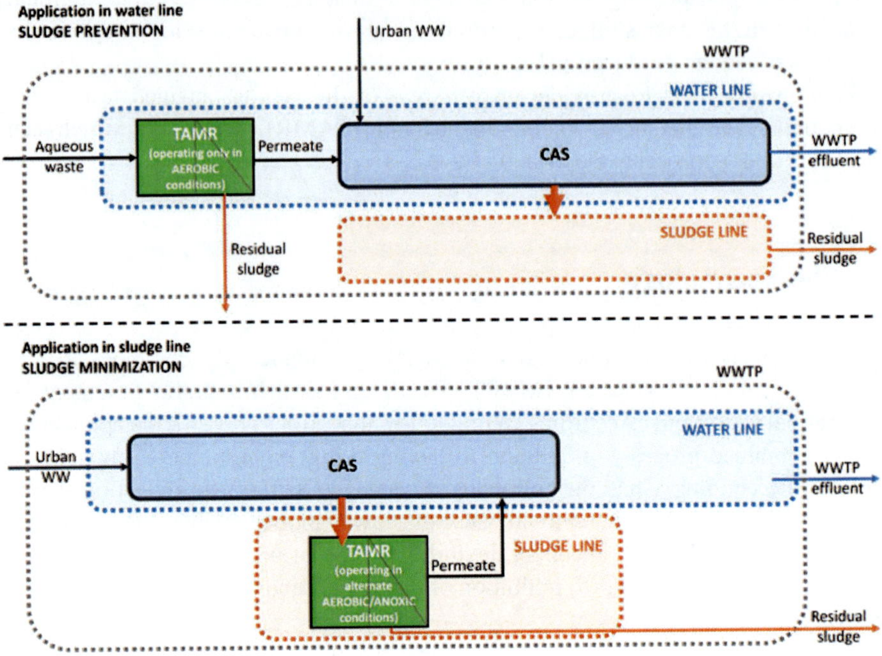

Fig. 1 Application of TAMR in water and sludge line. *CAS* conventional active sludge, *WW* wastewater

Table 1 Specific production of sludge in TAMR technology. *WW* wastewater, *R* real scale, *S* semi-industrial scale, *TP* total phosphorus, *TN* total nitrogen

Substrate	Scale	Specific sludge production ($kg_{VSS\ produced}\ kg_{COD\ removed}^{-1}$)	References
Aqueous waste			
Pharmaceuticals and detergents production WW and landfill leachate	R	0.092–0.101	[26]
Saline WW, neutral/acid/basic WW, landfill leachate, solvent WW, and slurries	R	0.08–0.09	[25]
WWs with highly recalcitrant pollutants (e.g., surfactants, solvents, pharmaceutical products)	R	0.052[a]	[27]
High strength WWs	S	0.09	[28]
High strength WW mainly containing dyes, surfactants, and solvents	S	0.04	[20]
WW with high concentrations of COD, TP, TN, chloride, acetic acid, methylene chloride, ethanol	S	0.016	[29]
Aqueous waste and sewage sludge[a]	S	0.04	[30]

[a] Expressed in $kg_{VS\ produced}\ kg_{COD\ removed}^{-1}$

[b] Mixture composed of 30% sewage sludge and of 70% of aqueous waste

during experiments at the semi-industrial scale of the TAMR technology both on diverse aqueous waste and on BSS. In the case of aqueous waste treatment, specific sludge production data monitored in full-scale plants are also available. These results are lower than those achievable with a mesophilic MBR (0.10–0.19 $kg_{VSS\ produced}\ kg_{COD\ removed}^{-1}$) [21, 31] and close to those reported in the literature for aerobic thermophilic processes (0.08 $kg_{VSS\ produced}\ kg_{COD\ removed}^{-1}$) [21], (VSS: volatile suspended solids; COD: chemical oxigen demand). Even the granular anaerobic processes have higher values than the TAMR technology: for example, the specific production of sludge in a UASB reactor that treats sewage sludge is equal to 0.1 $kg_{VSS\ produced}\ kg_{COD\ removed}^{-1}$ [32].

3.2 *Sludge Quality Improvement*

In general, the Italian legislation on the recovery of sludge in agriculture imposes some stricter limit values (such as on total chromium, lead, arsenic, agronomic parameters, and several organic contaminants) compared to other legislations, including the French and German ones. In particular, in the current legislation in Lombardy (Italy) [33], a distinction is required between "suitable sludge" and "high quality sludge". Sludge suitable for spreading in agricultural fields must comply with the limit values set by current Italian legislation, while "high-quality" sludge requires more stringent limit values than national ones.

Regarding the thermophilic sludge residue, the only criticality could be represented by the insufficiency of organic carbon, which can be solved by mixing other BSS with the thermophilic sludge normally with high concentrations of COD [30].

However, in an experiment involving the treatment of industrial wastewater with high concentrations of chlorides and perfluoroalkyl, although most of the COD introduced was oxidized in the TAMR, only a minor but still significant part (6–12%) remained in the thermophilic sludge [34].

A high concentration of phosphorus in the crystalline phase has been identified in the thermophilic sludge. In the thermophilic reactor, the chemical precipitation of total phosphorus takes place in the form of salts, such as vivianite and hydroxyapatite [28, 29]. In agreement with the scientific literature [35], these results could be related to the increase in pH induced by the aeration of the reactor which allowed the crystallization of phosphorus [28].

A significant amount of nitrogen in the thermophilic sludge was also observed due to (1) the absorption of nitrogen by the biomass, (2) adhesion to sludge, and (3) precipitation of ammoniacal nitrogen in the form of struvite [29, 34].

As regards the presence of pathogenic microorganisms, thermophilic processes generally guarantee greater safety than the mesophilic ones, thanks to higher process temperatures [23, 24]. Therefore, thermophilic extracted sludge could be suitable for spreading in agriculture thanks to the high content of carbon, nitrogen, phosphorus, and potassium, the excellent degree of humification and sanitation that guarantees a

Table 2 Qualitative characteristics of mixed liquor. *TN* total nitrogen, *TP* total phosphorus

Results on nutrients	References
Organic carbon accumulation (6–12% of COD fed)	[8, 30, 34]
Nitrogen presence (8–10% of VSS; 8–24% of TN fed)	[26, 29, 30, 34]
Accumulation by chemical precipitation of phosphorus as inorganic salts (70–80% of TP fed)	[28, 29, 34, 36]
Other results	
High concentration of total solids (up to 190 kg m^{-3} in the full-scale applications)	[25–27]
Absence of foaming phenomena during the treatment of liquid waste (real laundry wastewater rich in TAS e MBAS)	[37]
Sanitation thanks to high temperatures (>45°C)	[26, 29]

healthy and safe recovery of the sludge. Table 2 shows the main qualitative characteristics of the thermophilic sludge extracted from TAMR.

4 Possibility of Permeate Reuse

Among the residues, the permeate is the most significant from a quantitative point of view. The ultrafiltration membranes allow to keep all the biomass inside the biological reactor, obtaining a permeate totally solids-free substrate [8]. In addition, it is rich in ammoniacal nitrogen thanks to excellent ammonification activity by the thermophilic bacteria in TAMR [30, 38]. Despite the excellent performance of TAMR process (COD removals up to 90% [26, 29, 30]), permeate contains significant concentrations of well biodegradable COD by a mesophilic biomass, confirming the complementarity between mesophilic and thermophilic processes for the biodegradation of organic substances [20, 26, 28, 36].

Therefore, this substrate can first be subjected to a stripping treatment for the recovery of ammonia nitrogen in the form of ammonium sulphate and, considering the good biodegradability of the organic substance by mesophilic biomass, recirculated in the denitrification reactor of a CAS to improve the kinetics of nitrate removal, in place of external sources of carbon of synthetic origin [27, 29, 34, 38].

5 Tips for Future Research and Applications

Considering the depletion of world natural reserves of phosphorus, it would be interesting to investigate the bioavailability of this nutrient in the sludge extracted from TAMR to evaluate the direct assimilation by crops in case of agricultural reuse.

Another aspect that would be interesting to investigate is the application of the technology on BSS resulting from the treatment of industrial wastewater and aqueous waste. In this case, the authors suggest evaluating the performance of

TAMR to minimize BSS production considering feed with diverse characteristics and comparing the results with those obtained treating urban BSS. At the same time, examining a possible toxic and chronic effect of these substrates on the thermophilic sludge can represent an interesting point that should be further investigated.

The authors also suggest studying the up-grade of the process. For instance, the introduction into the reactor of a mobile support material for the development of attached biomass could be an aspect to be investigated. The traditional suspended biomass already present and the new adherent biomass developed on supports with a high specific surface would thus work simultaneously, guaranteeing a hybrid process. The support materials introduced into the thermophilic reactor could also be recovered through recycling operations according to a circular economy perspective applied to integrated urban water cycle.

6 Conclusions

The TAMR technology ensures the prevention/minimization of the production of BSS and guarantees the recovery of the residues produced. The excess sludge extracted from the thermophilic biological reactor could be destined for recovery in agriculture thanks to its content of nutrients (organic carbon, nitrogen, and phosphorus) and greater protection against the pathogenic load. At the same time, the permeate can be reused as an external carbon source in a post-denitrification process, after stripping to produce ammonium sulphate, thanks to the high content of ammonia nitrogen and well-biodegradable organic carbon by mesophilic biomass. In this way, both residues obtained from the TAMR acquire an economic value as products, guaranteeing the important possibility of closing the cycle linked to the management of wastewater and BSS in a circular economy perspective.

References

1. European Commission (1998) EUR-Lex commission directive 98/15/EC of 27 February 1998 amending council directive 91/271/EEC with respect to certain requirements established in annex I thereof. Off J Eur Communities 67:29–30
2. European Commission (1991) EUR-Lex council directive 91/271/EEC of 21 May 1991 concerning urban wastewater treatment. Off J Eur Communities 135:40–52
3. Mininni G, Blanch AR, Lucena F, Berselli S (2015) EU policy on sewage sludge utilization and perspectives on new approaches of sludge management. Environ Sci Pollut Res 22:7361–7374. https://doi.org/10.1007/s11356-014-3132-0
4. Collivignarelli MC, Abbà A, Carnevale Miino M, Torretta V (2019) What advanced treatments can be used to minimize the production of sewage sludge in WWTPs? Appl Sci 9:2650. https://doi.org/10.3390/app9132650
5. European Commission (2018) EUR-Lex directive EU/2018/851 of the European Parliament and of the council of 30 May 2018 amending directive 2008/98/EC on waste. Off J Eur Communities 150:109–140

6. Bertanza G, Canato M, Laera G, Tomei MC (2015) Methodology for technical and economic assessment of advanced routes for sludge processing and disposal. Environ Sci Pollut Res 22: 7190–7202. https://doi.org/10.1007/s11356-014-3088-0
7. Bertanza G, Papa M, Canato M et al (2014) How can sludge dewatering devices be assessed? Development of a new DSS and its application to real case studies. J Environ Manag 137:86–92. https://doi.org/10.1016/j.jenvman.2014.02.002
8. Collivignarelli MC, Castagnola F, Sordi M, Bertanza G (2015) Treatment of sewage sludge in a thermophilic membrane reactor (TMR) with alternate aeration cycles. J Environ Manag 162: 132–138. https://doi.org/10.1016/j.jenvman.2015.07.031
9. Zhao G, Garrido-Baserba M, Reifsnyder S et al (2019) Comparative energy and carbon footprint analysis of biosolids management strategies in water resource recovery facilities. Sci Total Environ 665:762–773. https://doi.org/10.1016/j.scitotenv.2019.02.024
10. Ispra (2019) Special waste report, 2019 edition
11. Mininni G, Mauro E, Piccioli B et al (2019) Production and characteristics of sewage sludge in Italy. Water Sci Technol 79:619–626. https://doi.org/10.2166/wst.2019.064
12. European Commission (1986) EUR_Lex council directive 86/278/EEC of 12 June 1986 on the protection of the environment, and in particular of the soil, when sewage sludge is used in agriculture. Off J Eur Communities 181:6–12
13. Liu M, Wang C, Bai Y, Xu G (2018) Effects of sintering temperature on the characteristics of lightweight aggregate made from sewage sludge and river sediment. J Alloys Compd 748:522–527. https://doi.org/10.1016/j.jallcom.2018.03.216
14. Zhang L, Xu C, Champagne P, Mabee W (2014) Overview of current biological and thermo-chemical treatment technologies for sustainable sludge management. Waste Manag Res 32: 586–600. https://doi.org/10.1177/0734242X14538303
15. Ashekuzzaman SM, Forrestal P, Richards K, Fenton O (2019) Dairy industry derived waste-water treatment sludge: generation, type and characterization of nutrients and metals for agricultural reuse. J Clean Prod 230:1266–1275. https://doi.org/10.1016/j.jclepro.2019.05.025
16. Kacprzak M, Neczaj E, Fijałkowski K et al (2017) Sewage sludge disposal strategies for sustainable development. Environ Res 156:39–46. https://doi.org/10.1016/j.envres.2017.03.010
17. European Commission (2020) EU circular economy action plan
18. Collivignarelli MC, Abbà A, Frattarola A et al (2019) Legislation for the reuse of biosolids on agricultural land in Europe: overview. Sustainability 11:6015. https://doi.org/10.3390/su11216015
19. Cornejo PK, Becker J, Pagilla K et al (2019) Sustainability metrics for assessing water resource recovery facilities of the future. Water Environ Res 91:45–53. https://doi.org/10.2175/106143017X15131012187980
20. Collivignarelli MC, Abbà A, Bertanza G, Barbieri G (2017) Treatment of high strength aqueous wastes in a thermophilic aerobic membrane reactor (TAMR): performance and resilience. Water Sci Technol 76:3236–3245. https://doi.org/10.2166/wst.2017.492
21. Kurian R, Acharya C, Nakhla G, Bassi A (2005) Conventional and thermophilic aerobic treatability of high strength oily pet food wastewater using membrane-coupled bioreactors. Water Res 39:4299–4308. https://doi.org/10.1016/j.watres.2005.08.030
22. LaPara TM, Alleman JE (1999) Thermophilic aerobic biological wastewater treatment. Water Res 33:895–908. https://doi.org/10.1016/S0043-1354(98)00282-6
23. Lloret E, Pastor L, Pradas P, Pascual JA (2013) Semi full-scale thermophilic anaerobic digestion (TAnD) for advanced treatment of sewage sludge: stabilization process and pathogen reduction. Chem Eng J 232:42–50. https://doi.org/10.1016/j.cej.2013.07.062
24. Ziemba C, Peccia J (2011) Net energy production associated with pathogen inactivation during mesophilic and thermophilic anaerobic digestion of sewage sludge. Water Res 45:4758–4768. https://doi.org/10.1016/j.watres.2011.06.014

25. Collivignarelli MC, Abbà A, Frattarola A et al (2019) Treatment of aqueous wastes by means of thermophilic aerobic membrane reactor (TAMR) and nanofiltration (NF): process auditing of a full-scale plant. Environ Monit Assess 191:708. https://doi.org/10.1007/s10661-019-7827-z
26. Collivignarelli MC, Bertanza G, Sordi M, Pedrazzani R (2015) High-strength wastewater treatment in a pure oxygen thermophilic process: 11-year operation and monitoring of different plant configurations. Water Sci Technol 71:588–596. https://doi.org/10.2166/wst.2015.008
27. Collivignarelli MC, Carnevale Miino M, Caccamo FM et al (2021) Performance of full-scale thermophilic membrane bioreactor and assessment of the effect of the aqueous residue on mesophilic biological activity. Water 13:1754. https://doi.org/10.3390/w13131754
28. Collivignarelli MC, Abbà A, Bertanza G et al (2018) Integrating novel (thermophilic aerobic membrane reactor-TAMR) and conventional (conventional activated sludge-CAS) biological processes for the treatment of high strength aqueous wastes. Bioresour Technol 255:213–219. https://doi.org/10.1016/j.biortech.2018.01.112
29. Collivignarelli MC, Abbà A, Bertanza G (2014) Treatment of high strength pharmaceutical wastewaters in a thermophilic aerobic membrane reactor (TAMR). Water Res 63:190–198. https://doi.org/10.1016/j.watres.2014.06.018
30. Collivignarelli MC, Abbà A, Bertanza G, Frattarola A (2019) Drastic reduction of sludge in wastewater treatment plants: co-digestion of sewage sludge and aqueous waste in a thermophilic membrane reactor. Environ Technol 1–10. https://doi.org/10.1080/09593330.2019.1575478
31. Lee W, Kang S, Shin H (2003) Sludge characteristics and their contribution to microfiltration in submerged membrane bioreactors. J Memb Sci 216:217–227. https://doi.org/10.1016/S0376-7388(03)00073-5
32. Chang FY, Lin CY (2004) Biohydrogen production using an up-flow anaerobic sludge blanket reactor. Int J Hydrog Energy 29:33–39. https://doi.org/10.1016/S0360-3199(03)00082-X
33. Lombardy Region (2019) Regional decree n. 6665 of 14 May 2019 (in Italian). https://www.regione.lombardia.it/wps/wcm/connect/19df87f4-0ac9-4fac-b171-084e37a5bc51/D.d.u.o.+6665_2019.pdf?MOD=AJPERES&CACHEID=ROOTWORKSPACE-19df87f4-0ac9-4fac-b171-084e37a5bc51-mH4MG8K. Accessed 9 Sep 2019
34. Collivignarelli MC, Abbà A, Bertanza G et al (2021) Treatment of high strength wastewater by thermophilic aerobic membrane reactor and possible valorisation of nutrients and organic carbon in its residues. J Clean Prod 280. https://doi.org/10.1016/j.jclepro.2020.124404
35. Liu S, Zhu N, Li LY (2011) The one-stage autothermal thermophilic aerobic digestion for sewage sludge treatment. Chem Eng J 174:564–570. https://doi.org/10.1016/j.cej.2011.09.043
36. Collivignarelli MC, Abbà A, Bertanza G (2015) Why use a thermophilic aerobic membrane reactor for the treatment of industrial wastewater/liquid waste? Environ Technol 36:2115–2124. https://doi.org/10.1080/09593330.2015.1021860
37. Collivignarelli MC, Carnevale Miino M, Baldi M et al (2019) Removal of non-ionic and anionic surfactants from real laundry wastewater by means of a full-scale treatment system. Process Saf Environ Prot 132:105–115. https://doi.org/10.1016/j.psep.2019.10.022
38. Collivignarelli MC, Abbà A, Castagnola F, Bertanza G (2017) Minimization of municipal sewage sludge by means of a thermophilic membrane bioreactor with intermittent aeration. J Clean Prod 143:369–376. https://doi.org/10.1016/j.jclepro.2016.12.101

Fate of Neonicotinoids in the Environment: Why Bees Are Threatened

Adeniyi K. Aseperi, Rosa Busquets, Philip C. W. Cheung, Peter S. Hooda, and James Barker

Contents

Abstract Pollinators are vital for ecosystems, agriculture and the economy. Their population has been declining over several decades and the current situation is serious and of great concern. Multiple stressors are likely to have contributed to this, and exposure to neonicotinoid pesticides is one possible causative factor. Pollinators, including bees, can encounter neonicotinoids when foraging

A. K. Aseperi, R. Busquets (✉), and J. Barker
School of Life Sciences, Pharmacy and Chemistry, Kingston University London, Kingston-upon-Thames, Surrey, UK
e-mail: R.Busquets@kingston.ac.uk

P. C. W. Cheung
School of Life Sciences, Pharmacy and Chemistry, Kingston University London, Kingston-upon-Thames, Surrey, UK

Department of Chemical Engineering, Imperial College, London, UK

P. S. Hooda
School of Engineering and the Environment, Kingston University London, Kingston-upon-Thames, Surrey, UK

Avelino Núñez-Delgado and Manuel Arias-Estévez (eds.),
Emerging Pollutants in Sewage Sludge and Soils,
Hdb Env Chem (2023) 114: 147–166, DOI 10.1007/698_2022_853,

Published online: 18 April 2022

contaminated flowers, and although they were not the target organism for such pesticides, neonicotinoids can be fatal for them (lethal dose that will reduce the insect population by 50% is ~2 mg/kg). The specific application of these pesticides plays an important role in their dispersion into the environment and application to the seed coating itself provides a more targeted way to release the pesticide with a reduced risk. Neonicotinoids can disperse via aerosols (when sprayed), bound to soil dust and dust abrasion from the seed coating, via the degradation of contaminated plants or run-off from crops treated with the pesticide. They have high solubility in water (e.g. 39.8 g dinotefuran/L, 4.1 g thiamethoxam/L) and this favours their spread, however natural factors such as sunlight, warm weather and microorganisms can degrade them and reduce their persistence. The time required for the concentration of neonicotinoids in soil to reach half of their initial concentration is varied (e.g. 3.4–7,000 days). Soils that are poor in organic matter will poorly retain neonicotinoids and as a result they will be very mobile in them and potentially pollute water systems. In contrast, soils rich in organic matter will have greater retention of neonicotinoids and once saturated, neonicotinoids will leach in a sustained manner. The benefits given by neonicotinoids explain why they are widely used across the globe. Their potential impact in the fields on bees and pollinators in general thus calls for a globally responsible and restricted use of neonicotinoids as well as innovation to reduce their ecotoxicity.

Keywords Colony collapse disorder, Pollinators, Seed coating, Systemic insecticide

1 Introduction

Perilous times have befallen bees. A combination of multiple stressors including pathogens such as varroa mites and virulent fungal strains, the deterioration of habitat, lack of forage, poor nutrition, lack of genetic diversity, over-exposure to pesticides and colony collapse disorder are threatening both cultivated honeybees and their bumble cousins in the wild to existential limits. This is a bad news for humans too; the process of pollination is agriculturally important to the environment and food production. According to the United Nations' Food and Agriculture Organization, better known as the FAO, 90% of the world's food supply comes from 100 crop species, of which 71 (principally fruits and vegetables) are pollinated by bees. For example, a substantial proportion of the global agricultural pollination services is attributed to the European honeybee (*Apis mellifera*) [1].

Agricultural practices have a myriad of effects on ecosystems, and pesticides used to increase crop productivity can have a knock-on effect on beneficial insects. The decline in the population of bees, which has been happening over several decades, has triggered the re-evaluation of the benefits of pesticides to the toxicity caused by them to non-target organisms. The link between bee decline and the use of systemic pesticides, when applied according to label directions, is still controversial (see CropLife America statement).

This chapter addresses the historical introduction of neonicotinoid (often abbreviated to *neonics*) insecticides. It introduces their modes of application in agriculture, and the phenomenon of colony collapse disorder. Factors affecting the environmental persistence, fate and distribution of neonicotinoids which could contribute to bee decline are discussed in this chapter and illustrated with recent data.

2 The Introduction of Neonicotinoids in the Market

Neonicotinoids are systemic insecticides used for the protection of agricultural crops. They were developed in response to evolved pest resistances and can target several pests in the Homoptera, Coleoptera and Lepidoptera families [2, 3]. The molecular structures of the main neonicotinoids are shown in Fig. 1.

In 1985, Bayer AG patented imidacloprid as the first commercial neonic and in 1994 this was registered by the U.S. Environmental Protection Agency (EPA) [5]. Soon afterwards, in 1995, acetamiprid and nitenpyram appeared on the market. That same year, Canada deployed its first neonic, imidacloprid, to combat insects known to damage apples, tomatoes, potatoes and lettuces [6]. Thiamethoxam became commercially available in 1998, clothianidin and thiacloprid in 2001, and dinotefuran in 2002 [7].

Overall, the demand for neonics has been on the rise for the past three decades. By the late 1990s, the efficacy of neonics as insecticides in farming was recognized worldwide and they were used to kill and control aphids, mirids, whiteflies, termites, beetles, thrips, locusts, leaf miners, stink bugs and root-feeding grubs. Veterinarians also discovered the effectiveness of neonics in eliminating fleas and ticks from animals [8]. Neonics can perform many functions, e.g., nitenpyram is a versatile neonicotinoid used to control sucking insects on rice, but is also used to eliminate external parasites in livestock and domestic pets.

Neonicotinoids arrived at a time of agrochemical development where organochlorines, organophosphates and carbamates amounted to over 90% of insecticides applied worldwide. Since then, neonicotinoids' usage has grown. In 1997, synthetic pyrethroids and other newer insecticides accounted for <5% of the quantity of insecticides used and only covered about one-third of farmland areas subjected to treatments. By 2013, almost all corn planted in the United States was treated with clothianidin or thiamethoxam. By 2014, a third of American soybean acreage was planted with seeds treated with imidacloprid or thiamethoxam. The economic activity of seed treatment alone grew from US$0.2 billion in the 1990s to approximately $1 billion in 2008 [9, 10]. By 2009, imidacloprid amounted to 41% of all neonics' sales, at a market value of $1.091 billion, when the total neonics market value was U.S. $2.63 billion [3]. In comparison, thiamethoxam was at $0.627 billion, and clothianidin at $0.439 billion [3]. By 2011, the global demand for thiamethoxam had risen to US $1 billion, and in 2012, to US $1.1 billion [11],

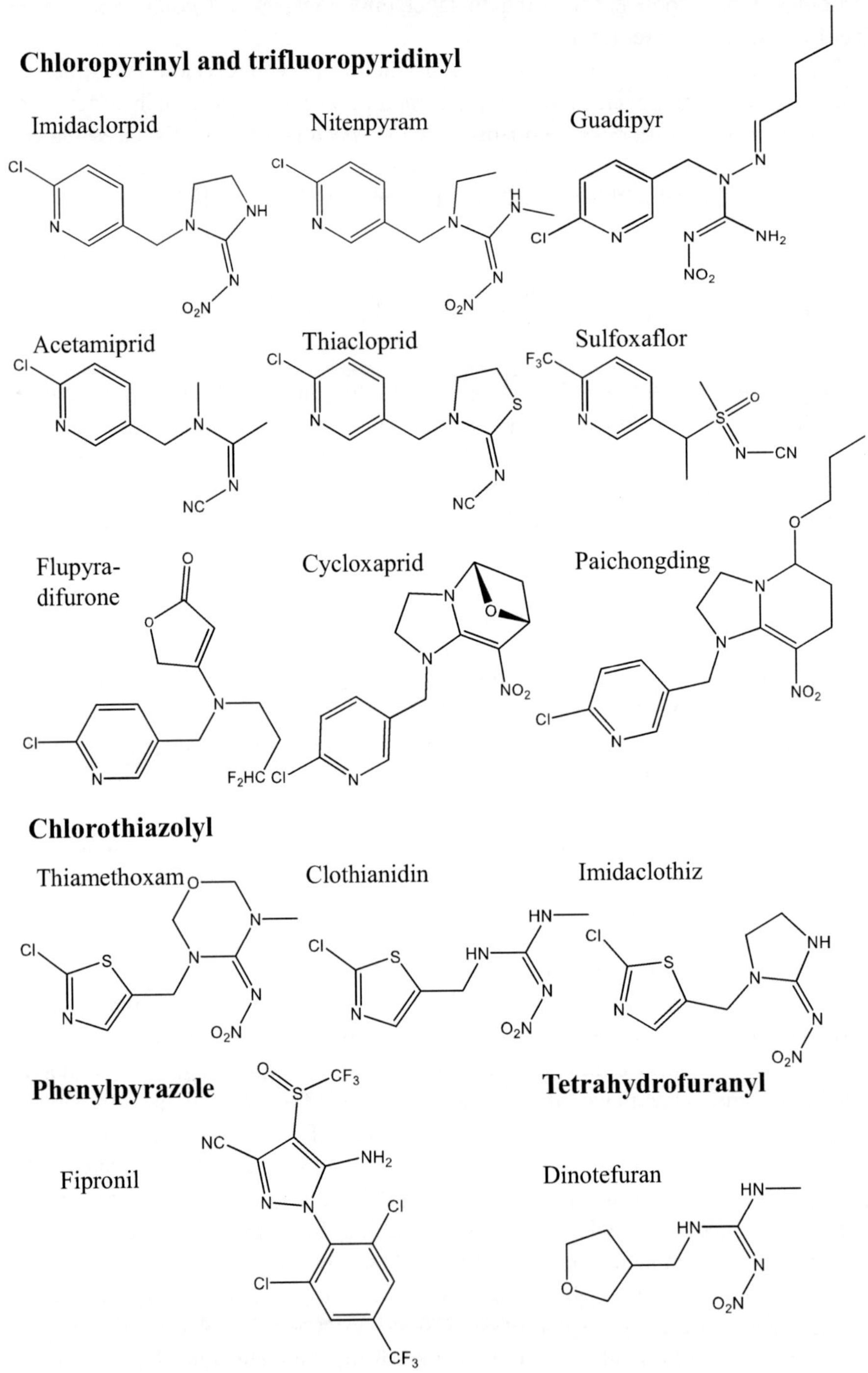

Fig. 1 Chemical structure and common name of neonicotinoids. Adapted from Giorio, [4]

therefore these were accounting for more than a quarter of the global neonics' sector, the sector at that time having a commercial value of around $3 billion. In 2012, imidacloprid, thiamethoxam and clothianidin together contributed to over 85% of all neonics' sales in the world, with the total sales of all neonicotinoids close to $2.7 billion [7].

Neonicotinoids were registered for use in over 120 countries by 2014, with 140 different crop applications; they became the most utilized group of insecticides in the world, representing a quarter of all pesticides sold in that year [12] and have been estimated to constitute 20% of all insecticides used [11, 13].

Thiamethoxam has gradually outperformed imidacloprid in sales and gained prominence in the marketplace. It is possible that this shift has stemmed from toxicity tests designed for various aquatic organisms sensitive to neonics. These experimental tests, conducted by independent research groups, have demonstrated consistently that imidacloprid is more toxic [14–17]. In 2019, the global market for crop protection chemicals dropped a little (0.8%) from U.S. $60.3 billion, however the insecticide market itself experienced a slight increase of 0.2% in the same period, to $15.1 billion, which represented approximately 25% of the crop protection market, the rest consisting of fungicides and herbicides. The small increase of 0.2% was due to improved soybean output in Latin America, and the need to treat infestations of a moth called Fall Armyworm (*Spodoptera frugiperda*), whose larvae can cause damage by consuming foliage – important in the Asia Pacific region [18, 19].

In some locations (e.g., European Union), a decline in the use of neonicotinoids has started as a consequence of regulation introduced. Imidacloprid and thiamethoxam are only allowed to be used in permanent greenhouses in the EU [20]. However, thiamethoxam was again allowed briefly for temporary emergency use to control a virus affecting sugar beet seeds in the UK in early 2021. Thiacloprid stopped being used in Europe in early 2020 and acetamiprid is, at the time of the preparation of this chapter (August 2021), approved for use outdoor and in greenhouses in the UK and in Europe [20].

In the European Union and the UK, the more restrictive use of neonicotinoids followed a previous moratorium (years 2013–2015), where there was limited use of such pesticides. As a result of this moratorium, for instance, in the UK, there was a drop of 90% in the area of land where imidacloprid and thiamethoxam were used and ~82% decrease in their total mass used (Fig. 2a, b) by the end of 2016. However, the amount of acetamiprid and thiacloprid usage during the same years remained relatively unchanged [21].

The use of neonicotinoids in Europe is expected to decrease further due to the limited uses of neonicotinoids permitted. Elsewhere, guidelines to mitigate the exposure of pollinators to neonicotinoids are also being investigated and adopted and plans for restoration of pollinators habitats were prioritized to increase the quality and abundance of habitat and forage [22].

However, the benefits of pesticides are also wanted by some nations: Switzerland voted to reject a ban on all synthetic pesticides on June 13th 2021 [23]. CropLife America insisted in the agricultural benefits of neonicotinoids, and according to

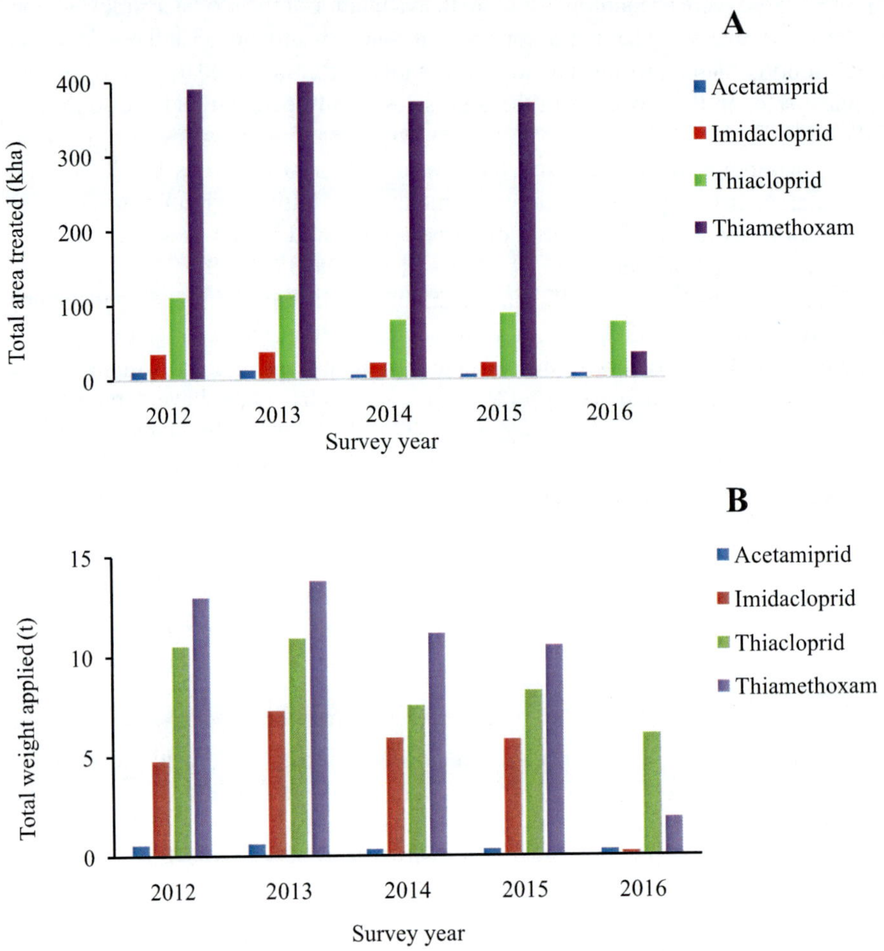

Fig. 2 Evolution of the use of a selection of neonicotinoids in the UK. The total area (ha) treated is shown in (**a**) and the mass, in tonnes, of the pesticides used is shown in (**b**). FERA [21]

them, these pesticides do not affect the health of bees when applied in the field following the label directions [24]. Nevertheless, companies producing neonicotinoids, in response to their effect on bees, and the EU enforcing their restricted use and also following the environmental principles of the UN Global Compact, are taking actions towards risk mitigation, such as measuring and minimizing dust abrasion from treated seeds to reduce unwanted impact [25].

3 Application Modes

Neonicotinoids are available in formulations that allow applications such as seed treatment (coating), foliar spraying or soil drenching. Efforts to reduce aerial spraying of pesticides (and drifts during and after spraying) have resulted in the shift towards strategies such as systematism, manifested in neonicotinoids as seed treatments, so that the insecticides can be transported to all parts of the crop plant. The idea of systemic insecticides is one in which pests die by chewing and sucking on any part of the crop plant and ingesting a significant dose of the neural disruptor. However, neonicotinoids will travel and be present in pollen and nectar that pollinators (bees included) will encounter when foraging. Implicit in this is that a substantial proportion of the applied product will be distributed within the crop plant itself, so that large percentages of wastage suffered by aerial spraying can be avoided.

Neonics applied through seed coatings function on the principle of a more targeted application of the pesticides [26]. This treatment may have made neonicotinoids more persistent as those sprayed require multiple applications. Seed coatings are used for a variety of crops including cotton, canola (oilseed rape), maize (corn), sunflowers and soybeans. However, nearly two decades of monitoring by researchers have shown that a maximum of only a fifth of neonics from seed coatings entered the plant body, the rest of the neonics stayed in soil or were degraded [27–30]. The loss of neonicotinoids from seed coating can lead to neonicotinoids (e.g. thiamethoxam) in soil leachate in concentrations acutely toxic to aquatic organisms [26, 31]. The loss of the seed coating itself as dust during planting appears to be low, in the order of 2% [32], therefore, it is not the main route of loss of neonics. The company Bayer®, which is a main producer of neonicotinoid-based pesticides, is currently working on innovations to minimize contaminated dust. Bayer is improving the adhesion of seed coatings to minimize the dust potentially released during planting by using lubricants to improve planting and attaching deflectors to pneumatic sowing machinery to reduce the release of dust particles from the seeder to the air [25].

Neonicotinoids can also be applied as granular formulations and as foliar sprays. In Canada, for example, imidacloprid is still being applied by all these methods to protect apples, potatoes, tomatoes, lettuce, sweet and sour cherries, highbush blueberries, Brussels sprouts and eggplants. Target pests include aphids, beetles, fruit flies, the Japanese beetle larvae and European chafer larvae.

One issue (amongst many) which renders the establishment of a long-term mass balance for neonics for a particular ecosystem of interest is that of the underreporting of the utilization of pesticides, as farmers genuinely do not know exactly the active ingredient of pesticides coated on the seeds they sowed. Hitaj et al. [33] reported that this happens if seeds are delivered to farmers after a "default" treatment which contains several active ingredients.

4 The Role of Neonicotinoids in Organisms and Colony Collapse Disorder (CDD)

In addition to bees, many other living organisms which play important roles in maintaining the health of ecosystems such as butterflies and earthworms together with other terrestrial and aquatic organism are adversely affected by neonics [34–36]. This effect on non-target organisms has triggered the re-evaluation of their permitted applications [6, 11].

The decline of bees recorded in many parts of the world follows a Colony Collapse Disorder (CCD) [37–40] by which there is loss of the adult bee population in hives without presence of dead bees in or in the proximity of the hives [41]. Bees are unable to return to their hives due to memory problems potentially after having been foraging in plants treated with neonicotinoids [39, 42, 43]. Other causes for the CCD phenomenon have also been proposed such as nutritional stress resulting from the loss of pollinator natural habitats [44]; dietary deficiency of pyrethrum, which are natural mite toxins that bees get from some flowering plants [45]; the presence of entombed pollen (bee bread covered by a sunken capping) which happens to be contaminated by the fungicide chlorothalonil [40] or contact with the Israeli acute paralysis virus, IAPV [46] and other pathogens [47], amongst other factors.

Colony decline has happened in the past: in 1869, between 1905–1919, and 1960–1975, and during mid-1999 [48]. The collapse of bee colonies is likely to be caused by a combination of factors: biological, chemical, environmental, and bee-keeping practices [49]. However, the European Food Safety Agency (EFSA) reports linked evidence from exposure to neonicotinoids to bee health and death. This led to a 2 years' moratorium restriction on neonicotinoids' use in Europe (2013–2015) [50] and, thereafter, the ban of three neonicotinoids use on all field crops and with limited use in greenhouses [51].

The insecticidal action of neonicotinoids results from their high affinity for the nicotinic acetylcholine receptors (*n*AChR) [52] acting as a nicotinergic neuronal pathway blocker. The receptors are mostly found at the junctions, called synapses, where neurons are positioned in proximity and signal to each other. When a molecular entity binds to the receptor, it stimulates the nerve and creates an electrical pulse. This pulse is information which tells the insect to crawl, fly, think or learn. Normally, an enzyme molecule comes along and deactivates or catabolizes the substance which is stimulating the nerve. Problems arise if the stimulant is a neonic. Neonic moieties are not catabolized or removed from the synaptic cleft easily. This high affinity to the receptor means that the nerve is continually stimulated. Normal neural functions are first impaired, but as neurons are over-stimulated at higher doses, insects can become hyperactive, resulting in epileptic seizures and incapacitation of nerve cells.

For instance, when imidacloprid was administered to bees at sub-lethal doses, they showed decreased capacity to communicate and their social behaviour declined [53]. Exposure to neonicotinoids leads to the accumulation of the neurotransmitter acetylcholine and this triggers paralysis and death [54]. The high affinity of

Table 1 Comparison of neonicotinoids' toxicity with other classes of insecticides. Adapted from Tomizawa and Casida [61]

Class	Nerve target	Potency (LD_{50}, mg/kg)		Selectivity factor[a]
		Insects	Rats	
Neonicotinoids	nAChR	2.0	912	456
Organophosphates	AChE	2.0	67	33
Methyl carbamates	AChE	2.8	45	16
Organochlorines	Na^+ or cl^+ channels	2.6	230	91
Pyrethroids	Na^+ channels	0.45	2000	4,500

[a] LD_{50} in rats/LD_{50} in insects

neonicotinoid with *n*AChR is attributed to the cyano and nitro groups in the pesticides (see Fig. 1) [42, 52, 54], where the interaction by the nitro groups appears to be more toxic, probably because of stronger interaction with the receptor via hydrogen bonding [55].

The dosages of neonics which bees encounter may or may not be high enough to kill them immediately, and it is possible that bee colonies collapse because they do not return to their hives. Perhaps, they cannot find their way home because they were flying under the strong influence of neonics, which causes depolarization of the postsynaptic membrane, thus blocking electrical signal transmission. The large amount of neonics-related bee research has revealed many interesting facts about the life of bees. For bees that survive to repeated exposure, it has been alleged that some develop preference for food contaminated by them [56]. Other studies showed that bees' gene expression, behavioural mechanisms and flight endurance can also be affected by neonics [57–60]. The challenge for melittologists and beekeepers is to find out how affected individuals disturb a bee colony, leading to migration *en masse*. Future work could address whether abnormal behaviour of affected bees and/or premature deaths alert the swarm to a dangerous situation so that the entire colony emigrates.

The generally high hydrophilicity and outstanding plant systemic activity, though shared by organophosphates and methylcarbamates, are advantages over the more lipophilic organochlorines, pyrethroids, organophosphates and methylcarbamates, which have a poor selectivity factor (given in Table 1). For neonics, the LD_{50} parameter, i.e., toxicity, that is measured as the dose that causes the death of half of the population of the organism under investigation, is roughly half of that of pyrethroids, but much higher than that of organophosphates, carbamates and organochlorines (see Table 1). The unique mode of action of neonicotinoids as an agonist at nAChR of insects differs from organochlorines and pyrethroids with Na^+ or Cl^- and Na^+ modulators, respectively (Table 1). This explains the preferential use of neonicotinoids over other insecticides available [61].

5 Physico-Chemical Properties of Neonicotinoids

The physico-chemical properties of neonicotinoids are related to their mobility and fate in the air/soil/water/plant compartments. Neonicotinoids have a low vapour pressure (3×10^{-7} mPa – 1.7×10^{-3} mPa at 25°C) and a low Henry's Law constant (2.9×10^{-16} (air) – 5.3×10^{-8} (water) atm·m^3/mol). Thus, the low values of these parameters indicate that neonicotinoids have very limited volatility. However they can still be found in aerosols after spraying land [31].

Neonicotinoids generally present a high solubility in water although large differences among them can be seen (see Table 2). This high solubility favours their systemic activity and facilitates their inclusion in the formulation of pesticides. The Log K_{ow} value, which is the ratio of the affinity of the pesticide for a hydrophobic environment with respect to water, indicates that whereas thiacloprid is quite hydrophobic, the rest have hydrophilic character. The values for Log K_{ow} collected in Table 2 correlate with their solubility in water and K_{oc} (or organic carbon/water partition coefficient). Large values of K_{oc} indicate strong sorption onto the organic matter of the soil [63].

The relatively high solubility of neocotinoids impacts their ecotoxicity. Their solubility values are greater than for substances such as lindane (7 mg lindane/L water) [64] or fipronil (4 mg fipronil/L), which is a broad-spectrum insecticide. Owing to their high solubilities in water, neonicotinoids can spread quickly in aqueous media and within plants.

Compared to other pesticides, the Log K_{ow} values for all neonicotinoids (−0.13 to 1.26) are relatively low. For example, Log K_{ow} of organophosphates, methyl carbamates, organochlorine and pyrethroid are typically 1 to 5.5; −1 to 3; 5.5 to 7.5 and 4 to 9, respectively [61]. The positive side of presenting low hydrophobicity and a high solubility in water is that their accumulation in biological sytems is relatively low.

6 Transport of Neonicotinoids in Air, Water and Soil

Now that the origins of neonics in soils have been established, we turn our attention to their fates in the natural environment. The way neonicotinoids are applied in agriculture, discussed in Sect. 2, and their high solubility in water have contributed

Table 2 Molecular properties that affect the distribution of a selection of neonicotinoids in the environment [31, 62]

Pesticide	Solubility in water (mg/L at 20°C)	p*K*a	Log K_{ow}	K_{OC}
Dinotefuran	39,830	12.6	0.549	26
Thiamethoxam	4,100	–	−0.13	33–117
Acetamiprid	2,950	0.7	0.8	200
Imidacloprid	610	–	0.57	156–960
Thiacloprid	180	–	1.26	261–870

to their high spread in the environment. Moreover, other factors such as soil type, rain regime, wind, degradation by sunlight or microorganisms can affect their pathways.

Neonicotinoids have been found in water [65], soil [66], plants [67], insects [68], birds [69], aquatic organisms [70], mammals [71, 72] and food [73–75]. An interesting case that illustrates the entrance of neonicotinoids to the food web was the finding of clothianidin and thiamethoxam in the carcasses of free-ranging wild turkeys in the Province of Ontario, Canada [76]. It is possible that the turkeys might have ingested agricultural seeds coated with these pesticides. The same researchers also discovered neonics-coated corn and soybean seeds in the gastro-intestinal tracts of some other bird species.

They envision the outcome of their research as baseline data for Southern Ontario wild turkeys and provide context for reference values in future analyses. Eng et al. [77] hypothesized that the delay in migration of some songbirds was related to the consumption of materials containing neonicotinoids. Adegun et al. [78] quantified thiacloprid, acetamiprid and thiamethoxam (used on cocoa farms) in six cultivable fish species in the Owena River Basin in Nigeria. Of the six fish species, it was found that the four insectivores had accumulated more neonicotinoids than the two plankton-eaters [78]. These examples evince the ubiquity of neonicotinoids in the environment.

6.1 Neonicotinoids in Air

Neonicotinoids can stay in air temporarily subsequent to being sprayed onto leaves. This process makes them particularly available to non-target organisms, such as honeybees, if these are within reach of the site being sprayed. Neonicotinoids could also be part of contaminated particles (soil or dust), which can impact foraging non-target insects [37, 79, 80] and can affect land beyond the crop being sprayed, finding their ways to run-offs and natural bodies of water.

6.2 Neonicotinoids in Aquatic Systems

The high usage and solubility of neonicotinoids are major factors that make them likely to be present in natural water. The deposition of contaminated dust or plants and their decay once in water will also affect the presence of the pesticides in water. Indeed, neonicotinoids have been detected in lakes [16, 65], rivers [78, 81] and estuaries [82]. The characteristics of the water (pH, suspended solids, organic matter content, temperature), exposure to sunlight and microbial presence can affect the stability of neonicotinoids in water. The extent of the exposure to factors contributing to their degradation will affect neocotinoids' stability in water. Photolytic degradation can play a major role in the stability of neonicotinoids; for example, imidacloprid, dinotefuran and thiamethoxam have half-lives of 2.3, 3.6 and 3.8 h, respectively, in water under natural light conditions [83]. Interestingly, the nitroguanidine group in thiamethoxam and imidacloprid can be photolysed, but the cyanoimine group of acetamiprid and thiacloprid was found to be stable under UV irradiation [84]. There is a scarcity of studies addressing the photodegradation mechanisms of neonicotinoids and also their degradation in different types of surface water.

Substances in aquatic media may sorb these pesticides or hinder their photolysis by blocking their exposition to sunlight (e.g. suspended solids and organic matter); or may attenuate or promote the degradation of neonicotinoids via scavenging or release of free radicals. All these physico-chemical phenomena will affect the level of neonicotinoids over time, and this will impact on their ecotoxicity and bioavailability. The presence of neonicotinoids in surface water has been widely reported and they are regarded as a threat to aquatic animals [81]. The European Commission has now listed some neonicotinoids in the Watch List of EU (Decision 2015/495) as substances to be monitored.

6.3 Neonicotinoids in Soil

Major routes that introduce neonicotinoids in soil are planting seeds coated with them or their direct deposition when neonicotinoids are sprayed on plants. Indirect pollution pathways can be through contaminated dust, water or leaves transported from other crops. For instance, in a monitoring study carried out in 291 fields in France during a moratorium on some neonicotinoids (2014–2018), nectar from rapeseed oil flowers was frequently found to contain imidacloprid with levels up to 45 ppb. Transport of imidacloprid from non-neighbouring crops to the study sites via dust drift or contaminated run-off were suggested as likely pollution pathways [85]. From soil, plant uptake of the neonicotinoids' bioavailable fraction occurs, which as with other contaminants will vary with soil type and plant species. For instance, cabbage uptakes acetamiprid and imidacloprid more than cotton [31].

Degradation of neonicotinoids in soil however is an important natural attenuation process. As a result, different persistence levels will be observed for different neonicotinoids. For instance, their stability can be very short (of less than a week) or very long (e.g. several years). Neonicotinoids can also affect the diversity and activity of bacteria that could take part in their biodegradation in soil, and this will change the persistence of the pesticide [86].

Table 3 presents the period required for the concentration of neonicotinoids in soil to reach half of their initial value (DT_{50}). Such variable stability is affected by factors such as the soil type and its conditions (moisture, presence of organic matter); the extent to which UV irradiation reaches the pesticides; pH, temperature [31], bacteria present [87] and concentration of neonicotinoid [88]. However, to the best of our knowledge, there are no studies focusing on the degradation of pesticides based on their dose. Temperature has an important effect on the degradation of

Table 3 Degradation rates (DT_{50}) of commonly used neonicotinoids in soil. Information compiled from IUPAC PPDB [62], Goulson [79]; Bonmatin et al. [31]

Insecticides	Range of DT_{50} (days)
Thiacloprid	3.4–>1,000
Thiamethoxam	7–335
Nitenpyram	8
Acetamiprid	31–450
Dinotefuran	75–82
Imidacloprid	28–1,250
Clothianidin	121–7,000

neonicotinoids, for instance, because it affects their biological degradation. The information available from imidacloprid indicates that its half-life decreased (547, 153 and 85 days) with increasing temperature (5, 15 and 85°C), respectively. Hence, it is relevant to consider the climate or geographical location when formulating regulations for pesticide usage.

Mechanisms driving the movement of neonicotinoids in soil are still largely unknown and research is needed in this direction. A number of factors related to soil and water, including soil pH [89], ionic strength [90], organic matter (amount and its functional groups) [91, 92] have been reported to affect the mobility of neonicotinoids in soil, however, scarce information is available for such highly soluble pesticides.

The authors of this chapter carried out research to find answers related to the migration of neonicotinoids in soil [93]. When incubating contrasting soils with the major neonicotinoids in batch mode, using a concentration of neonicotinoids (2.5 μg/mL), which is lower than what is used to spray plants, it became apparent that the soil with the lowest amount of organic carbon (SOC) presented the least uptake of every neonicotinoid assayed (Fig. 3). Thiamethoxam presented very distinctive sorption onto soil and its uptake was very limited in all kinds of soils. In contrast, the other neonicotinoids studied, with greater K_{ow} than thiamethoxam, presented significantly greater sorption onto soils.

The mobility of the same neonicotinoids (imidacloprid, acetamiprid, thiacloprid and thiamethoxam) was further studied with a flow-through study incorporating a soil column with the two most different soils which had 0.8% and 12.5% SOC. The soil (sieved at <2 mm) was packed in 15 cm length columns and the packing of the columns achieved a uniform bulk density of 1.1 g soil/mL. Neonicotinoids were spiked by adding 1 mL of 1 mg neonicotinoid/g solution on the top layer of the column. Water was pumped through the top of the column at 0.8 mL/min, and about a 10 cm water-head was left constant to avoid affecting the surface of the topsoil. Fraction leachates were collected and were used to establish breakthrough curves of contrasting neonicotinoids when moving in contrasting soils (show in Fig. 4). The soil with 0.8% SOC leached both neonicotinoids within the first bed volume, whereas the mobility and subsequent leaching of both neonicotinoids in the soil with 12.5% SOC was much more restrained, and it took over nine bed volumes for thiacloprid to elute. Crucially, the content of organic matter in the soil was found to be a very important factor for the retention of the study neonicotinoids: low organic

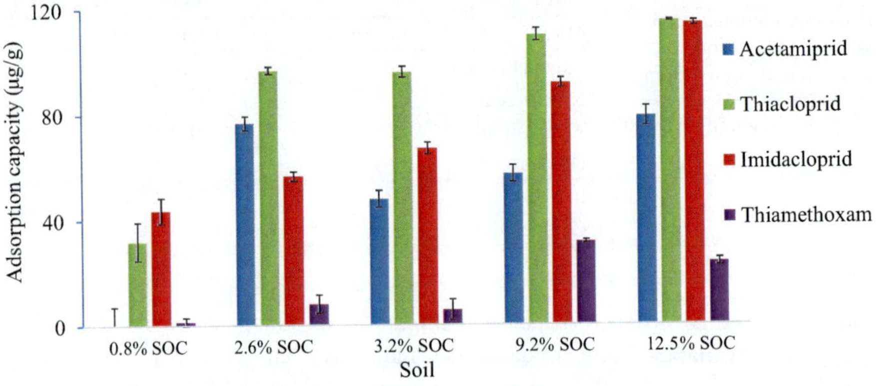

Fig. 3 Comparison between soils sorption capacity for neonicotinoids based on their different soil organic carbon (SOC). These results were obtained from a batch study with 1:5 soil/solution, where individual pesticides were in water contaminated at 2.5 μg neonicotinoid/mL water. The incubation of the soil samples with the pesticides in solution took place in an orbital shaker at 100 rpm for 48 h at 25°C. Results given as average ($n = 3$) ± SD. Adapted from Aseperi et al. [93]

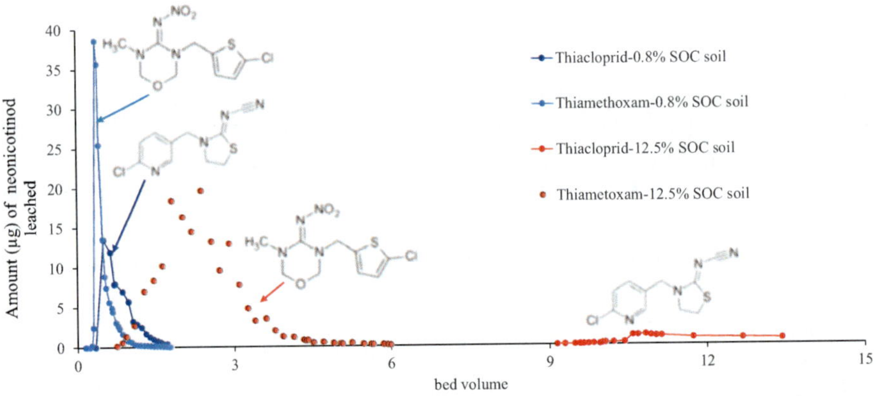

Fig. 4 Neonicotinoids with differentiated K_{ow} leaching from a soil column. The pesticides were applied at the top of a soil column (192 g soil, 4 cm i.d. × 15 cm length) under a constant flow of water at 0.8 mL/min. Adapted from Aseperi et al. [93]

matter in soil led to greater mobility of the pesticides and this has a greater risk of water contamination.

Thiacloprid, which was the most hydrophobic neonicotinoid studied (see Table 2) presented the lowest tendency to migrate through soil columns [93]. Hence, soil organisms could encounter the retained insecticides in the soil. In contrast, thiamethoxam, the most hydrophilic of all, was the least sorbed in both soils, hence it was the most mobile and had the greatest potential to contaminate groundwater. Furthermore, the content of organic matter in the soil appeared to affect the region in the soil column where neonicotinoids were retained. However, further details on the distribution of organic matter within the soil column would be needed to predict the location of the contamination; although, importantly, soil rich in organic matter had more prolonged leaching of the pesticides than soil with low organic matter content. Taking these findings into consideration, it would be useful to account for the soil properties of crops when selecting the neonicotinoid that should be used as the phytopharmaceutical product, in order to minimize the dispersion of the pesticides in the environment.

7 Conclusions and Future Trends

Neonicotinoids are pesticides that were initially introduced as advantageous alternatives to more toxic pesticides and their use expanded to all continents and many different types of crops. Neonicotinoids present a high binding affinity for the nAChR receptors of some organisms and this causes neurological problems. After neonicotinoids' introduction, there has been increasing evidence that non-target insects are being severely affected by them and they have been linked to CCD in honeybees. In response to this, regulatory bodies have started to restrict their use and producers are innovating to minimize ecotoxicity. Honeybees become exposed to neonicotinoids because these pesticides are highly soluble in water and spread through the different parts of the plants easily. Some evidence suggests that foragers actually prefer plants contaminated with neonicotinoids (thiamethoxam) once they have encountered such pesticides.

The stability of neonicotinoids in water and soil is variable. Higher temperature, sun irradiation and presence of certain microorganisms will contribute to their degradation. The soil organic content is a factor that determines the mobility of the neonicotinoids; soils richer in organic matter retain more these pesticides and leaching will be prolonged. The hydrophobic character of some of the neonicotinoids, such as thiacloprid, will favour the retention in soil, and those highly hydrophilic, such as thiamethoxam, readily leach from soil, can disperse more and potentially contaminate groundwater. It would be desirable to carry out a more targeted prescription of neonicotinoids considering conditions that would affect the stability of the molecule (sun irradiation, temperature) and take into consideration the content of organic matter in the soil to control their presence. The suitability of neonicotinoids for controlling pests is being re-evaluated because their risks may not outweigh their benefits and research is being carried out to minimize the diffusion of neonicotinoids to the environment.

References

1. Breeze TD, Bailey AP, Balcombe KG, Potts SG (2011) Pollination services in the UK: how important are honeybees? Agric Ecosyst Environ 142:137–143

2. Jeschke P, Nauen R (2008) Neonicotinoids—from zero to hero in insecticide chemistry. Pest Manag Sci 64(11):1084–1098
3. Jeschke P, Nauen R, Schindler M, Elbert A (2011) Overview of the status and global strategy for neonicotinoids. J Agric Food Chem 59(7):2897–2908
4. Giorio C, Safer A, Sánchez-Bayo F, Tapparo A, Lentola A, Girolami V, van Lexmond MB, Bonmatin JM (2021) An update of the worldwide integrated assessment (WIA) on systemic insecticides. Part 1: new molecules, metabolism, fate, and transport. Environ Sci Pollut Res Int 28(10):11716–11748
5. US EPA (2020) Imidacloprid. Proposed interim registration review decision. U.-S. Environmental Protection Agency, Docket Number EPA-HQ-OPP-2008-0844, pp 5 and 32. Imidacloprid proposed interim registration review decision case number 7605 (epa.gov)
6. Anderson JC, Dubetz C, Palace VP (2015) Neonicotinoids in the Canadian aquatic environment: a literature review on current use products with a focus on fate, exposure, and biological effects. Sci Total Environ 505:409–422
7. Bass C, Denholm I, Williamson MS, Nauen R (2015) The global status of insect resistance to neonicotinoid insecticides. Pestic Biochem Physiol 121:78–87
8. Hovda LR, Hooser SB (2002) Toxicology of newer pesticides for use in dogs and cats. Vet Clin Small Anim 32:455–467
9. Stokstad E (2013) Pesticides under fire for risks to pollinators. Science 340(6133):674–676. https://doi.org/10.1126/science.340.6133.674
10. US EPA (2014) Benefits of neonicotinoid seed treatments to soybean production. https://www.epa.gov/sites/default/files/2014-10/documents/benefits_of_neonicotinoid_seed_treatments_to_soybean_production_2.pdf
11. Simon-Delso N, Amaral-Rogers V, Belzunces LP, Bonmatin JM, Chagnon M, Downs C, Furlan L, Gibbons DW, Giorio C, Girolami V, Goulson D, Kreutzweiser DP, Krupke CH, Liess M, Long E, McField M, Mineau P, Mitchell EAD, Morrissey CA, Noome DA, Pisa L, Settele J, Stark JD, Tapparo A, Van Dyck H, Van Praagh J, Van der Sluijs JP, Whitehorn PR, Wiemers M (2014) Systemic insecticides (neonicotinoids and fipronil): trends, uses, mode of action and metabolites. Environ Sci Pollut Res 22:5–34
12. Lundin O, Rundlöf M, Smith HG, Fries I, Bommarco R (2015) Neonicotinoid insecticides and their impacts on bees: a systematic review of research approaches and identification of knowledge gaps. PLoS One 10(8):e0136928
13. Merck KGaA (2021) Neonicotinoids. https://www.sigmaaldrich.com/technical-documents/articles/analytical/neonicotinoids-bee-pesticides.html. Accessed Jul 2021
14. Cavallaro MC, Morrissey CA, Headley JV, Peru KM, Liber K (2017) Comparative chronic toxicity of imidacloprid, clothianidin, and thiamethoxam to Chironomus dilutus and estimation of toxic equivalency factors. Environ Toxicol Chem 36(2):372–382
15. Craddock HA, Huang D, Turner PC, Quirós-Alcalá L, Payne-Sturges DC (2019) Trends in neonicotinoid pesticide residues in food and water in the United States, 1999–2015. Environ Heal Glob Access Sci Source 18:1–16
16. Morrissey CA, Mineau P, Devries JH, Sanchez-Bayo F, Liess M, Cavallaro MC, Liber K (2015) Neonicotinoid contamination of global surface waters and associated risk to aquatic invertebrates: a review. Environ Int 74:291–303
17. Raby M, Zhao X, Hao C, Poirier DG, Sibley PK (2018) Chronic toxicity of 6 neonicotinoid insecticides to Chironomus dilutus and Neocloeon triangulifer. Environ Toxicol Chem 37(10):2727–2739
18. IHS Markit (2019) Global crop protection market down 1% in 2019. https://ihsmarkit.com/research-analysis/global-crop-protection-market-down-1-in-2019.html. Accessed Jul 2021
19. Phillips MWA (2020) Agrochemical industry development, trends in R&D and the impact of regulation. Pest Manag Sci 76:3348–3356
20. European Commission (2021) Neonicotinoids Food Safety (europa.eu). Neonicotinoids (europa.eu). Accessed Jul 2021

21. Food and Environment Research Agency (FERA) (2019) Pesticide usage survey reports in UK. https://secure.fera.defra.gov.uk/pusstats/. Accessed Jul 2021
22. The White House (2014) Presidential memorandum - creating a federal strategy to promote the health of honey bees and other pollinators. https://obamawhitehouse.archives.gov/the-press-office/2014/06/20/presidential-memorandum-creating-federal-strategy-promote-health-honey-b. Accessed Jul 2021
23. Illien N (2021) Swiss voters reject proposal to ban synthetic pesticides. The New York Times. https://www.nytimes.com/2021/06/13/world/swiss-pesticide-referendum-ban.html?smid=url-share. Accessed Jul 2021
24. Glenn (2014) CropLife America statement on TFSP report on systemic pesticides. CropLife America. https://www.croplifeamerica.org/news-releases/oggeb7e6f76txh68ehaxjixzh5rwo2?rq=neonicotinoid. Accessed Jul 2021
25. Bayer (2021) Neonicotinoid insecticides @Bayer. Systematic risk mitigation & portfolio evolution towards minimized risk for the environment, pp 1–11. https://www.bayer.com/sites/default/files/Report%20Neonicotinoid%20Insecticides%40Bayer.pdf. Accessed Jul 2021
26. Struger J, Grabuski J, Cagampan S, Sverko E, McGoldrick D, Marvin CH (2017) Factors influencing the occurrence and distribution of neonicotinoid insecticides in surface waters of southern Ontario, Canada. Chemosphere 169:516–523
27. Jones A, Harrington P, Turnbull G (2014) Neonicotinoid concentrations in arable soils after seed treatment applications in preceding years. Pest Manag Sci 70(12):1780–1784
28. Sur R, Stork A (2003) Uptake, translocation and metabolism of imidacloprid in plants. Bull Insectol 56:35–40
29. Goulson D (2013) An overview of the environmental risks posed by neonicotinoid insecticides. J Appl Ecol 50(4):977–987. https://doi.org/10.1111/1365-2664.12111
30. Bonmatin JM, Giorio C, Girolami V, Goulson D, Kreutzweiser DP, Krupke C, Liess M, Long E, Marzaro M, Mitchell EAD, Noome DA, Simon-Delso N, Tapparo A (2014) Environmental fate and exposure; neonicotinoids and fipronil. Environ Sci Pollut Res 22:35–67. https://doi.org/10.1007/s11356-014-3332-7
31. Bonmatin JM, Giorio C, Girolami V, Goulson D, Kreutzweiser DP, Krupke C, Liess M, Long E, Marzaro M, Mitchell EAD, Noome DA, Simon-Delso N, Tapparo A (2014) Environmental fate and exposure; neonicotinoids and fipronil. Environ Sci Pollut Res:35–67
32. Tapparo A, Marton D, Giorio C, Zanella A, Soldà L, Marzaro M, Vivan L, Girolami V (2012) Environ Sci Technol 46(5):2592–2599
33. Hitaj C, Smith DJ, Code A, Wechsler S, Esker PD, Douglas MR (2020) Sowing uncertainty: what we do and don't know about the planting of pesticide-treated seed. Bioscience 70(5): 390–403
34. Chevillot F, Convert Y, Desrosiers M, Cadoret N, Veilleux É, Cabana H, Bellenger JP (2017) Selective bioaccumulation of neonicotinoids and sub-lethal effects in the earthworm Eisenia andrei exposed to environmental concentrations in an artificial soil. Chemosphere 186:839–847. https://doi.org/10.1016/j.chemosphere.2017.08.046
35. Butcherine P, Benkendorff K, Kelaher B, Barkla BJ (2018) The risk of neonicotinoid exposure to shrimp aquaculture. Chemosphere 217:329–348
36. Hernando MD, Gámiz V, Gil-Lebrero S, Rodríguez I, García-Valcárcel AI, Cutillas V, Fernández-Alba AR, Flores JM (2018) Viability of honeybee colonies exposed to sunflowers grown from seeds treated with the neonicotinoids thiamethoxam and clothianidin. Chemosphere 202:609–617
37. Cresswell JE (2011) A meta-analysis of experiments testing the effects of a neonicotinoid insecticide (imidacloprid) on honey bees. Ecotoxicology 20:149–157
38. Evans JD, Schwarz RS (2011) Bees brought to their knees: microbes affecting honey bee health. Trends Microbiol 19:614–620
39. Paradis D, Bérail G, Bonmatin JM, Belzunces LP (2014) Sensitive analytical methods for 22 relevant insecticides of 3 chemical families in honey by GC-MS/MS and LC-MS/MS. Anal Bioanal Chem 406:621–633

40. vanEngelsdorp D, Evans JD, Donovall L, Mullin C, Frazier M, Frazier J, Tarpy DR, Hayes J, Pettis JS (2009) Entombed pollen: a new condition in honey bee colonies associated with increased risk of colony mortality. J Invertebr Pathol 101:147–149
41. Cox-Foster DL, Conlan S, Holmes EC, Palacios G, Evans JD, Moran NA, Quan PL, Briese T, Hornig M, Geiser DM, Martinson V, vanEngelsdorp D, Kalkstein AL, Drysdale A, Hui J, Zhai J, Cui L, Hutchison SK, Simons JF, Egholm M, Pettis JS, Lipkin WI (2007) A metagenomic survey of microbes in honey bee colony collapse disorder. Science 318:283–286
42. Farooqui T (2013) A potential link among biogenic amines-based pesticides, learning and memory, and colony collapse disorder: a unique hypothesis. Neurochem Int 62:122–136
43. Tapparo A, Giorio C, Marzaro M, Marton D, Soldà L, Girolami V (2011) Rapid analysis of neonicotinoid insecticides in guttation drops of corn seedlings obtained from coated seeds. J Environ Monit 13:1564–1568
44. Naug D (2009) Nutritional stress due to habitat loss may explain recent honeybee colony collapses. Biol Conserv 142:2369–2372
45. Sharpe RJ, Heyden LC (2009) Honey bee colony collapse disorder is possibly caused by a dietary pyrethrum deficiency. Biosci Hypotheses 2:439–440
46. Di Prisco G, Pennacchio F, Caprio E, Boncristiani HF, Evans JD, Chen Y (2011) Varroa destructor is an effective vector of Israeli acute paralysis virus in the honeybee, Apis mellifera. J Gen Virol 92:151–155
47. Genersch E (2010) Honey bee pathology: current threats to honey bees and beekeeping. Appl Microbiol Biotechnol 87:87–97
48. Underwood R, VanEngelsdorp D (2007) Colony collapse disorder: have we seen this before? Bee Cult 17110:13–18
49. EFSA (2008) Bee mortality and bee surveillance in Europe (EFSA-Q-2008-428). EFSA J:1–28. https://doi.org/10.2903/j.efsa.2008.154r
50. EU Directives (2013) Commission implementing regulation (EU) no. 485/2013. Conclusion on the peer review of the pesticide risk assessment for bees for the active substance thiamethoxam. Off J Eur Union 56:12–26. https://doi.org/10.2903/j.efsa.2013.3067
51. EU Directives (2018) Commision regulation (EU) 2018/781. Off J Eur Union 61
52. Tomizawa M, Casida JE, Mu P (2009) Molecular recognition of neonicotinoid insecticides: the determinants of life or death. Acc Chem Res 42:260–269
53. García MDG, Galera MM, Valverde RS, Galanti A, Girotti S (2007) Column switching liquid chromatography and post-column photochemically fluorescence detection to determine imidacloprid and 6-chloronicotinic acid in honeybees. J Chromatogr A 1147:17–23
54. Pandey G, Dorrian SJ, Russell RJ, Oakeshott JG (2009) Biotransformation of the neonicotinoid insecticides imidacloprid and thiamethoxam by pseudomonas sp. 1G. Biochem Biophys Res Commun 380:710–714
55. Le Questel J, Graton J, Cerón-carrasco JP, Jacquemin D, Planchat A, Thany SH (2011) Bioorganic & medicinal chemistry new insights on the molecular features and electrophysiological properties of dinotefuran, imidacloprid and acetamiprid neonicotinoid insecticides. Bioorg Med Chem 19:7623–7634. https://doi.org/10.1016/j.bmc.2011.10.019
56. Arce AN, Rodriguez Ramos A, Jiaju Y, Colgan TJ, Wurm Y, Gill RJ (2018) Foraging bumblebees acquire a preference for neonicotinoid-treated food with prolonged exposure. Proc R Soc B 285(1885)
57. Colgan TJ, Fletcher IK, Gill RJ, Ramos Rodriguez A, Stolle E, Chittka L, Wurm Y (2019) Caste-and-pesticide-specific effects of neonicotinoid pesticide exposure on gene expression in bumblebees. Mol Ecol 28:1964–1974
58. Crall JD, Switzer CM, Oppenheimer RL, For Versypt AN, Dey B, Brown A, Eyster M, Guérin C, Pierce NE, Combe SA, de Bivort BL (2018) Neonicotinoid exposure disrupts bumblebee nest behavior, social networks, and thermoregulation. Science 362(6415):683–686
59. Evans M (2019) Neonics hinder bees' grooming ability: study. The Western Producer, May 23, 2019. https://www.producer.com/2019/05/neonics-hinder-bees-grooming-ability-study/. Accessed Jul 2021

60. Kenna D, Cooley H, Pretelli I, Ramos Rodrigues A, Gill SD, Gill RG (2019) Pesticide exposure affects flight dynamics and reduces flight endurance in bumblebees. Ecol Evol 9:5637–5650
61. Tomizawa M, Casida JE (2005) Neonicotinoid insecticide toxicology: mechanisms of selective action. Annu Rev Pharmacol Toxicol 45:247–268
62. IUPAC Pesticide Properties Database, AERU-University of Hertfordshire. PPDB A to Z Index (herts.ac.uk). Accessed Jul 2021
63. Franco A, Trapp S (2008) Estimation of the soil-water partition coefficient normalized organic carbon for ionizable organic chemicals. Environ Toxicol Chem 27:1995–2004
64. Fliedner A (1997) Ecotoxicology of poorly water-solube substances. Chemosphere 35 (97):295–305
65. Hladik ML, Corsi SR, Kolpin DW, Baldwin AK, Blackwell BR, Cavallin JE (2018) Year-round presence of neonicotinoid insecticides in tributaries to the Great Lakes, USA. Environ Pollut 235:1022–1029
66. Dankyi E, Gordon C, Carboo D, Fomsgaard IS (2014) Quantification of neonicotinoid insecticide residues in soils from cocoa plantations using a QuEChERS extraction procedure and LC-MS/MS. Sci Total Environ 499:276–283
67. Botías C, David A, Hill EM, Goulson D (2016) Contamination of wild plants near neonicotinoid seed-treated crops, and implications for non-target insects. Sci Total Environ 566–567:269–278
68. Gbylik-Sikorska M, Sniegocki T, Posyniak A (2015) Determination of neonicotinoid insecticides and their metabolites in honey bee and honey by liquid chromatography tandem mass spectrometry. J. Chromatogr B Anal Technol Biomed Life Sci 990:132–140
69. Humann-Guilleminot S, Clément S, Desprat J, Binkowski ŁJ, Glauser G, Helfenstein F (2019) A large-scale survey of house sparrows feathers reveals ubiquitous presence of neonicotinoids in farmlands. Sci Total Environ 660:1091–1097
70. Zhiming et al (2013) Determination of neonicotinoid insecticides residues in eels using subcritical water extraction and ultra-performance liquid chromatography–tandem mass spectrometry. Anal Chim Acta 777:32–40
71. Kavvalakis MP, Tzatzarakis MN, Theodoropoulou EP, Barbounis EG, Tsakalof AK, Tsatsakis AM (2013) Development and application of LC–APCI–MS method for biomonitoring of animal and human exposure to imidacloprid. Chemosphere 93:2612–2620
72. Taira K, Fujioka K, Aoyama Y (2013) Qualitative profiling and quantification of neonicotinoid metabolites in human urine by liquid chromatography coupled with mass spectrometry. PLoS One 8:1–13
73. Gaweł M, Kiljanek T, Niewiadowska A, Semeniuk S, Goliszek M, Burek O, Posyniak A (2019) Determination of neonicotinoids and 199 other pesticide residues in honey by liquid and gas chromatography coupled with tandem mass spectrometry. Food Chem 282:36–47
74. Golge O, Kabak B (2015) Determination of 115 pesticide residues in oranges by high-performance liquid chromatography-triple-quadrupole mass spectrometry in combination with QuEChERS method. J Food Compos Anal 41:86–97
75. Montiel-León JM, Duy SV, Munoz G, Verner MA, Hendawi MY, Moya H, Amyot M, Sauvé S (2019) Occurrence of pesticides in fruits and vegetables from organic and conventional agriculture by QuEChERS extraction liquid chromatography tandem mass spectrometry. Food Control 104:74–82
76. MacDonald AM, Jardine CM, Thomas PJ, Nemeth NM (2018) Detection in wild turkeys (Meleagris gallopavo silvestris) in Ontario, Canada. Environ Sci Pollut Res 25:16254–16260
77. Eng ML, Stutchbury BJM, Morrissey CA (2019) A neonicotinoid insecticide reduces fueling and delays migration in songbirds. Science 365(6458):1177–1180
78. Adegun AO, Akinnifesi TA, Ololade IA, Busquets R, Hooda PS, Cheung PCW, Aseperi AK, Barker J (2020) Quantification of neonicotinoid pesticides in six cultivable fish species from the river Owena in Nigeria and a template for food safety assessment. Water 12:2422
79. Goulson D (2013) An overview of the environmental risks posed by neonicotinoid insecticides. J Appl Ecol 50:977–987

80. Sanchez-Bayo F, Goka K (2014) Pesticide residues and bees – a risk assessment. PLoS One 9 (4):e94482. https://doi.org/10.1371/journal.pone.0094482
81. Sánchez-Bayo F, Hyne RV (2014) Detection and analysis of neonicotinoids in river waters – development of a passive sampler for three commonly used insecticides. Chemosphere 99:143–151
82. Hano T, Ito K, Ohkubo N, Sakaji H, Watanabe A, Takashima K, Sato T, Sugaya T, Matsuki K, Onduka T, Ito M, Somiya R, Mochida K (2019) Occurrence of neonicotinoids and fipronil in estuaries and their potential risks to aquatic invertebrates. Environ Pollut 252:205–215
83. Kurwadkar S, Evans A, DeWinne D, White P, Mitchell F (2016) Modeling photodegradation kinetics of three systemic neonicotinoids—dinotefuran, imidacloprid, and thiamethoxam—in aqueous and soil environment. Environ Toxicol Chem 35:1718–1726
84. Banić ND, Šojić DV, Krstić JB, Abramović BF (2014) Photodegradation of neonicotinoid active ingredients and their commercial formulations in water by different advanced oxidation processes. Water Air Soil Pollut 225:1955
85. Wintermantel D, Odoux J-F, Decourtye A, Henry M, Allier F, Bretagnolle V (2020) Neonicotinoid-induced mortality risk for bees foraging on oilseed rape nectar persists despite EU moratorium. Sci Total Environ 707:135400
86. Zhang H, Zhang Z, Song J, Mei J, Fang H, Gui W (2021) Reduced bacterial network complexity in agricultural soils after application of the neonicotinoid insecticide thiamethoxam. Environ Pollut 274:116540
87. Hussain S, Hartley CJ, Shettigar M, Pandey G (2016) Bacterial biodegradation of neonicotinoid pesticides in soil and water systems. FEMS Microbiol Lett 363:1–13
88. Chopra I, Chauhan R, Kumari B, Dahiya KK (2011) Fate of fipronil in cotton and soil under tropical climatic conditions. Bull Environ Contam Toxicol 86:242–245
89. Cao J, Guo H, Zhu HM, Jiang L, Yang H (2008) Effects of SOM, surfactant and pH on the sorption-desorption and mobility of prometryne in soils. Chemosphere 70:2127–2134
90. Peng X, Wang J, Fan B, Luan Z (2009) Sorption of endrin to montmorillonite and kaolinite clays. J Hazard Mater 168:210–214
91. Kasozi GN, Nkedi-Kizza P, Li Y, Zimmerman AR (2012) Sorption of atrazine and ametryn by carbonatic and non-carbonatic soils of varied origin. Environ Pollut 169:12–19
92. Zhang P, Ren C, Sun H, Min L (2018) Sorption, desorption and degradation of neonicotinoids in four agricultural soils and their effects on soil microorganisms. Sci Total Environ 615:59–69
93. Aseperi AAK, Busquets R, Hooda PS, Cheung PCW, Barker J (2020) Behaviour of neonicotinoids in contrasting soils. J Environ Manag 276:111329

Current Progress of Microplastics in Sewage Sludge

Anqi Huang, Dan Zhi, and Yaoyu Zhou

Contents

Abstract Microplastics (MPs), an emerging global pollutant, are widely present in different ecosystems due to the overuse of plastic worldwide, causing potential risks to the health of humans and other animals. Wastewater contains high levels of MPs, and most of them (≈99%) remain in sewage sludge even after highly efficient wastewater treatment processes. Part of these MPs finally enters the soil ecosystem as sludge is spread on soil. Many studies have focused on MPs in marine ecosystems or other aquatic environments, while limited research has focused on MPs in sewage sludge. Therefore, we strive to comprehensively describe the current and latest state of MPs in sewage sludge. Several main sources of MPs in sludge are summarized. The traditional pretreatment, extraction, and identification methods as well as some latest new methods are illustrated. The influence of MPs in sludge on sewage treatment, sludge digestion, and soil (including the effects on fauna, plants, and microorganisms in soil) are summarized. Based on this information, we also offer several proposals involving MPs in sewage sludge for future research.

A. Huang, D. Zhi, and Y. Zhou (✉)
Hunan International Scientific and Technological Cooperation Base of Agricultural Typical Pollution Remediation and Wetland Protection, College of Resources and Environment, Hunan Agricultural University, Changsha, China
e-mail: zhouyy@hunau.edu.cn

Avelino Núñez-Delgado and Manuel Arias-Estévez (eds.),
Emerging Pollutants in Sewage Sludge and Soils,
Hdb Env Chem (2023) 114: 167–188, DOI 10.1007/698_2022_854,

Published online: 16 April 2022

Graphical Abstract

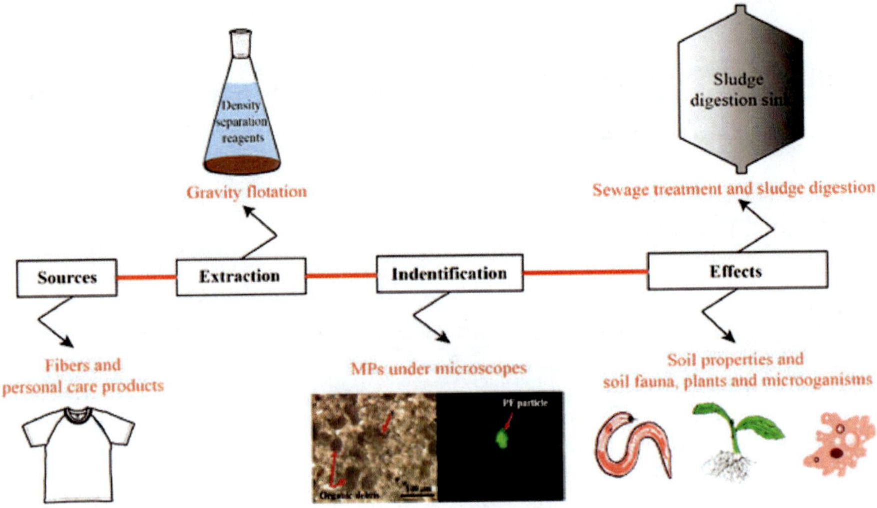

Keywords Ecosystems, Effluent, Microplastics, Sewage, Sludge

1 Introduction

Plastic products have been widely used worldwide because of their convenience and low price [1], and a large amount of plastic has been discarded in the environment and split into microplastics (MPs) due to the low recycling rate [2]. MPs are plastic particles with diameters less than 5 mm that enter the ecological environment after long-term physical, chemical, and biological processes [3]. They are present in marine ecosystems [4, 5], freshwater ecosystems (rivers and lakes) [6–9], soil and deposit sediment [10–12], and even in drinking water [13] and polar regions [14]. In addition, it is possible for MPs to adsorb persistent organic pollutants and metals and then release them as they migrate from their initial environment to other ecosystems [15]; MPs can even enter the human body through breathing and food consumption [16].

Over the last few decades, MPs have received much attention. However, most of these studies mainly focused on the investigation of the distribution and effects of MPs in marine or other aquatic ecosystems. However, humans are exposed to terrestrial MPs more frequently, as land is the main source of MP production and an important pollution sink [3, 4]. Quite a few MPs in land are released through human activities (e.g., use of personal care products, fabrication of textiles, rubbers of automotive tires, etc.) and finally remain in sewage sludge after wastewater treatment. Studies have reported that the amount of sewage sludge produced by wastewater treatment plants (WWTPs) in the European Union (EU) in 2010 was over 10 million tons [17]. In China, nearly 40 million tons of sewage sludge were produced by WWTPs in 2015, and the total amount of sludge production has grown by 13% every year [18]. WWTPs, a significant component of urban water systems, cause the accumulation of MPs in sludge [17] because they have a strong ability to remove MPs from municipal effluents with a high removal rate of 98.41% [19]. The species of MPs in sludge are very abundant, including polyamide (PA), polyethylene (PE), polyester (PES), polypropylene (PP), polyvinyl chloride (PVC), polyethylene terephthalate (PET), polystyrene (PS), and so on [20–23]. It has been pointed out that plastic consumption around the world follows the order PE > PP > PVC > PET > PS [24, 25].

To comprehensively understand the current status of MPs in sewage sludge, this study aims to summarize their main sources, analytic methods of extraction, and identification and effects on sewage treatment, sludge digestion, and soil. To further understand and investigate MP pollution, some key factors and future prospects are discussed as examples.

2 Main Sources of MPs in Sewage Sludge

To date, the sources of MPs in the environment are not fully understood [19]. In general, they can be broadly divided into two categories: primary MPs and secondary MPs [19, 26–28]. Primary MPs are original small-scale plastic products that are used for commercial applications; such MPs include plastic particles in personal care and cosmetic products [26], plastic fibers in laundry effluent [29], and even tiny particles in air blasting media [30]. Secondary MPs originate from the decomposition of large plastic/fiber products, and decomposition may occur because of wind, light, water, and other pressures [31, 32]. These secondary MPs are abundant, and their composition is complex; they may come from different sources, such as fishing nets, a variety of fibers, films, industrial raw materials, and household supplies [6].

There is no doubt that most of the MPs in sewage sludge come from raw domestic and industrial wastewater, because MPs remain in sludge even after highly efficient wastewater treatment [18, 19]. In other words, the MPs in municipal effluents are the sources of MPs in sludge. Therefore, several main specific sources of MPs in effluents are further illustrated:

1. Washing processes of textiles. The textile industry is a major source of MP pollution [3]. Studies have reported that over 42 million tons of synthetic fibers are produced every year, and they generate over 170% more MPs upon disposal than natural fibers [3, 33]. Microfibers are mainly produced both in the process of domestic and industrial synthetic cloth washing. A garment wash can release over 1,900 fibers one time, and all garments can release 100 fibers per liter of effluent [34]; a 6-kg wash load of acrylic fabric can produce more than 700,000 microfibers under domestic washing conditions [35]. The number of microfibers released is affected by many factors, including wash cycles [36], types of washing

agent [36], wash parameters [33], and textile constructions [37]. As an example, PES materials shed the greatest amounts of microfibers, 7,360 fibers/m^2/L on average, which is higher than the number of fibers shed by acrylic and nylon materials. Additionally, loose and worn textiles shed more microfibers than high twist yarns [37].

2. Personal care products (PPSPs). Microbeads were firstly identified in PPSPs by Zitko and Hanlon [38]. Microbeads are plastic that have been made of different sizes and shapes, and they have been widely used in PPCPs (e.g., facial scrubs, shampoos, toothpaste, skin cream, etc.) as abrasive agents [39]. According to the report of the European Cosmetic Industry Association, 4,130 tons of microbeads are used in soap every year in Norway, Switzerland, and European Union countries [40]. In mainland China, microbeads in nine facial scrubs were identified with a mean density of 20,860 particles/g and diameters ranging between 85 and 186 μm. Nearly 209.7 trillion microbeads (approximately 306.9 tons) are discharged into aquatic environments every year [39]. In Malaysia, the microbeads in toothpaste and facial cleaner were investigated; they were found to have diameters ranging from 3 to 178 μm, and 0.199 trillion PPCPs were emitted per year [41]. These microbeads are usually emitted into WWTPs with domestic sewage, and most of them finally remain in sludge.
3. Other sources. In addition to the two main sources mentioned above, others also need to be mentioned. For example, some plastic processing industries generate MPs through abrasive blast cleaning, molding, and other operations, and the generated MPs are finally discharged into sewage [42, 43]. In addition, rubber from automotive tires can produce a large amount of debris on the nanometer or micrometer scale [44]. Some of the debris accumulate in rainwater and may run off into sewage [45]. Additionally, it is possible for leachates containing MPs to be transported into sludge after various plastics are broken up in solid waste landfills [46].

3 Analytical Methods for Microplastics in Sewage Sludge

3.1 *Pretreatment and Extraction Methods*

It is important to first purify and separate MPs from the original sludge matrix. Because they are made of a viscous matrix of organics, microorganisms, and inorganics bound by biopolymers, sludge samples have a high affinity for many polymer surfaces [47]. Additionally, sludge contains high ratios of organic impurities (30%–50%), which may prevent the identification of MPs [48]. Usually, strong inorganic acids and bases are used to eliminate organics from the matrix; such substances include 1 and 5 mol/L nitric acid (HNO_3), 1 and 5 mol/L hydrochloric acid (HCl), 10 mol/L sodium hydroxide (NaOH), and 10% potassium hydroxide (KOH) at 60°C per 30–60 g of sludge sample [22, 49, 50]. However, these purification methods have certain effects on the quantitative analysis of MPs. For example, some MPs melt due to heating methods (110–120°C) combined with acid treatment or microwaves [20]. Additionally, the surface of PET is damaged during alkali treatment [50]. Studies have further found that Fenton's reagent has a lesser effect on MPs but a positive effect on organic decomposition in a short time during sludge pretreatment [51]. This method is also called wet peroxide oxidation (WPO), and Fenton's reagent consists of H_2O_2 and ferrous sulfate heptahydrate ($FeSO_4$). The process of H_2O_2 decomposition in WPO can produce many hydroxyl radicals that are able to oxidize most natural organics but not MPs [47, 52, 53]. $FeSO_4$ is a catalyst that can accelerate the digestion of organic matter under mild conditions including heating at 70°C and stirring for 30 min or longer [52]. Hurley et al. [49] used an ice bath at intervals to maintain temperatures less than 40°C, and these steps can reduce the decomposition of H_2O_2 and better protect the MPs.

At present, the extraction of MPs from sludge is primarily performed using density-based extraction techniques, which separate MPs from sludge by utilizing their density differences. The MP samples can be suspended on the surface of the solution after being placed in a gravity flotation solution such as a high-density salt solution, and the high-density sludge components stay at the bottom of the solution [17]. The solution usually consists of high-density salts such as NaCl, $ZnCl_2$, and NaI, and these salts are also able to improve the recovery of MPs [22, 54]. Li et al. [55] pointed out that the pretreatment step before flotation can improve the efficiency of MP extraction; such pretreatment involved putting sludge samples in 30% H_2O_2 and heating at 70°C. In general, the next step after the flotation mixture settles is filtration through a polycarbonate membrane, and these steps should be repeated several times [47].

Other methods for MP extraction in sludge have also been investigated in recent years. For example, the oil extraction protocol (OEP), a high-recovery, economical, unharmful, and Fourier transform infrared (FTIR)-compatible method, is very suitable for MP extraction [52]. The principle of the OEP is utilizing the lipophilicity and detectability of MPs to distinguish dense and heavy plastics from organics. In addition, Talvitie et al. [23] suggested that direct filtration was effective for sludge sample extraction. The samples were filtered through a stacked sieve device after dilution with water, which could prevent damage from harsh chemicals in the next steps. However, the drawback was that the size of the processable samples was limited.

In general, current methods of extraction of MPs in sludge are time-consuming and costly (especially density separation reagents and organic removal), and chemicals easily affect the identification of MPs in sludge samples.

3.2 Detection and Identification Methods

It is a challenge to detect and identify MPs accurately, as every microplastic species has its own morphology and properties. Traditional methods for MP identification include visual inspection, FTIR spectroscopy, and Raman spectroscopy. Visual methods are usually carried out through a microscope, and the steps for MP detection include size measurement, morphology characterization, and enumeration. This method has strict requirements when measuring objects such as a lack of cell or organic structure, a unified thickness and length of fibers, and a unified, clear color of particles [56]. Additionally, the sole use of visual observations may result in a deviation of quantity because it is easy to have subjective determination and confuse natural and synthetic materials. For example, a study reported that only 50% of fibers were treated as synthetic fibers in wastewater through visual methods [22]. Therefore, spectroscopic techniques and analysis are recommended in addition to visual observation to ensure the consistency of the results and avoid deviation in chemical composition [57].

In recent years, other methods for MP identification have been investigated; for example, FTIR and Raman spectrometers coupled with microscopes, portable spectrometers, fluorescent dyeing, and some thermal analysis methods are also worthy of attention. These methods are able to couple with each other or existing conventional instruments (e.g., scanning electron microscopy (SEM), energy dispersive spectrometry (EDS), gas chromatography-mass spectrometry (GC-MS), etc.) to obtain more accurate quantitative and qualitative analysis for MP identification. The combination mode of different technologies should be determined by the specific situation of samples and the variety of MPs. There are several methods for MP identification in sludge, and four of the most common methods are summarized in Table 1.

The combination of FTIR technology and microscopes is very common in the field of MP detection, especially for precise and qualitative identification. A spectral chart can be obtained when MP samples are exposed to infrared radiation, and the sample components are identified by referencing the spectral library [59]. Normal micro-FTIR/micro-Raman faces a major challenge in the identification of small microparticles and complex components, especially Raman spectroscopy, which is time-consuming [58, 60]. Some authors reported that atomic force microscopy (AFM) coupled with infrared (IR) or Raman spectroscopy may be able to detect nanoplastics (NPs) [64], and coupled AFM-IR has successfully produced spectra for polystyrene (PS) MPs [65].

In addition, Song et al. [66] found that using FTIR or Raman spectroscopy in advance and then utilizing stereomicroscopy to count the number of MPs can reduce the error of MP counts. Figure 1 shows an image of six typical MPs extracted from sewage sludge through stereomicroscopy [18].

The use of portable spectrometers for environmental MP samples has overcome the disadvantages of normal techniques that are time-consuming, and such spectrometers avoid the need for chemical pretreatment and extraction. As a reference, a visible near-infrared spectrometer (Vis-NIR) has been successfully used in the direct detection and quantification of soil MPs. The reflected light of MP surfaces and the reflection percentage at certain wavelengths were measured by Vis-NIR spectroscopy, which can predict the chemical composition of the sample [67]. In addition, some other methods can also shorten the testing time of MP detection. For example, MPs can be directly and rapidly dyed with a 5-mg/L Nile red (NR, a fluorescent dying) solution. It is easy to observe the dyed particles, which exhibit green

Table 1 Four of the most common analytical methods in MPs identification in sludge

Analytical methods	Conditions	Advantages	Deficiencies	References
Visual methods	Pretreatment No cells/organics, unified length, and clear and unified color of sample	Direct; simple Not chemically destructive	Too subjective Deviation in results Unable to test chemical composition	[56, 22]
Micro-FTIR Micro-Raman TGA	Pretreatment Need a stereo microscope Dry samples Pretreatment Heating Inert or air atmosphere	Nondestructive Small amount of sample Simple sample pretreatment Chemical composition analysis Particle size distribution information Nondestructive Small amount of sample Eco-friendly Simple sample pretreatment Can detect small particles (1–2 μm) Can provide particle size distribution information Quick, simple, easy and direct Does not require pretreatment	Expensive instruments Laborious work Time-consuming identification Unable to test small-size plastics Interferes with fluorescence Expensive instruments Laborious work Time-consuming identification Disable in testing NPs Limited range of application Unable to test chemical composition dependent on MP size and shape	[58–60] [61, 62] [25, 63]

fluorescence under a microscope after staining (Fig. 2). NR has been proven to successfully dye MPs in various materials (e.g., PVC, PES, PA, PS) [68]. It is worth noting that rose red dyeing is nontoxic and can dye materials other than plastics [69].

Several thermal analysis methods have been investigated in the characterization and quantitation of MPs in soil. Thermogravimetry (TGA) can monitor most of the samples while heating to carry out a quantitative analysis without pretreatment. The temperature can be programmed to maintain an isothermal system or ramp up the temperature under conditions of an inert or air atmosphere [25]. A universal modeling method related to TGA has been proposed for the quantitative and qualitative analysis of some MPs, including PE, PS, PVC, and PET [70]. This method is simple, cost-saving, and time-saving, and the most important advantage is that it does not require pretreatment, even for a complex soil sample. However, only one type of soil was tested, and the types of MPs analyzed were limited in this

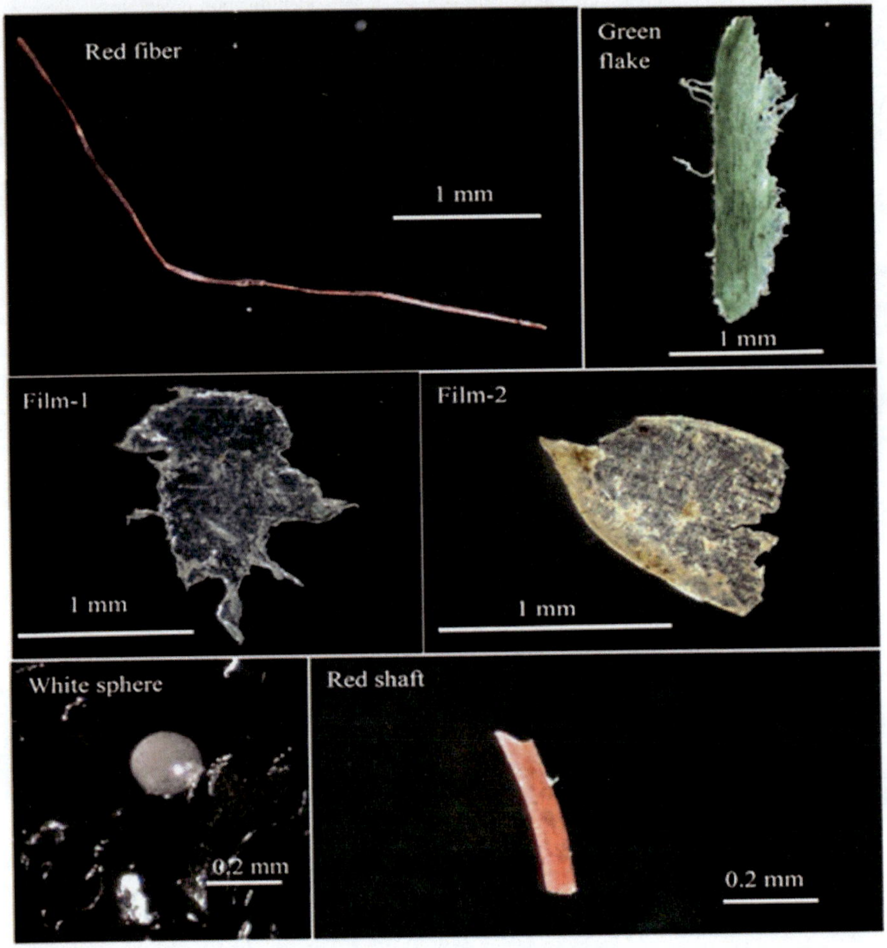

Fig. 1 Stereomicrograph of six typical MPs extracted from sludge [18]

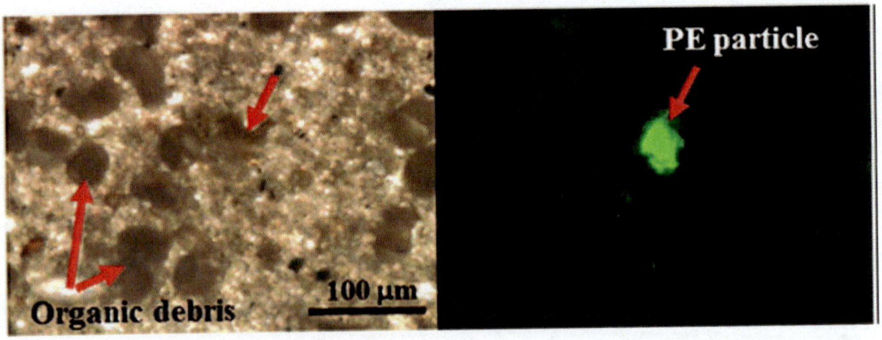

Fig. 2 Identification of PE MPs among organic debris under a microscope (left) and a fluorescence microscope (right) after Nile red staining [68]

research. The approach provides a worthy reference for future studies, but more considerations need to be taken. A method of alkali-assisted thermal hydrolysis has been used to decompose MPs of PE and PET in pentanol or butanol phases. The amounts of PE and PET were confirmed by the determination of the concentrations of bisphenol A and *p*-phthalic acid [63]. Solid samples, including sludge samples, can be directly measured by this method after physical separation.

4 Effects of Microplastics on Sewage Sludge

4.1 Effects of Microplastics on Sewage Treatment and Sludge Digestion

Due to the strong purification ability of WWTPs, most MPs enter sewage sludge during wastewater treatment processes incrementally. In Fig. 3, the specific percentages of MPs with various forms and different colors in sludge and sewage are clearly exhibited. This result indicates that fibrous MPs and those that are white in color are most prevalent in sludge, while foam MPs and those that are red in color are more prevalent in sewage [18, 19]. In previous studies, the retention rate of MPs remaining in sewage sludge from raw wastewater reached 99% [42, 71], and the concentration of MPs in sludge was between 1.5×10^3 MPs/kg and 2.4×10^4 MPs/kg

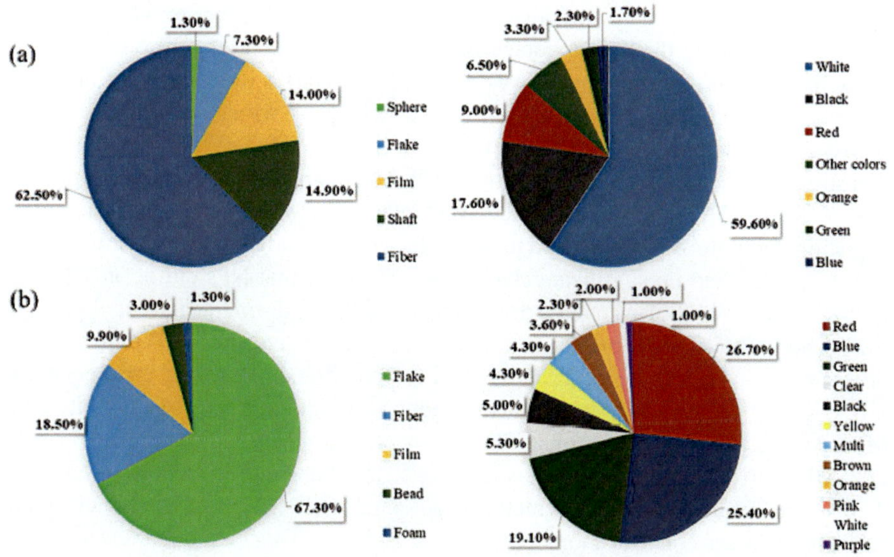

Fig. 3 (**a**) Average percentage of different MP types and colors in sludge samples; (**b**) Average percentage of different MP types and colors in sewage samples [18, 19, 47]

[17, 22]. During the entire process of WWTPs, MPs may affect both wastewater treatment and sewage sludge treatment.

MPs have a certain degree of influence on wastewater treatment. The efficiency of primary treatment can be affected by MPs. The abundant MPs in wastewater are able to block the fine grille, as the holes of the fine grille usually have a size of 3–10 mm. Additionally, MPs can adsorb chemical agents during oxidation and neutralization steps and even adsorb toxic matter due to their large surface area and hydrophobicity [72]. During the process of secondary treatment, MPs (especially at the nanoscale) decrease the ammonia-nitrogen conversion efficiency since they may restrain denitrification by changing the microbial-mediated process [47, 73, 74]. In addition, there is a positive correlation of biochemical oxygen demand (BOD), dissolved oxygen (DO), total nitrogen (TN), and total phosphorus (TP) with MP concentrations [75]. MPs can also offer surfaces for microorganisms to stick and grow in the process of biologically active filtration (BFA) [23]. The cost of sludge disposal increases if a high level and long-term processing of MPs are present. This is because MPs can lead to a decrease in the destruction of volatile solids, and thus, more sludge (approximately 9.1%) is produced [76, 77]. In the process of tertiary treatment, the efficiency of the coagulation process decreases, as MPs easily interact with flocculants [78]. MPs also affect air flotation, as they adsorb other pollutants and agglomerate during the process; thus, the primary bubbles cannot bring pollutants with a larger size and density to the top of the water, and the membranes used in microfiltration or reverse osmosis are easily worn and fouled by irregular MPs [79, 80].

MPs also have a significant influence on sludge digestion. In a usual sludge stabilization method, the particulate organics in sludge can be converted into a soluble state during the primary stage of anaerobic digestion [81]. The results of a study indicated that the chemical oxygen demand (COD) is directly proportional to the concentration of polyvinyl chloride (PVC) MPs [82], which is likely due to the lipids and nucleic acids from wastewater activated sludge (WAS) in anaerobic digestion; in other words, the dissolution of WAS is enhanced after a long exposure to PVC MPs [47]. Figure 4a shows the SEM image of sewage sludge after anaerobic digestion. Numerous NPs attached to the surface of the sludge, while the structure of sewage sludge had no significant change, which meant that NPs did not result in a higher release of cellular contents for biogas production during anaerobic digestion [83]. The MPs in WAS have a negative effect on the hydrolysis of proteins and polysaccharides and thus decrease the validity of acidified matter and even generate less gas [81]. In addition, the process of methane production is easily inhibited by MPs. Wei et al. [76] found that PE MPs with high concentrations of 100 and 200 MPs/g (total solids) obviously decreased methane production by 12.4% to 27.5% and resulted in a lower methane potential and hydrolysis coefficient than PE MPs at lower concentrations (i.e., 10, 30 and 60 MPs/g). Another study found that PVC MPs at high concentrations also decreased methane production and WAS hydrolysis, while low concentrations of PVC MPs benefited methane production in WAS [82]. Additionally, cationic nanoscale PS exhibited a greater influence on methane production than anionic PS [84], and nanoscale PS temporarily inhibited

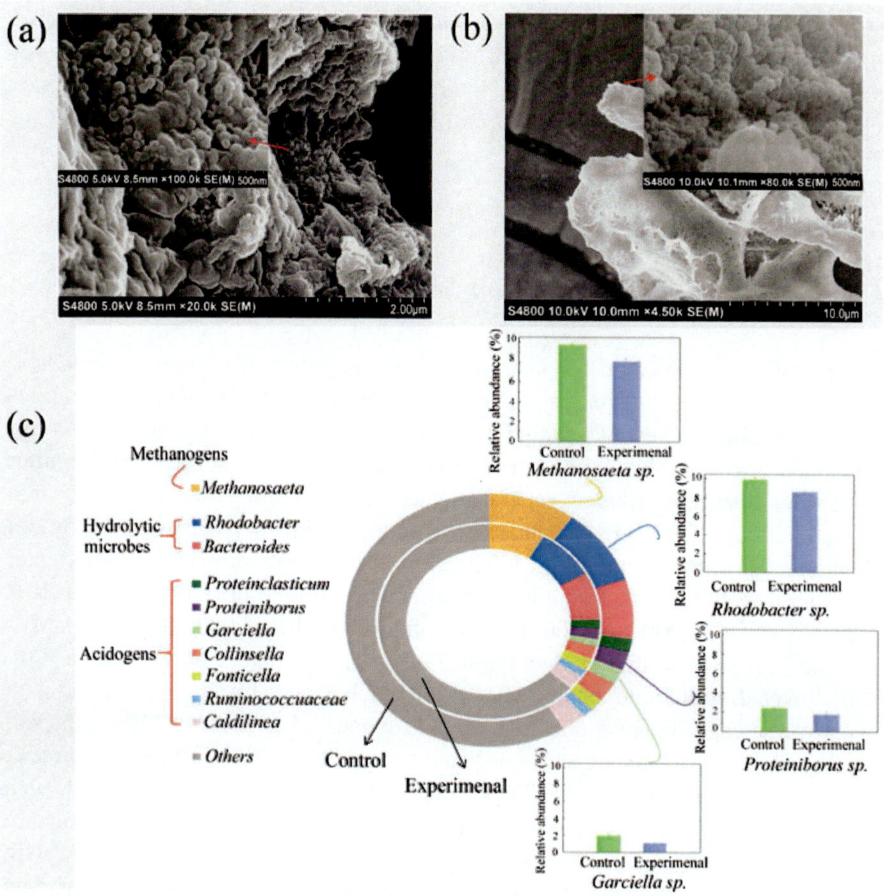

Fig. 4 SEM images of (**a**) sewage sludge and (**b**) *Acetobacteroides hydrogenigenes* in the presence of NPs [83]; (**c**) Distribution of microbial population at the genus level exposed to MPs in the anaerobic digestion system [82]

the gaseous products of aerobic granular sludge (AGS) after brief contact. While volatile fatty acids (VFAs) were an intermediate product, their concentration was not affected by nano-PS since methanogens may be more sensitive to chemical toxicities than acid-forming bacteria [85].

In addition, MPs affect microbial community diversity and abundance. Zettler et al. [86] mentioned that the structure and metabolism of microflora on the external surface of MPs is different from others. It is possible to affect other species by altering their microflora in the presence of MPs, and thus isotopes and elements were altered and growth and reproduction changed [86]. Figure 4b shows SEM images of NPs attached to the cytomembrane of *Acetobacteroides hydrogenigenes*. Additionally, many nanosized pores were present on the cytomembrane of *Acetobacteroides hydrogenigenes* due to the diapirism of NPs [83]. Figure 4c shows that the abundance of the five main bacteria in sludge digestion, *Chloroflexi*, *Bacteroidetes*,

Actinobacteria, *Proteobacteria*, and *Firmicutes,* declined due to the influence of MPs [83]. The relative abundance of other kinds of bacteria also decreased; for example, *Rhodobacter* sp. (perform hydrolysis), *Proteiniclasticum* sp. (produce acetic acid from proteins), *Methanothrix* (decarboxylate acetate to CH_4 and CO_2), and so on [84, 87].

4.2 *Effects of Microplastics on Soil*

MPs pollution in soil could originate from different potential sources, such as land utilization of sewage sludge, irrigation, organic fertilizer, film mulching, and atmospheric deposition. Sewage sludge is a crucial source of MPs in soil. As we mentioned above, a large number of MPs accumulate in sewage sludge after wastewater treatment. Sludge generally enters soil after disposal, especially into farmland through fertilizer, and sludge is permitted to enter soil accompanied by soil amendment in most countries [88]. Therefore, remarkable amounts of MPs enter agricultural soils through the utilization of sewage sludge on farmland [89]. It is estimated that 630,000 to 4,300,000 tons and 440,000 to 3,000,000 tons of MPs translate into soil through sludge application in Europe and North America [90]. Corradini et al. [91] found that the MPs in soil in Chile showed an upward trend with increasing sewage sludge application, and the average abundance of MPs in the soil (200 tons/hm^2 with a dry weight) reached 3.5 p/g after applying sludge five times.

The presence of MPs could influence the physiochemical properties of soil, such as hydraulic characteristics, soil aggregates, soil bulk density, and soil structure [92, 93]. They are incorporated into soil by biota or plowing and are mixed with other pieces of organic matter, so they are easily insulated inside soil aggregate [94]. Different types of MPs have different influences on soil; for example, PES can significantly decrease the content of water-stable aggregates in soil, while PE has the opposite effect [93]. PES is able to enhance soil moisture retention to keep the degree of water saturation always at a high level, and large aggregates are proportional to the amount of PES [92]. MPs may also affect the soil bulk density because they are less dense than soil minerals [93]. Polyethylene films can significantly accelerate soil water evaporation through building channels for water shifts and even destroy the integrity of the soil structure, resulting in desiccation cracking. This phenomenon is increasingly obvious with increasing MP abundance and decreasing particle size [95]. The large water-stable macroaggregates (>2 mm) and size of macropores (> 30 μm) in clayey soil obviously increased after PES microfibers were added [96]. In addition, MPs may also affect pedological mass circulation in soil. For instance, low-density PE and biodegradable plastics have a great influence on the pH, electrical conductivity (EC), and C/N ratio, and biodegradable plastics have significantly affected the release of volatile organic compounds (VOC) near the wheat rhizosphere [97]. As another example, the amount of dissolved organic carbon (DOC), nitrogen and phosphorus in dissolved organic matter (DOM) is proportional to the amount of added PP MPs to soil after 30 days [98]. This suggests that MPs have hidden contributions to carbon storage in soil [99].

In addition, MPs can affect fauna, plant traits, and microorganisms in soil. Rillig and Bonkowski [100] reported that MPs influence the growth and reproduction of invertebrates in soil. Earthworms and collembolans are the most widely investigated species in soil, and their traits, such as growth inhibition, weight, gut damage, and mortality, have been studied after exposure to MPs [10, 101, 102]. For example, Rodriguez-Seijo et al. [101] found that a concentration of PE less than 0.1% (w/w in dry soil) resulted in gut damage but had no effect on the weight or reproduction of earthworms, while growth was inhibited and mortality increased when the concentration was higher than 1% [103]. Another study found that PE at a concentration of 0.1% inhibited the reproduction of collembolans; a concentration of 0.5% obviously changed the microflora in the gut of collembolans; and a concentration of 1% decreased the reproductive rate by 70% compared with the control group [104]. MPs not only affect fauna in soil through the ingestion of small particles but also relate to changes in the surrounding environment or physical harm outside organisms [105]. Additionally, further studies should pay more attention to the impact of MPs on more kinds of soil fauna as well as the mechanism of the process.

Some studies have also focused on the terrestrial plants affected by MPs. The presence of MPs changed some soil properties that we mentioned above (e.g., soil aggregates, soil bulk density, and soil structure), which may further affect the root traits, growth status, and nutrient uptake process of plants [106]. Wheat, a popular research object, was first investigated by Qi et al. [107]. They added 1% (w/w) microplastic film residues to dry soil, and the results showed that the growth (vegetative and reproductive stages) of wheat (*Triticum aestivum*) was inhibited. The reason was attributed to the changes in soil properties after plastics were added. Biodegradable plastics also affected the growth of wheat more substantially than low-density PE [107]. MPs of PS (0 to 100 mg/kg) had a negative effect on the photosynthesis of wheat leaves, and protein synthesis was inhibited [108]. Studies proved that MPs affected other terrestrial plants; for example, the photosynthesis, growth, and antioxidant defense system of lettuce were affected [109]; the pores of cress were blocked, and thus, water absorption was restrained, which prolonged the germination and growth of roots [110]; and the elemental composition and root traits of spring onion were altered [111]. Exploring the transmission mechanisms of MPs in the soil–plant interaction system and the risks of MPs entering the human body through the food chain requires further effort.

Moreover, MPs can impact microflora by changing the enzymatic activities. PE films significantly inhibited the activity of soil urease for long-term use and further changed the abundance of genes related to the carbon and nitrogen cycles in soil [112]. Nanoscale PS inhibited the activity of dehydrogenase and other enzymes participating in the carbon and nitrogen cycles [113]. MPs of PP at concentrations of 7% and 28% (w/w) promoted the activity of fluorescein diacetate hydrolase in soil [98]. MPs also affect bacterial transport and deposition in soil [114] and provide adsorption sites to microorganisms to form a biological membrane layer on the surface of the MPs, which may change the functional properties of soil [115]. Until now, most research has focused on the changes and activities of soil enzymes exposed to MPs, and other aspects should be considered, such as how to protect the biological membrane on the surface of MPs when observing microorganisms or MPs after density flotation [106].

It is worth noting that MPs can adsorb a large number of other pollutants, such as antibiotics or other organic pollutants and heavy metals, in sludge or soil [116–118]. These chemical pollutants carried on the surface of MPs easily move with MP migration in soil due to wet–dry cycles [119, 120], soil management practices [121] and bioturbation [122] and the life activities of earthworms [123, 124], such transfer of pollutants may pose ecological and health risks [125]. Ramos et al. [126] found that sorption by MPs can also slow the degradation rate of organic pollutants in soil. Few studies have investigated the impact of MPs on the transport and degradation of these toxic chemical pollutants [106]. Thus, the combined effect of MPs and those pollutants in soil requires more attention, and systematic and scientific regulars need to be determined.

5 Conclusion

A large number of MPs gradually enter sewage sludge during highly efficient wastewater treatment processes and are finally released into soil. The sources of MPs in sewage sludge are abundant, and they include textiles, personal care products, plastic processing, rubber from automotive tires, and others. Pretreatment or extraction must be performed before identification to separate MPs from the original sludge, though this step is time-consuming, expensive, and easily affects the identification of MPs. Visual observation, a combination of FTIR or Raman spectroscopy and microscopy are very common methods of identifying MPs in sludge; some new methods, such as TGA, Vis-NIR, and fluorescent dying, have proven effective in quantitative and qualitative analyses. MPs can affect sewage treatment and sludge digestion and even threaten the soil environment. Until now, limited research has focused on MPs in sludge; thus, some future research directions are suggested:

1. Establish a unified and standard approach for quick extraction, portable identification, and highly effective analysis of MPs in sewage sludge.
2. Adjust wastewater and sludge treatment process to reduce the influence of MPs, and design a specific and suitable unit for MP treatment to reduce their amount before they enter sludge.
3. Investigate the adsorption and release mechanisms and influencing factors between MPs and other pollutants in sludge and soil, as well as the risks of MPs to the human body due to accumulation through the food chain.
4. Centralize high-density MP sludge treatment and limit their discharge to agricultural soil to reduce damage to fauna, plants, microorganisms, and even human health.

References

1. Sharma MD, Elanjickal AI, Mankar JS, Krupadam RJ (2020) Assessment of cancer risk of microplastics enriched with polycyclic aromatic hydrocarbons. J Hazard Mater 398:122994
2. Chae Y, An Y-J (2017) Effects of micro- and nanoplastics on aquatic ecosystems: current research trends and perspectives. Mar Pollut Bull 124:624–632
3. Xu C, Zhang B, Gu C, Shen C, Yin S, Aamir M, Li F (2020) Are we underestimating the sources of microplastic pollution in terrestrial environment? J Hazard Mater 400:123228
4. Andrady AL (2011) Microplastics in the marine environment. Mar Pollut Bull 62:1596–1605
5. Cole M, Lindeque P, Halsband C, Galloway TS (2011) Microplastics as contaminants in the marine environment: a review. Mar Pollut Bull 62:2588–2597
6. Free CM, Jensen OP, Mason SA, Eriksen M, Williamson NJ, Boldgiv B (2014) High-levels of microplastic pollution in a large, remote, mountain lake. Mar Pollut Bull 85:156–163
7. Mani T, Hauk A, Walter U, Burkhardt-Holm P (2015) Microplastics profile along the Rhine River. Sci Rep 5:17988
8. McCormick A, Hoellein TJ, Mason SA, Schluep J, Kelly JJ (2014) Microplastic is an abundant and distinct microbial habitat in an Urban River. Environ Sci Technol 48:11863–11871
9. Su L, Xue Y, Li L, Yang D, Kolandhasamy P, Li D, Shi H (2016) Microplastics in Taihu Lake, China. Environ Pollut 216:711–719
10. Huerta Lwanga E, Gertsen H, Gooren H, Peters P, Salánki T, van der Ploeg M, Besseling E, Koelmans AA, Geissen V (2016) Microplastics in the terrestrial ecosystem: implications for Lumbricus terrestris (Oligochaeta, Lumbricidae). Environ Sci Technol 50:2685–2691
11. Nizzetto L, Langaas S, Futter M (2016) Pollution: do microplastics spill on to farm soils? Nature 537:488–488
12. Peng G, Zhu B, Yang D, Su L, Shi H, Li D (2017) Microplastics in sediments of the Changjiang estuary, China. Environ Pollut 225:283–290
13. Pivokonsky M, Cermakova L, Novotna K, Peer P, Cajthaml T, Janda V (2018) Occurrence of microplastics in raw and treated drinking water. Sci Total Environ 643:1644–1651
14. Kanhai LDK, Gardfeldt K, Krumpen T, Thompson RC, O'Connor I (2020) Microplastics in sea ice and seawater beneath ice floes from the Arctic Ocean. Sci Rep 10:5004
15. Engler RE (2012) The complex interaction between marine debris and toxic chemicals in the ocean. Environ Sci Technol 46:12302–12315
16. Prata JC, da Costa JP, Lopes I, Duarte AC, Rocha-Santos T (2020) Environmental exposure to microplastics: an overview on possible human health effects. Sci Total Environ 702:134455
17. Mahon AM, O'Connell B, Healy MG, O'Connor I, Officer R, Nash R, Morrison L (2017) Microplastics in sewage sludge: effects of treatment. Environ Sci Technol 51:810–818
18. Li X, Chen L, Mei Q, Dong B, Dai X, Ding G, Zeng EY (2018) Microplastics in sewage sludge from the wastewater treatment plants in China. Water Res 142:75–85
19. Murphy F, Ewins C, Carbonnier F, Quinn B (2016) Wastewater treatment works (WwTW) as a source of microplastics in the aquatic environment. Environ Sci Technol 50:5800–5808
20. Carr SA, Liu J, Tesoro AG (2016) Transport and fate of microplastic particles in wastewater treatment plants. Water Res 91:174–182
21. Lares M, Ncibi MC, Sillanpää M, Sillanpää M (2018) Occurrence, identification and removal of microplastic particles and fibers in conventional activated sludge process and advanced MBR technology. Water Res 133:236–246
22. Mintenig SM, Int-Veen I, Löder MGJ, Primpke S, Gerdts G (2017) Identification of microplastic in effluents of waste water treatment plants using focal plane array-based micro-Fourier-transform infrared imaging. Water Res 108:365–372
23. Talvitie J, Mikola A, Setälä O, Heinonen M, Koistinen A (2017) How well is microlitter purified from wastewater? – a detailed study on the stepwise removal of microlitter in a tertiary level wastewater treatment plant. Water Res 109:164–172

24. Bond T, Ferrandiz-Mas V, Felipe-Sotelo M, van Sebille E (2018) The occurrence and degradation of aquatic plastic litter based on polymer physicochemical properties: a review. Crit Rev Environ Sci Technol 48:685–722
25. Peñalver R, Arroyo-Manzanares N, López-García I, Hernández-Córdoba M (2020) An overview of microplastics characterization by thermal analysis. Chemosphere 242:125170
26. Fendall LS, Sewell MA (2009) Contributing to marine pollution by washing your face: microplastics in facial cleansers. Mar Pollut Bull 58:1225–1228
27. Magni S, Binelli A, Pittura L, Avio CG, Della Torre C, Parenti CC, Gorbi S, Regoli F (2019) The fate of microplastics in an Italian wastewater treatment plant. Sci Total Environ 652:602–610
28. Salvador Cesa F, Turra A, Baruque-Ramos J (2017) Synthetic fibers as microplastics in the marine environment: a review from textile perspective with a focus on domestic washings. Sci Total Environ 598:1116–1129
29. Belzagui F, Crespi M, Álvarez A, Gutiérrez-Bouzán C, Vilaseca M (2019) Microplastics' emissions: microfibers' detachment from textile garments. Environ Pollut 248:1028–1035
30. Gregory MR (2009) Environmental implications of plastic debris in marine settings—entanglement, ingestion, smothering, hangers-on, hitch-hiking and alien invasions. Philosophical Transact R Soc B Biol Sci 364:2013–2025
31. Kontrick AV (2018) Microplastics and human health: our great future to think about now. J Med Toxicol 14:117–119
32. Rezaei M, Riksen MJPM, Sirjani E, Sameni A, Geissen V (2019) Wind erosion as a driver for transport of light density microplastics. Sci Total Environ 669:273–281
33. Kelly MR, Lant NJ, Kurr M, Burgess JG (2019) Importance of water-volume on the release of microplastic fibers from laundry. Environ Sci Technol 53:11735–11744
34. Browne MA, Crump P, Niven SJ, Teuten E, Tonkin A, Galloway T, Thompson R (2011) Accumulation of microplastic on shorelines worldwide: sources and sinks. Environ Sci Technol 45:9175–9179
35. Napper IE, Thompson RC (2016) Release of synthetic microplastic plastic fibres from domestic washing machines: effects of fabric type and washing conditions. Mar Pollut Bull 112:39–45
36. De Falco F, Gullo MP, Gentile G, Di Pace E, Cocca M, Gelabert L, Brouta-Agnésa M, Rovira A, Escudero R, Villalba R, Mossotti R, Montarsolo A, Gavignano S, Tonin C, Avella M (2018) Evaluation of microplastic release caused by textile washing processes of synthetic fabrics. Environ Pollut 236:916–925
37. Carney Almroth BM, Åström L, Roslund S, Petersson H, Johansson M, Persson N-K (2018) Quantifying shedding of synthetic fibers from textiles; a source of microplastics released into the environment. Environ Sci Pollut Res 25:1191–1199
38. Zitko V, Hanlon M (1991) Another source of pollution by plastics: skin cleaners with plastic scrubbers. Mar Pollut Bull 22:41–42
39. Cheung PK, Fok L (2017) Characterisation of plastic microbeads in facial scrubs and their estimated emissions in mainland China. Water Res 122:53–61
40. Gouin T, Avalos J, Brunning I, Brzuska K, Wolf T (2015) Use of micro-plastic beads in cosmetic products in Europe and their estimated emissions to the North Sea environment. SOFW-J 141:1–33
41. Praveena SM, Shaifuddin SNM, Akizuki S (2018) Exploration of microplastics from personal care and cosmetic products and its estimated emissions to marine environment: an evidence from Malaysia. Mar Pollut Bull 136:135–140
42. Gies EA, LeNoble JL, Noël M, Etemadifar A, Bishay F, Hall ER, Ross PS (2018) Retention of microplastics in a major secondary wastewater treatment plant in Vancouver, Canada. Mar Pollut Bull 133:553–561
43. Long Z, Pan Z, Wang W, Ren J, Yu X, Lin L, Lin H, Chen H, Jin X (2019) Microplastic abundance, characteristics, and removal in wastewater treatment plants in a coastal city of China. Water Res 155:255–265

44. Galafassi S, Nizzetto L, Volta P (2019) Plastic sources: a survey across scientific and grey literature for their inventory and relative contribution to microplastics pollution in natural environments, with an emphasis on surface water. Sci Total Environ 693:133499
45. Kreider ML, Panko JM, McAtee BL, Sweet LI, Finley BL (2010) Physical and chemical characterization of tire-related particles: comparison of particles generated using different methodologies. Sci Total Environ 408:652–659
46. He P, Chen L, Shao L, Zhang H, Lü F (2019) Municipal solid waste (MSW) landfill: a source of microplastics? -evidence of microplastics in landfill leachate. Water Res 159:38–45
47. Zhang Z, Chen Y (2020) Effects of microplastics on wastewater and sewage sludge treatment and their removal: a review. Chem Eng J 382:122955
48. Ke Y, Ning X-A, Liang J, Zou H, Sun J, Cai H, Lin M, Li R, Zhang Y (2018) Sludge treatment by integrated ultrasound-Fenton process: characterization of sludge organic matter and its impact on PAHs removal. J Hazard Mater 343:191–199
49. Hurley RR, Lusher AL, Olsen M, Nizzetto L (2018) Validation of a method for extracting microplastics from complex, organic-rich, environmental matrices. Environ Sci Technol 52: 7409–7417
50. Li X, Chen L, Ji Y, Li M, Dong B, Qian G, Zhou J, Dai X (2020) Effects of chemical pretreatments on microplastic extraction in sewage sludge and their physicochemical characteristics. Water Res 171:115379
51. Tagg AS, Harrison JP, Ju-Nam Y, Sapp M, Bradley EL, Sinclair CJ, Ojeda JJ (2017) Fenton's reagent for the rapid and efficient isolation of microplastics from wastewater. Chem Commun 53:372–375
52. Crichton EM, Noël M, Gies EA, Ross PS (2017) A novel, density-independent and FTIR-compatible approach for the rapid extraction of microplastics from aquatic sediments. Anal Methods 9:1419–1428
53. Huang A, Zhi D, Tang H, Jiang L, Luo S, Zhou Y (2020) Effect of Fe2+, Mn2+ catalysts on the performance of electro-Fenton degradation of antibiotic ciprofloxacin, and expanding the utilizing of acid mine drainage. Sci Total Environ 720:137560
54. Ziajahromi S, Neale PA, Rintoul L, Leusch FDL (2017) Wastewater treatment plants as a pathway for microplastics: development of a new approach to sample wastewater-based microplastics. Water Res 112:93–99
55. Li Q, Wu J, Zhao X, Gu X, Ji R (2019) Separation and identification of microplastics from soil and sewage sludge. Environ Pollut 254:113076
56. Hidalgo-Ruz V, Gutow L, Thompson RC, Thiel M (2012) Microplastics in the marine environment: a review of the methods used for identification and quantification. Environ Sci Technol 46:3060–3075
57. Gatidou G, Arvaniti OS, Stasinakis AS (2019) Review on the occurrence and fate of microplastics in sewage treatment plants. J Hazard Mater 367:504–512
58. Gong J, Xie P (2020) Research progress in sources, analytical methods, eco-environmental effects, and control measures of microplastics. Chemosphere 254:126790
59. Sun J, Dai X, Wang Q, van Loosdrecht MCM, Ni B-J (2019) Microplastics in wastewater treatment plants: detection, occurrence and removal. Water Res 152:21–37
60. Tagg AS, Sapp M, Harrison JP, Ojeda JJ (2015) Identification and quantification of microplastics in wastewater using focal plane array-based reflectance micro-FT-IR imaging. Anal Chem 87:6032–6040
61. He D, Luo Y, Lu S, Liu M, Song Y, Lei L (2018) Microplastics in soils: analytical methods, pollution characteristics and ecological risks. TrAC Trends Anal Chem 109:163–172
62. Rocha-Santos T, Duarte AC (2015) A critical overview of the analytical approaches to the occurrence, the fate and the behavior of microplastics in the environment. TrAC Trends Anal Chem 65:47–53
63. Wang L, Zhang J, Hou S, Sun H (2017) A simple method for quantifying polycarbonate and polyethylene terephthalate microplastics in environmental samples by liquid chromatography–tandem mass spectrometry. Environ Sci Technol Lett 4:530–534

64. Ando T, Uchihashi T, Kodera N, Yamamoto D, Miyagi A, Taniguchi M, Yamashita H (2008) High-speed AFM and nano-visualization of biomolecular processes. Pflugers Arch - Eur J Physiol 456:211–225
65. Dazzi A, Prater CB, Hu Q, Chase DB, Rabolt JF, Marcott C (2012) AFM–IR: combining atomic force microscopy and infrared spectroscopy for nanoscale chemical characterization. Appl Spectrosc 66:1365–1384
66. Song YK, Hong SH, Jang M, Han GM, Rani M, Lee J, Shim WJ (2015) A comparison of microscopic and spectroscopic identification methods for analysis of microplastics in environmental samples. Mar Pollut Bull 93:202–209
67. Corradini F, Bartholomeus H, Huerta Lwanga E, Gertsen H, Geissen V (2019) Predicting soil microplastic concentration using vis-NIR spectroscopy. Sci Total Environ 650:922–932
68. Shim WJ, Song YK, Hong SH, Jang M (2016) Identification and quantification of microplastics using Nile red staining. Mar Pollut Bull 113:469–476
69. Lares M, Ncibi MC, Sillanpää M, Sillanpää M (2019) Intercomparison study on commonly used methods to determine microplastics in wastewater and sludge samples. Environ Sci Pollut Res 26:12109–12122
70. David J, Weissmannová HD, Steinmetz Z, Kabelíková L, Demyan MS, Šimečková J, Tokarski D, Siewert C, Schaumann GE, Kučerík J (2019) Introducing a soil universal model method (SUMM) and its application for qualitative and quantitative determination of poly (ethylene), poly(styrene), poly(vinyl chloride) and poly(ethylene terephthalate) microplastics in a model soil. Chemosphere 225:810–819
71. Mason SA, Garneau D, Sutton R, Chu Y, Ehmann K, Barnes J, Fink P, Papazissimos D, Rogers DL (2016) Microplastic pollution is widely detected in US municipal wastewater treatment plant effluent. Environ Pollut 218:1045–1054
72. Bakir A, Rowland SJ, Thompson RC (2014) Transport of persistent organic pollutants by microplastics in estuarine conditions. Estuar Coast Shelf Sci 140:14–21
73. Ling SD, Sinclair M, Levi CJ, Reeves SE, Edgar GJ (2017) Ubiquity of microplastics in coastal seafloor sediments. Mar Pollut Bull 121:104–110
74. Sun X, Chen B, Li Q, Liu N, Xia B, Zhu L, Qu K (2018) Toxicities of polystyrene nano- and microplastics toward marine bacterium Halomonas alkaliphila. Sci Total Environ 642:1378–1385
75. Kataoka T, Nihei Y, Kudou K, Hinata H (2019) Assessment of the sources and inflow processes of microplastics in the river environments of Japan. Environ Pollut 244:958–965
76. Wei W, Huang Q-S, Sun J, Dai X, Ni B-J (2019) Revealing the mechanisms of polyethylene microplastics affecting anaerobic digestion of waste activated sludge. Environ Sci Technol 53: 9604–9613
77. Wei W, Zhang Y-T, Huang Q-S, Ni B-J (2019) Polyethylene terephthalate microplastics affect hydrogen production from alkaline anaerobic fermentation of waste activated sludge through altering viability and activity of anaerobic microorganisms. Water Res 163:114881
78. Perren W, Wojtasik A, Cai Q (2018) Removal of microbeads from wastewater using electrocoagulation. ACS Omega 3:3357–3364
79. Abdelrasoul A, Doan H, Lohi A (2013) A mechanistic model for ultrafiltration membrane fouling by latex. J Membr Sci 433:88–99
80. Lai CY, Groth A, Gray S, Duke M (2014) Enhanced abrasion resistant PVDF/nanoclay hollow fibre composite membranes for water treatment. J Membr Sci 449:146–157
81. Ak MS, Muz M, Komesli OT, Gökçay CF (2013) Enhancement of bio-gas production and xenobiotics degradation during anaerobic sludge digestion by ozone treated feed sludge. Chem Eng J 230:499–505
82. Wei W, Huang Q-S, Sun J, Wang J-Y, Wu S-L, Ni B-J (2019) Polyvinyl chloride microplastics affect methane production from the anaerobic digestion of waste activated sludge through leaching toxic bisphenol-a. Environ Sci Technol 53:2509–2517
83. Fu S-F, Ding J-N, Zhang Y, Li Y-F, Zhu R, Yuan X-Z, Zou H (2018) Exposure to polystyrene nanoplastic leads to inhibition of anaerobic digestion system. Sci Total Environ 625:64–70

84. Feng Y, Feng L-J, Liu S-C, Duan J-L, Zhang Y-B, Li S-C, Sun X-D, Wang S-G, Yuan X-Z (2018) Emerging investigator series: inhibition and recovery of anaerobic granular sludge performance in response to short-term polystyrene nanoparticle exposure. Environ Sci Water Res Technol 4:1902–1911
85. Mu H, Zheng X, Chen Y, Chen H, Liu K (2012) Response of anaerobic granular sludge to a shock load of zinc oxide nanoparticles during biological wastewater treatment. Environ Sci Technol 46:5997–6003
86. Zettler ER, Mincer TJ, Amaral-Zettler LA (2013) Life in the “Plastisphere”: microbial communities on plastic marine debris. Environ Sci Technol 47:7137–7146
87. Wang Y, Zhao J, Wang D, Liu Y, Wang Q, Ni B-J, Chen F, Yang Q, Li X, Zeng G, Yuan Z (2018) Free nitrous acid promotes hydrogen production from dark fermentation of waste activated sludge. Water Res 145:113–124
88. Gao D, Li X-Y, Liu H-T (2020) Source, occurrence, migration and potential environmental risk of microplastics in sewage sludge and during sludge amendment to soil. Sci Total Environ 742:140355
89. Nizzetto L, Futter M, Langaas S (2016) Are agricultural soils dumps for microplastics of urban origin? Environ Sci Technol 50:10777–10779
90. Ng E-L, Huerta Lwanga E, Eldridge SM, Johnston P, Hu H-W, Geissen V, Chen D (2018) An overview of microplastic and nanoplastic pollution in agroecosystems. Sci Total Environ 627: 1377–1388
91. Corradini F, Meza P, Eguiluz R, Casado F, Huerta-Lwanga E, Geissen V (2019) Evidence of microplastic accumulation in agricultural soils from sewage sludge disposal. Sci Total Environ 671:411–420
92. de Souza Machado AA, Lau CW, Kloas W, Bergmann J, Bachelier JB, Faltin E, Becker R, Görlich AS, Rillig MC (2019) Microplastics can change soil properties and affect plant performance. Environ Sci Technol 53:6044–6052
93. de Souza Machado AA, Lau CW, Till J, Kloas W, Lehmann A, Becker R, Rillig MC (2018) Impacts of microplastics on the soil biophysical environment. Environ Sci Technol 52:9656–9665
94. Rillig MC (2018) Microplastic disguising as soil carbon storage. Environ Sci Technol 52: 6079–6080
95. Van Cauwenberghe L, Claessens M, Vandegehuchte MB, Janssen CR (2015) Microplastics are taken up by mussels (Mytilus edulis) and lugworms (Arenicola marina) living in natural habitats. Environ Pollut 199:10–17
96. Zhang GS, Zhang FX, Li XT (2019) Effects of polyester microfibers on soil physical properties: perception from a field and a pot experiment. Sci Total Environ 670:1–7
97. Qi Y, Ossowicki A, Yang X, Huerta Lwanga E, Dini-Andreote F, Geissen V, Garbeva P (2020) Effects of plastic mulch film residues on wheat rhizosphere and soil properties. J Hazard Mater 387:121711
98. Liu H, Yang X, Liu G, Liang C, Xue S, Chen H, Ritsema CJ, Geissen V (2017) Response of soil dissolved organic matter to microplastic addition in Chinese loess soil. Chemosphere 185: 907–917
99. Rillig MC, Lehmann A (2020) Microplastic in terrestrial ecosystems. Science 368:1430–1431
100. Rillig MC, Bonkowski M (2018) Microplastic and soil protists: a call for research. Environ Pollut 241:1128–1131
101. Rodriguez-Seijo A, Lourenço J, Rocha-Santos TAP, da Costa J, Duarte AC, Vala H, Pereira R (2017) Histopathological and molecular effects of microplastics in Eisenia andrei Bouché. Environ Pollut 220:495–503
102. Zhu B-K, Fang Y-M, Zhu D, Christie P, Ke X, Zhu Y-G (2018) Exposure to nanoplastics disturbs the gut microbiome in the soil oligochaete Enchytraeus crypticus. Environ Pollut 239: 408–415
103. Cao D, Wang X, Luo X, Liu G, Zheng H (2017) Effects of polystyrene microplastics on the fitness of earthworms in an agricultural soil. IOP Conf Ser Earth Environ Sci 61:012148

104. Ju H, Zhu D, Qiao M (2019) Effects of polyethylene microplastics on the gut microbial community, reproduction and avoidance behaviors of the soil springtail, Folsomia candida. Environ Pollut 247:890–897
105. Selonen S, Dolar A, Kokalj AJ, Skalar T, Dolcet LP, Hurley R, Gestel CV (2020) Exploring the impacts of plastics in soil - the effects of polyester textile fibers on soil invertebrates. Sci Total Environ 700:134451.134451–134451.134412
106. Wang W, Ge J, Yu X, Li H (2020) Environmental fate and impacts of microplastics in soil ecosystems: progress and perspective. Sci Total Environ 708:134841
107. Qi Y, Yang X, Pelaez AM, Huerta Lwanga E, Beriot N, Gertsen H, Garbeva P, Geissen V (2018) Macro- and micro- plastics in soil-plant system: effects of plastic mulch film residues on wheat (Triticum aestivum) growth. Sci Total Environ 645:1048–1056
108. Yuan-Chen L, Jahitbek N, Mei L, Xiao-Lin W, Li-Juan J (2019) Effects of microplastics on the growth, physiology, and biochemical characteristics of wheat (Triticum aestivum). Huan jing ke xue= Huanjing kexue 40:4661–4667
109. Gao M, Liu Y, Song Z (2019) Effects of polyethylene microplastic on the phytotoxicity of di-n-butyl phthalate in lettuce (Lactuca sativa L. var. ramosa Hort). Chemosphere 237:124482
110. Bosker T, Bouwman LJ, Brun NR, Behrens P, Vijver MG (2019) Microplastics accumulate on pores in seed capsule and delay germination and root growth of the terrestrial vascular plant Lepidium sativum. Chemosphere 226:774–781
111. Machado A, Lau CW, Kloas W, Bergmann J, Bacheher JB, Faltin E, Becker R, Goerlich AS, Rillig MC (2019) Microplastics can change soil properties and affect plant performance. Environ Sci Technol 53:6044–6052
112. Qian H, Zhang M, Liu G, Lu T, Qu Q, Du B, Pan X (2018) Effects of soil residual plastic film on soil microbial community structure and fertility. Water Air Soil Pollut 229:261
113. Awet TT, Kohl Y, Meier F, Straskraba S, Grün AL, Ruf T, Jost C, Drexel R, Tunc E, Emmerling C (2018) Effects of polystyrene nanoparticles on the microbiota and functional diversity of enzymes in soil. Environ Sci Eur 30:11
114. He L, Wu D, Rong H, Li M, Tong M, Kim H (2018) Influence of nano- and microplastic particles on the transport and deposition behaviors of bacteria in quartz sand. Environ Sci Technol 52:11555–11563
115. Huang Y, Zhao Y, Wang J, Zhang M, Jia W, Qin X (2019) LDPE microplastic films alter microbial community composition and enzymatic activities in soil. Environ Pollut 254:112983
116. Groh KJ, Backhaus T, Carney-Almroth B, Geueke B, Inostroza PA, Lennquist A, Leslie HA, Maffini M, Slunge D, Trasande L, Warhurst AM, Muncke J (2019) Overview of known plastic packaging-associated chemicals and their hazards. Sci Total Environ 651:3253–3268
117. Hahladakis JN, Velis CA, Weber R, Iacovidou E, Purnell P (2018) An overview of chemical additives present in plastics: migration, release, fate and environmental impact during their use, disposal and recycling. J Hazard Mater 344:179–199
118. Holmes LA, Turner A, Thompson RC (2012) Adsorption of trace metals to plastic resin pellets in the marine environment. Environ Pollut 160:42–48
119. Hüffer T, Metzelder F, Sigmund G, Slawek S, Schmidt TC, Hofmann T (2019) Polyethylene microplastics influence the transport of organic contaminants in soil. Sci Total Environ 657: 242–247
120. O'Connor D, Pan S, Shen Z, Song Y, Jin Y, Wu W-M, Hou D (2019) Microplastics undergo accelerated vertical migration in sand soil due to small size and wet-dry cycles. Environ Pollut 249:527–534
121. Steinmetz Z, Wollmann C, Schaefer M, Buchmann C, David J, Tröger J, Muñoz K, Frör O, Schaumann GE (2016) Plastic mulching in agriculture. Trading short-term agronomic benefits for long-term soil degradation? Sci Total Environ 550:690–705
122. Huerta Lwanga E, Gertsen H, Gooren H, Peters P, Salánki T, van der Ploeg M, Besseling E, Koelmans AA, Geissen V (2017) Incorporation of microplastics from litter into burrows of Lumbricus terrestris. Environ Pollut 220:523–531

123. Rillig MC, Ingraffia R, de Souza Machado AA (2017) Microplastic incorporation into soil in agroecosystems. Front Plant Sci 8:1805
124. Rillig MC, Muller LAH, Lehmann A (2017) Soil aggregates as massively concurrent evolutionary incubators. ISME J 11:1943–1948
125. Guo J-J, Huang X-P, Xiang L, Wang Y-Z, Li Y-W, Li H, Cai Q-Y, Mo C-H, Wong M-H (2020) Source, migration and toxicology of microplastics in soil. Environ Int 137:105263
126. Ramos L, Berenstein G, Hughes EA, Zalts A, Montserrat JM (2015) Polyethylene film incorporation into the horticultural soil of small periurban production units in Argentina. Sci Total Environ 523:74–81

Revision of the Most Harmful Organic Compounds Present in Sewage and Sludge

Javier Canas Jiménez, Juan García Rodríguez, Blanca Hermana Mendioroz, V. Ismael Águeda Maté, and Silvia Álvarez-Torrellas

Contents

Abstract Water treatment is a vital process to ensure the development and sustainability of today's society, both from an environmental and public health point of view. Wastewater treatment has a huge variety of studies and operations. In these processes, which are usually carried out in wastewater treatment plants (WWTPs), large quantities of toxic and highly heterogeneous sludge are generated, the proper management of which is a major challenge.

The overexploitation of limited natural resources and the enormous consumption of energy by modern society mean that substantial changes are needed in water and

J. C. Jiménez
Catalysis and Separation Processes Group (CyPS), Chemical Engineering and Materials Department, Chemistry Sciences Faculty, Complutense University, Madrid, Spain

ECOLOTUM, Energía Recuperable, S.L., Madrid, Spain

J. G. Rodríguez, V. Ismael Águeda Maté, and S. Álvarez-Torrellas (✉)
Catalysis and Separation Processes Group (CyPS), Chemical Engineering and Materials Department, Chemistry Sciences Faculty, Complutense University, Madrid, Spain
e-mail: satorrellas@ucm.es

B. H. Mendioroz
ECOLOTUM, Energía Recuperable, S.L., Madrid, Spain

Avelino Núñez-Delgado and Manuel Arias-Estévez (eds.),
Emerging Pollutants in Sewage Sludge and Soils,
Hdb Env Chem (2023) 114: 189–208, DOI 10.1007/698_2022_855,

Published online: 16 April 2022

sludge treatment and purification systems. In this sense, it has been estimated that in Europe the generation of sludge in WWTPs will exceed 13 million tons/year in 2021. The detection of new contaminants in sewage sludge, as well as the significant increase in its production and its limited usefulness in agricultural applications, makes it necessary to invest in research and development of technological solutions that respond to the demanding restrictions established by the European legislation.

To aid in the knowledge on the presence and concentration of organic chemicals in sewage sludge, peer-reviewed literature and official government reports have been examined in this chapter.

Keywords Industrial sludge, Organic compounds, Sewage sludge, Toxic compounds, Wastewater

Abbreviations

EPA	United States Environmental Protection Agency
ISS	Industrial sewage sludge
OPs	Organophosphate pesticides
PAH	Polycyclic aromatic hydrocarbons
PBB	Polybrominated biphenyls
PBDE	Polybrominated diphenyl ethers
PCB	Polychlorinated biphenyls
PCDDs	Polychlorinated dibenzo-p-dioxins
PCDFs	Polychlorinated dibenzo-p-furans
PFC	Perfluorinated aliphatic compounds
SS	Sewage sludge
TCDD	Tetrachlorodibenzo-p-dioxin

1 Introduction

Sewage sludge is a by-product generated after the municipal and industrial wastewater treatment. So, the concept of sewage sludge is usually understood as the waste generated in wastewater treatment plants, where the wastewater is mainly urban or municipal in nature, although, depending on the location, it usually incorporates more or less quantity of industrial wastewater. Therefore, in order to distinguish between the two origins of sludge, sewage sludge (SS) will refer to waste generated in wastewater treatment plants (WWTP), whereas when referring to sludge of strictly industrial origin we will name it as industrial sewage sludge (ISS).

Regarding the management of sewage sludge, this is one of the most complex environmental problems associated with the treatment of wastewater. The fundamental problem related to sewage sludge stems from the fact that, in most cases, it is a waste that needs to be managed. This implies an obligation to treat sludge from

WWTPs, which is now at least partly integrated in the water treatment facilities themselves. This strategy is one of the key elements of the Circular Economy [1].

Generally, SS is that which is generated at WWTP; the wastewater treated at WWTP comes from different sources. The first and main source is the domestic wastewater, resulting from human activity and generated around population centers. The other source of wastewater is industrial wastewater, which is discharged into the environment, generally after its treatment, as well as rainwater, and any wastewater discharge that may occur. The proportions in which the different sources of wastewater are found vary greatly, depending on the geographical location of each WWTP, the size of nearby population centers and the presence and type of surrounding industries, making it very difficult to group them in a uniform way [2]. However, considering urban waters, especially domestic waters, it could be considered that they present characteristics that are quite similar to each other, so an overview of them is presented.

On the other hand, ISS has characteristics completely different depending on the specific industry. It is impossible to try to group them in a homogeneous way, so this work will highlight some specific industrial sludge, either because of their toxicity or because of the volume generated, such as those coming from the petrochemical, pharmaceutical, agricultural, food, textile, or paper industries.

Sewage sludge is, in general, a mixture of solids in an aqueous matrix with a very high content of compounds of various kinds and in variable percentage depending on the treatment and the type of wastewater. Generally, the presence of organic compounds stands out, but we can also find inorganic solids, heavy metals, minerals, etc.

All this makes sewage sludge a heterogeneous substance, with a large number of compounds, many of which are toxic for humans and the environment, both from a chemical and biological point of view. For this reason, various treatments have been developed to reduce the hazardousness and toxicity of the sludge, as well as to take advantage of the recoverable compounds.

As a result of the development of legislation associated with the sludge treatment, which is constantly being updated, more and more compounds present in sludge are being classified as toxic or hazardous, depending on their concentration, and therefore require an appropriate treatment. Moreover, considering that the growing world population is causing an increase in the generation of sludge, the importance of good sludge management, both in terms of its treatment and its continuous study and research, is a challenge for the current and future society [3].

2 Organic Compounds in Sludge

Sewage sludge is obtained after the various wastewater treatments in the WWTP, which are generated at different points of the plant. In general, the incoming wastewater is subjected to a physical-chemical treatment, from which the primary sludge is obtained after the first settling. Afterwards, the water undergoes biological treatment and, after subsequent settling, the secondary or biological sludge is produced. Often, the wastewater requires tertiary or chemical treatments, which may produce a tertiary sludge [4].

A common treatment of this sludge is to mix it and subject it to anaerobic digestion, to stabilize it and obtain biogas and, generally, after undergoing dewatering processes, it is available for agriculture, incineration, landfill, thermal treatments, composting, etc. [2].

In this sense, the organic matter content of sewage sludge is tremendously diverse [5] and is influenced by both the origin of wastewater and the type of treatment to which the wastewater is subjected, and from which the sludge is obtained. In this work, the assessment of organic compounds present in sludge is evaluated along three points:

- Firstly, the aim is to provide a complete and global view of sludge as it is usually characterized and, more specifically, of the organic fraction present.
- The organic matter of biological origin or character will be described, with the focus on the various populations of bacteria and microorganisms that are present in the sludge, as they are mainly responsible for the biological treatments that are carried out in the WWTP.
- Finally, some organic compounds are described, grouped into families, which stand out for their toxicity to humans and/or the environment, which persist after treatment in the WWTP, or which are even generated in the WWTP or in subsequent treatments.

2.1 General Composition

In this section, the aim is to give an overview of the presence of organic compounds in the sewage sludge, focusing on its origin and treatment, i.e., whether it comes from primary, secondary, mixed, or digested treatment of a sludge generated in a WWTP, for example, of a municipal nature.

Regarding the organic matter content, it has been estimated that the equivalent chemical oxygen demand (COD) load per capita –or inhabitant equivalent (IE)– discharged in municipal wastewater streams is of 100 g COD/IE day [6] while referring to biological oxygen demand –measured within 5 days– (BOD_5) is of 60 g BOD_5/IE day [7]. These values, depending on the treatment to which the wastewater is subjected, generate a greater or lower amount of solids in the sludge, as well as different concentrations of organic matter.

Table 1 shows typical values as volatile solids (VS) associated with the organic matrix present in the solid, and other groups of organic compounds present in the sewage sludge [2, 5, 8]. It is assumed that the amount of total solids (TS) corresponds almost absolutely to the suspended solids, or dried solids. The table shows the ranges and proportions where the described parameters are normally found, depending on the type of sludge.

Table 1 General composition of sewage sludge [2, 5, 8]

	Primary sludge	Secondary sludge	Digested mixed sludge
Total solids (TS, %)	1–8	0.4–1.5	3–6
Volatile solids (% TS)	60–85	60–88	50–75
C (% VS)	50–51	52–53	50
Grease and fats (% TS)	6–35	8–10	10
Proteins (% TS)	20–30	34–36	18
Fibers (cellulose) (% TS)	8–16	7–10	10
Organic acids (g/L as acetic acid)	0.2–2	1.1–1.7	–

Primary sludge is obtained after primary or physical-chemical treatment, followed by primary settling; the objective of which is to remove suspended matter, solids, grease, etc. from the water line. Dates in 1.1 kg of primary sludge/per capita (IE) by day is generated, in which between 50–70% of the total solids and 25–40% of the biodegradable organic matter (BOD_5) entering the WWTP are removed [9, 10].

This is the sludge with the highest amount of solids, mainly of volatile character, as shown in Table 1, concentrating the highest organic load of the WWTP and, being mainly biodegradable, causing that it decomposes easily and becomes septic, producing bad odors. Numerous pathogenic microorganisms and bacteria are present, but there is also a significant presence of fats and fibers [2].

Secondary sludge, also known as biological sludge, is produced by biological treatments in WWTP as activated sludge, membrane bioreactors, or rotating biological contactors. Bacteria and microorganisms act by reducing the organic matter, or nutrients, in the wastewater, and the solid residue generated contains mostly a wide variety of bacterial and pathogens cells, with a flocculus appearance and a quick decomposition. Between 90 and 99% of the pathogens involved in biological sludge are concentrated in the secondary sludge. Its total solids content is lower than in the primary sludge, as shown in Table 1, but the proportion of volatile solids is similar. Thus, the amount generated differs greatly depending on the type of treatment [2, 5].

Digested mixed sludge is the result of combining primary and secondary sludge and subjecting it to anaerobic digestion. The treatment stabilizes the organic matter and reduces odors and pathogens content [9].

This sludge, after undergoing dewatering or drying to reduce its moisture content to different extents (usually, up to 20–25% of dried solids), is evacuated from WWTP toward its final disposal, which may be in direct agricultural application (for example, in Spain, it must follow the guidelines of Order AAA/1072/2013 of 7 June on the use of sewage sludge in the agricultural sector), or it may be treated by composting, incineration, etc.

As for the ISS sludge generated in industrial processes, its composition and characteristics differ greatly depending on the type of industry that generates it. In this way, in the petrochemical industry, the sludge generated, known as oily sludge, is the result of the treatment of different contaminated streams generated in the oil refining process, constituting a complex mixture of compounds of different origins, such as residues from oil/water separation (API separator), cleaning sludge, sediments at the bottom of tanks or sludge from flocculation-flotation units, many of them highly pollutant and toxic [11]. They are notable for their high organic content, mostly non-biodegradable, and, therefore, have a high COD/BOD ratio. In terms of composition, this kind of sludge may contain between 50 and 85% of water, 5–20% of solids, with a high content of heavy metals, and from 10 to 30% of hydrocarbon content [12], including heavier petroleum hydrocarbons, which are mostly composed by C20+ hydrocarbons molecules [13].

There are also a multitude of organic compounds in water and sludge generated in the pharmaceutical industry. Both the method of production of pharmaceuticals, by chemical, extractive or fermentative means, and the wastewater treatment carried out, by biological means or by advanced wastewater treatments, produce a very different quantity and composition of sludge, but they usually have common characteristics. Generally, they present high amounts of organic matter, mostly highly biodegradable, on the one hand, with the presence of various solvents commonly used in the industry, such as methanol, ethanol, acetone, isopropanol, benzene, or chloroform, and on the other hand, of the drugs themselves, compounds with great influence and effect on living organisms, pollutants and even toxic in low doses, such as hormones, antibiotics, steroids, etc. [14].

The textile industry, one of the oldest in the world, uses enormous quantities of dyes and pigments, the presence of which is found in sludge, after the wastewater treatment, both in the industry itself and in WWTP where the contaminated water arrives. Sludge is characterized by its high organic matter, much of which is readily biodegradable, and by the presence of numerous toxic organic compounds, such as dyes, a complex matrix of aromatic compounds, with several characteristics that make them toxic due to their reactivity, acidity, or hydrophobicity [15, 16]. In addition, the solids content of sludge produced from textile industry is much higher than in conventional WWTP [17].

Therefore, the pulp and paper industry generates large quantities of sludge of varied composition, generally categorized as non-toxic [18]. The high organic matter content stands out – between 50 and 75 wt.% of the sludge – in which a distinction can be made between primary sludge, which is not very biodegradable due to the abundance of cellulose fibers (15 wt.%), and biological sludge, with a greater presence of microorganisms and nutrients [19]. Other sludges of enormous generation in terms of volume are from the food and agricultural industry. This sludge is very similar to that of a strictly urban nature, highlighting the presence of some certainly dangerous and toxic organic compounds such as pesticides and herbicides, which have been detailed elsewhere [20, 21].

2.2 *Microbiological Composition of Sludge*

This section aims to provide an overview of the microbiological composition of sludge, depending on its municipal or industrial origin [22]. It is intended to highlight the presence of microorganisms as the major and fundamental organic components in sewage sludge, whether originating from the wastewater source from which the sludge originates, formed after the various biological treatments to which wastewater is normally subjected, or a combination of both.

The categories of microorganisms include bacteria, archaea, fungi, algae, protozoa, helminthes, and viruses. It is necessary to highlight the family of pathogenic microorganisms, as they are capable of producing infection and disease in the living organisms in which they coexist, some of the most common being *Eschericia coli, Legionella, Salmonella,* or *Vibrio cholera* in the case of bacteria, *Balantidium coli* in protozoa, *Ascaris lumbricoides* for helminths and *Adenovirus*, *Enterovirus,* and *Rotavirus* in the case of viruses [5].

Certain of these microorganisms are continuously measured, especially in the water line, as they are indicators of fecal contamination, associated with municipal WWTP. In sludge, continuous monitoring is less frequent, as the subsequent treatments contemplate their presence in the majority of cases and, therefore, design treatments for their elimination (incineration) or make use of them (composting). The solid waste derived from the various biological treatments that are carried out in WWTP, and that form part of the sludge, is mainly made up of the microorganisms present in the process, and therefore depends on the treatment that is carried out, although the main difference derives from the type of wastewater from which it originates. The most widely used treatment involves the presence of activated sludge, with bacteria having the major role [5].

Activated sludge clumps in flocs composed mainly by aggregates of bacteria and other particles, organic or inorganic, blended in a polymeric matrix, known as extracellular polymeric substance (EPS), substances secreted by microorganisms into their environment. The bacterial cell fraction contributes up to 20% of the total organic matter content of the floc, while the EPS constitutes more than 50%. In other words, more than 70% of the organic matter present in the sludge is either microorganisms or compounds produced by them [22].

The microbial composition of the municipal sludge consists mainly of bacteria, of which different phyla can be distinguished. After evaluating several studies with different types of sludge both in their origin and in the biological treatment from which they are originated, the bacterial profile can be evaluated and summarized as shown in Fig. 1.

Each municipal sewage sludge studied showed high bacterial diversity and different from each other. However, their source or biological treatment conditions do not affect the bacterial structures; thus, the various types of municipal sludge have a similar bacterial phyla profile.

Proteobacteria and *Bacteroidetes* constitute about 60% of the bacterial phyla present in the sludge, being practically half of each of them [23–25]. A large number of different minority of bacterial phyla, such as *Chloroflexi*, *Verrucomicrobia*, *Spirochaetes*, *Synergistetes*, *Nitrospirae*, *Gemmatimonadetes* or *Cyanobacteria*, are grouped under the label "Others."

Furthermore, the classes mostly present in both bacterial phyla are also analyzed in Fig. 2. Alfa and beta classes are the most abundant in proteobacteria phylum, while in *Bacteroidetes*, *Sphingobacteria* constitutes more than 60% of phylum.

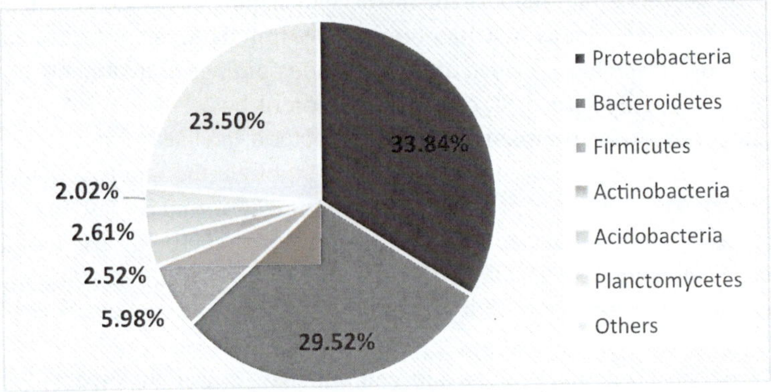

Fig. 1 Bacterial phyla composition in different municipal sewage sludge [23–25]

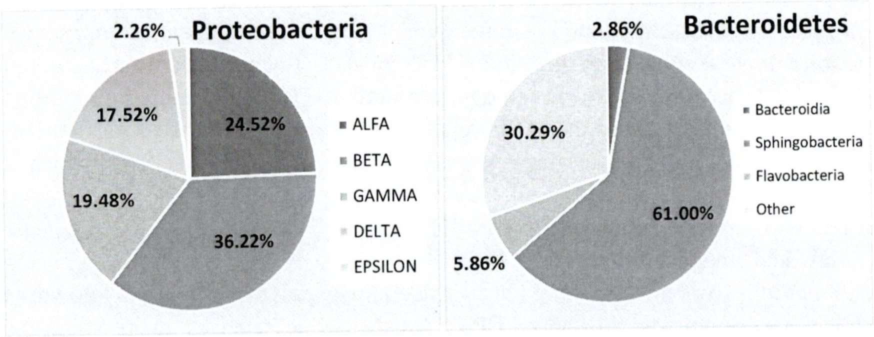

Fig. 2 Classes of bacterial phyla in different municipal sewage sludge [23, 24]

Regarding the analysis of the bacterial phyla, the study is extended to several types of sludge of industrial origin, showing the profiles in Fig. 3. For some of the different industrial sludge evaluated, multiple analyses of the bacterial profile carried out by numerous studies over different periods of time are summarized, showing consistency in the same [26, 27].

In all of them it can be observed that the phyla *Proteobacteria* and *Bacteroidetes* are the majority, making up from 50 to 90% of the bacterial presence in the sludge, with *Proteobacteria* being the majority in practically all of them. However, the difference in the proportions between industrial sludge and other sludge is much more important than in the case of municipal sludge. In addition, the bacterial diversity found in industrial sludge is substantially lower. For each sludge a specific bacterial profile is found, which is relatively undiverse compared to the municipal sludge and quite different for sludge from different industries.

Therefore, it can be considered that the bacterial composition of sludge depends more on the wastewater from which it originates than on the type of biological treatment applied or the conditions of the treatment. Therefore, municipal sludge has

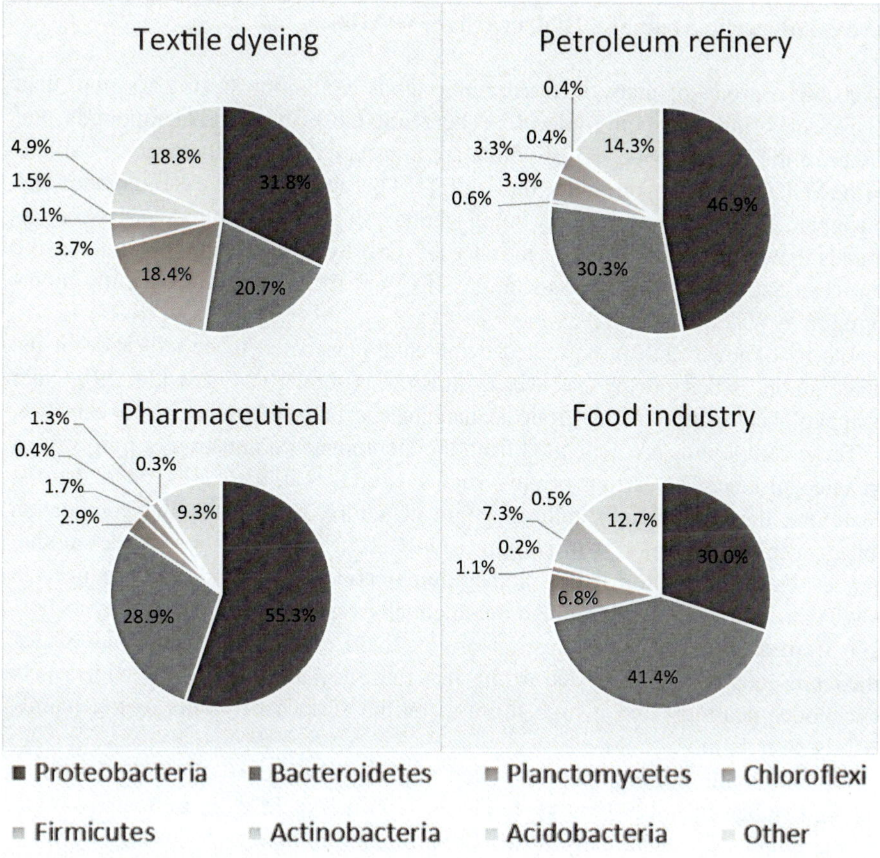

Fig. 3 Bacteria phyla composition in different industrial sewage sludge [26, 27]

a more similar profile, while sludge from each industry shows more specific and variable bacterial compositions.

2.3 *Toxic Organic Compounds Families*

This section describes the families of organic compounds that are important because of their toxicity. They are found in low concentrations (ppm, ppb, etc.) but even so they are extremely dangerous for the environment and living organisms. These groups of compounds have been categorized as hazardous in recent years, in which research on the measurement, detection, monitoring, and categorization of these compounds, known in many cases as emerging pollutants, has been multiplying. This research continues, and the list of emerging organic pollutants is expected to grow year by year. Thus, because of their size and characteristics, they are known as organic micropollutant compounds (OMPs).

2.3.1 Polycyclic Aromatic Hydrocarbons (PAHs)

PAHs are a group of many different compounds containing from 2 to more than 10 benzene rings in their molecule. There are more than 200 compounds that comprise the PAHs group, 33 of which are considered especially toxic to humans by the SCF (Scientific Committee on Food) [28] because they have carcinogenic and mutagenic properties, as well as being genotoxic and embryotoxic. These compounds have been included in the list of priority organic compounds of the European and United States Associations (EU and EPA) and are frequently monitoring in WWTP.

The best-known and most studied compound, considered to be indicative of the whole group, is benzo(α)pyrene, whose structure is shown in Fig. 4 [28, 29]. Other examples of compounds in this group are naphthalene, benzo(α)anthracene, or chrysene.

These compounds are generated from the incomplete combustion or pyrolysis of organic substances of various origins, such as wood, coal, fuels, oils, or foodstuffs. Therefore, they are mainly anthropogenic in nature, i.e., due to the combustion boilers, vehicle traffic or food cooking. Fires are a natural source, while another source of their formation is the incineration treatment of dry sludge produced in WWTP, a common and widespread treatment all over the world. All this favors that, both from a domestic and industrial origin, these compounds reach wastewater. Their presence in sludge is due to the fact that they are compounds with a very low biodegradability and a high affinity for the surface of solids, being mainly adsorbed on them.

While several studies have reported maximum concentrations of PAHs in digested sludge up to 200 mg/kg D.M. and 100 mg/kg D.M. in untreated sludge, in Spain, the Spanish Ministry of the Environment (*Ministry of Medio Ambiente y Medio Rural y Marino*) and the Ministry of Ecologic Transition (*Ministry of Transición Ecológica y el Reto Demográfico*) have published concentrations of these compounds of less than half of the legal limit concentration in sludge applicable to agriculture, which is 6 mg/kg D.M [28].

2.3.2 Dioxins and Furans (PCDDs/PCDFs)

This group of contaminants represents a mixture of approximately 210 different compounds [28] that are characterized as organochlorine aromatics and are divided in two families: the poly chlorinated dibenzo-p-dioxins (PCDDs) and the poly chlorinated dibenzo-p-furans (PCDFs).

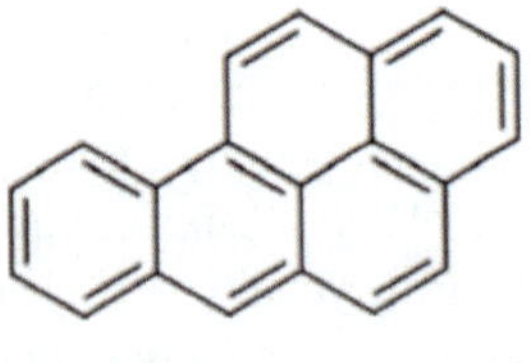

Fig. 4 Molecular structure of benzo (α) pyrene

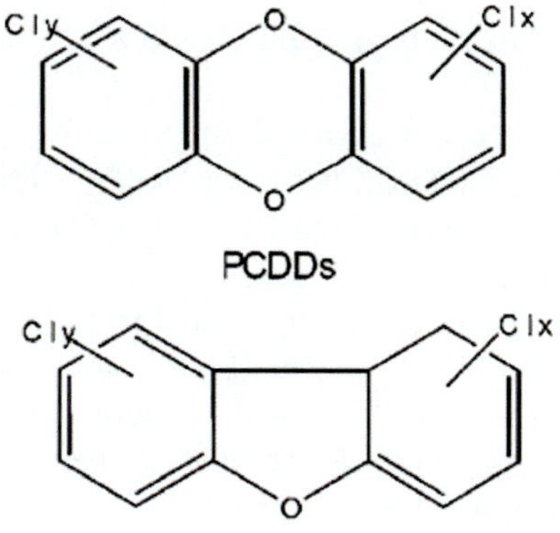

Fig. 5 Molecular structures of PCDDs and PCDFs

Chlorine substitution can occur at any of the free linkage vertices of the benzene rings, allowing the formation of numerous isomers of each family (Fig. 5). This factor will determine the toxicity of the compound formed. In this regard, 17 compounds (7 dioxins and 10 furans) are particularly harmful to humans.

The best-known and most dangerous compound, in terms of toxicity, of this family is 2,3,7,8 tetrachlorodibenzo-p-dioxin (2,3,7,8-TCDD), and the concentration of dioxins is usually expressed as the sum of equivalent toxicity with respect to the toxicity of the most toxic dioxin, TCDD [29].

These compounds are harmful because they accumulate in adipose tissue, due to their low water solubility and high fat solubility. Because of this, and because of their long half-life in living organisms, approximately 8 years, they accumulate and interfere in the food chain. They are very stable compounds, with high thermal and chemical resistance in oxidation processes [28].

Dioxins and furans are originated as undesirable by-products in the production or use of organochlorine compounds or in the presence of organic compounds and chlorine. They can be produced naturally in forest fires or by photochemical and metabolic reactions in crop soils in the presence of organochlorine pesticides. However, they are mostly generated in industries where combustion of organic matter with the presence of chlorinated compounds takes place. The paper industry, metallurgy, and all those related to the synthesis of chlorinated compounds or chlorine gas itself, as well as waste incineration, are the most important sources of dioxins and furans emissions. In addition, the generation of dioxins and furans derives from the use of dishwashers, vehicle traffic and rainwater run-off from soil, streets and roads.

Their presence in sewage sludge is mainly due to their resistance to the conventional treatment in WWTP and even their own generation in WWTP. In addition, the formation of dioxins and furans has been observed in some enzymatic reactions, such as the oxidation of chlorophenols, as well as in dechlorination treatments. If we add to this their formation during the thermal incineration treatments of the sludge, generating atmospheric pollution that returns in the form of run-off to the WWTP, these compounds can accumulate in the sludge.

2.3.3 Polychlorinated Biphenyls (PCBs)

In terms of chemical structure, this group of compounds is characterized by two benzene molecules connected by a bridge (the structure of the biphenyl compound) in which chlorine is substituted in multiple and diverse ring positions, as can be seen in Fig. 6.

There are more than 209 isomers forming the PCB family, divided in ten homologous groups (from mono to deca-chlorobiphenyl compounds) [28], all of them depending on the chlorine substitution in the different available positions, which defines their spatial configuration, their behavior in the environment as well as their toxicity.

The origin of these compounds can be centered on two routes. Firstly, the industrial production process of PCBs, since these compounds are used in various applications, such as hydraulic fluids, lubricants, impregnating agents or as additives in paints and insecticides. The presence of these compounds in products with a long useful life and everyday use allows their penetration into the natural human environment, which can occur in the form of evaporation or spills. On the other hand, they are compounds that are generated in incomplete combustion processes, such as in waste incineration treatment [29].

These compounds enter the human body via the dietary pathway, through respiration or through the skin. The major source of human contamination of PCBs is through the ingestion of animal fats. They quickly enter the circulatory system and accumulate in various tissues, mainly fat.

According to the EPA, when the main adverse effects of PCBs on humans are highlighted, they are classified as compounds with very probable carcinogenic effects, that weaken the immune system, and that affect the reproductive, nervous and endocrine systems [30].

They are very poorly soluble in water, less as higher is the chlorine substitution, so they are very soluble in fats. This facilitates their sorption on polar surfaces of solids, which is why they are present in sludge rather than in water. This may be due to their presence in the incoming wastewater, wet and dry deposition in the atmospheric air, and in wastewater treatment processes. The concentration values of PCBs in sludge vary from 0.003 to 0.65 mg/kg D.M. No differences were found in PCB concentrations in sewage sludge from municipal or industrial WWTP [28]. In Spain, the limit imposed in 2009 set at 0.8 mg/kg D.M., with the sum of seven compounds of the group, highlighting in some cases the exceeding of the limit in some WWTP that received point discharges of industrial nature.

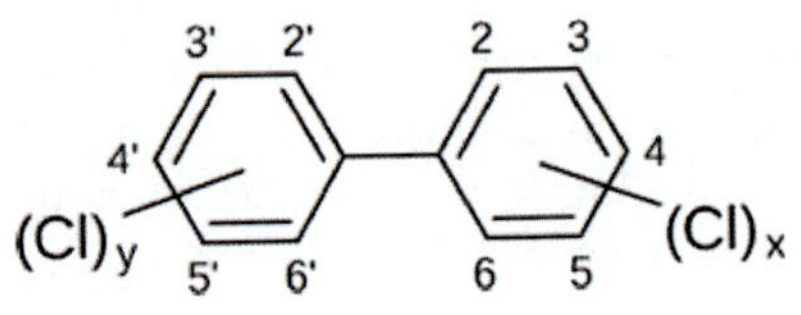

Fig. 6 Molecular structure of PCBs

2.3.4 Polybrominated Compounds (PBBs and PBDEs)

In terms of their molecular structure, PBBs (PolyBrominated Biphenyls) are similar compounds to PCBs, with the difference of the substituted group, Br instead of Cl. On the other hand, PBDEs (PolyBrominated Diphenyl Ethers) are compounds equally substituted with Br in the benzene rings, which in this case are joined by a bridge that forms the ether structure. Their structures are shown in Fig. 7.

These compounds are mainly used as flame retardants in the incineration of plastics. Therefore, they are present in electronic devices such as televisions or computers, as well as in textile articles, such as carpets or upholstery. Thus, the sources of emissions of these compounds are centered both in the industries in charge of producing them and in those where they are used, both in the manufacture of plastic compounds and textiles, as well as in the management of the waste that contains them. They can be released into the air, water and soil [28].

PBBs reach humans by ingestion, inhalation or through the skin. Since they have low solubility in water and high solubility in fats, they accumulate in these tissues. According to the EPA, they are considered possibly carcinogenic, immune-weakening, and toxic to the liver, kidney and thyroid systems [31]. Therefore, their use was banned a few decades ago, and PBDEs started to be produced as an industrial substitute. Currently, PBDEs are considered emerging pollutants, so their toxicity is under study, and a similar danger to PBBs can be observed, standing out for being endocrine disruptors [32].

Their presence in sewage sludge is due to their low solubility in water, ineffective treatments, and their accumulation in solids, being found in concentrations of up to 1 mg/g D.M. [33]. In Spain, most of the WWTP analyzed presented PBDEs concentrations that did not exceed the value of 1 mg/g D.M.

2.3.5 Perfluorinated Aliphatic Compounds (PFCs)

PFCs (perfluorinated aliphatic compounds) are compounds derived from hydrocarbons in which one or some hydrogen atoms are replaced by fluorine atoms. They include carboxylic derivatives (PFOA, perfluorooctanoic acid), sulfonates (PFOS, perfluorooctane sulfonate), and sulfonamides (PFOSA, perfluorooctane sulfonamide), as well as their esters, salts, and fluorides. The strong C-F bond formed with respect to the C-H bond gives this group its fundamental characteristics: high chain stiffness, high chemical and thermal stability, hydrophobicity, high density, and low biodegradability in the environment.

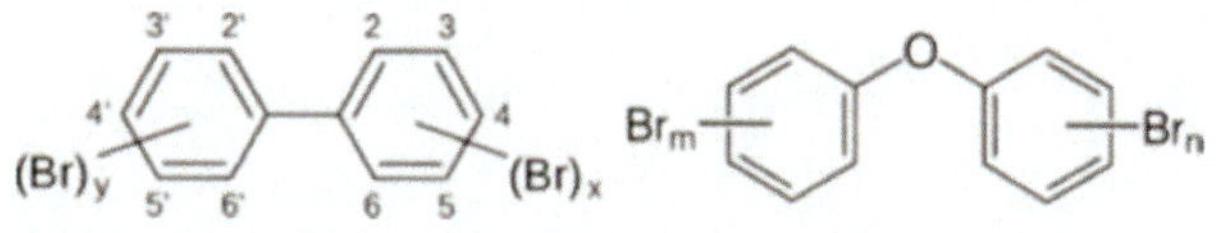

Fig. 7 Molecular structures of PBBs (left) and PBDEs (right)

These compounds are used in industry as flame retardants, industrial lubricants and in household products, such as pans and pots (Teflon), impregnated textiles, food packaging, cleaning and personal care products, shampoos or waxes.

Due to their high stability, these compounds are very persistent, both in the human body and in the environment. Contamination of water or ingestion by contaminated animals may cause them to bioaccumulate and eventually lead to adverse health effects. Studies on their potential toxicity have only been carried out for a few years, and they are therefore classified as emerging pollutants [28].

The EPA, after several studies, considers that PFOA and PFOS can affect the reproductive, hepatic, renal, immunological, and tumor-causing systems in animals. Human studies showed an increase in cholesterol levels, with more limited results regarding immune system disorders, cancer (for PFOA), and thyroid hormone disruption (for PFOS) [34].

The presence of these compounds in sewage sludge is variable, and they have been found up to 40 ng/g D.M. in European WWTPs (Spain and Germany), up to 6.6 mg/g D.M. in Chinese WWTPs and up to 1 mg/g D.M. in the USA. Their concentration seems to be closely related to equivalent inhabitants. These compounds are not removed by conventional treatments intended for agricultural application, and therefore end up being an environmental and health problem [35].

2.3.6 Pesticides

These are multiple and diverse synthetic compounds formulated in order to being applied directly to crops, either as herbicides, fungicides or insecticides, to increase the production or improve the crops quality.

In the case of insecticides, the most numerous within the family, organochlorine compounds are the most important; compounds that are halogen derivatives of multi-ring cycloparaffins and combinations of dienes, terpenes or benzene itself. Among them, the most toxic and best-known compound is DDT (dichloro-diphenyl-trichloroethane), widely used since its discovery and subsequently banned, and its analogues such as lindane, hexachlorobenzene, aldrin, or endrin (Fig. 8).

These compounds are highly lipid soluble and therefore tend to accumulate in the tissues of living organisms. In addition, they are characterized by their high resistance to detoxifying agents and their durability in the environment. They are toxic because they damage the cellular structures of nerve tissue and disrupt its basic functions.

Fig. 8 Molecular structure of DDT

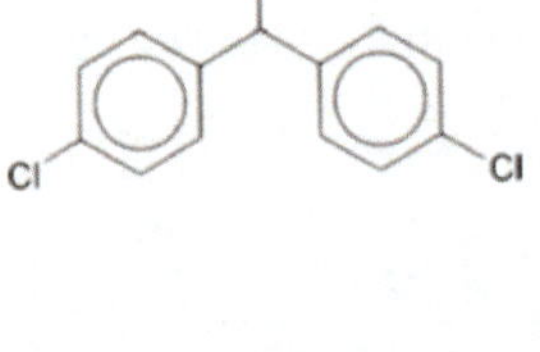

A new group belonging to this family and similar to the organochlorines is the one of the organophosphate pesticides (OPs). They are the most widely used at present, both in agriculture and in gardens or domestic use. Some examples of these compounds are diazinon or fenthion. Chronic exposure to OPs can cause toxicity (in addition to insects) to other living beings such as humans. It is known that OPs inhibit acetylcholinesterase activity, causing respiratory, reproductive, nervous, hepatic, and renal difficulties [36].

Pesticides amounts of up to 8–10 mg/kg D.M. of compounds such as hexachlorobenzene or lindane have been found in sludge at the end of the past century; however, in recent years, the presence of pesticides in sewage sludge seems to have been reduced to values below to the established limits [28].

2.3.7 Other Compounds

A number of other organic compounds stand out in sewage sludge, either because of their toxicity, their increased presence in recent years or their potential to cause environmental problems. Recent research is leading to the detection and treatment of an increasing number of emerging pollutants, as mentioned above. In turn, the development of industry generates new organic compounds that have to be evaluated, not only for their presence, but also for their potential risk to health or the environment. Examples include numerous drugs and hormones, substances that end up in wastewater and sludge, and are not removed in the conventional treatments. One of the best-known and most recently studied examples is a compound from the family of the endocrine disrupting compounds (EDCs), Bisphenol-A (BPA), whose structure can be seen in Fig. 9.

They are compounds capable of altering the normal operation or function of the endocrine system, causing excess or deficiency in the hormones production. They are found in many daily products such as plastic bottles, detergents, and cosmetics, although recent research is allowing to reduce their use [37].

Other pollutants in water and sewage sludge are microplastics (MPs). The widespread and extensive use of plastics in everyday life and their low biodegradability make these materials important potential pollutants for the environment. In the processing of plastics, as well as in their different uses, smaller plastic particles are generated. These particles, with a size lower than 5 mm, are known as microplastics.

In recent years, much effort has been focused on developing methods to detect and analyze microplastics, as well as to investigate their toxicity and their potential to pollute the environment. Thus, it has been reported their potential to adsorb other micropollutants (PAHs, PCBs or metals), in addition to the toxicity of the MPs themselves, which is still under investigation, but which warns of the risk to humans

Fig. 9 Molecular structure of BPA

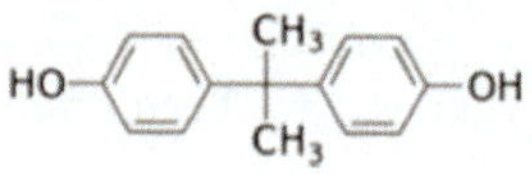

due to their position in the trophic chain. Whether by ingestion, inhalation or contact, MPs are genotoxic compounds and can cause inflammation in the respiratory tract [38, 39].

Both during production and after use, MPs are concentrated in wastewater and wastewater sludge. In WWTP, MPs affect the development of biological nutrient removal treatments, as they reduce the bacterial population responsible for treating both nitrogen and phosphorus in the biological processes.

Evidence of microplastics presence has been found in numerous samples of sewage sludge from multiple WWTPs and may be a source of environmental pollution, especially in the case of agricultural application or composting [40].

2.4 *Influence of the Treatment on OMPs*

The presence of most of the organic micropollutants (OMPs) present in sewage sludge went unnoticed until the 1980s without being considered a risk to health or the environment, but it was not until the 1990s that dioxins, furans, PCBs or halogenated organic compounds began to be detected in various European countries and legislation was introduced.

With the aim of removing OMPs from sludge, numerous studies have been carried out in order to evaluate the sludge treatment method and how it affects the elimination of these compounds.

According to some reports [28, 41, 42], these organic micropollutants can be biodegraded by themselves. Thus, the degradation of dioxins and furans, PCBs, and PAHs has been observed both in the aerobic process and during sludge stabilization by anaerobic digestion. In the case of PCBs, reduction values during methane fermentation up to 40% in mesophilic conditions (35°C) and up to 80% in thermophilic conditions (55°C) have been reported [28]. In other cases, pretreatments prior to anaerobic digestion, such as ultrasounds or thermal hydrolysis, have been performed, reporting the improvement in the degradation of certain OMPs [41, 42].

However, these processes are still subject to research, just as new pollutants belonging to the OMPs family are being reported every year.

2.5 *Toxic Organic Compounds in Sewage Sludge*

In the annually revised U.S. Code of Federal Regulations (CFR), in the 2020 report, named as 40-Environmental Protection, it was presented a list of priority pollutants found in wastewater [43]. Most of them are organic compounds. The concentration values of any of these compounds in sewage sludge, after the review of several studies, have been summarized in Table 2.

Table 2 Priority pollutants in sewage sludge [28, 29, 44, 45]

Priority pollutants	Concentration (mg/kg DM)	Compounds family
Acenaphthene	6.6	PAH
Acenaphthylene	0.3	PAH
Anthracene	44	PAH
Benzidine	12.7	PAH
Benzo(a)anthracene	99	PAH
Benzo(a)pyrene	24.7	PAH
Benzo(ghi)perylene	12.9	PAH
Chrysene	32.4	PAH
Dibenz(a,h)anthracene	13	PAH
Fluoranthene	60	PAH
Fluorene	8.1	PAH
Indeno(1,2,3-cd)pyrene	9.5	PAH
Naphthalene	6.61	PAH
Phenanthrene	44	PAH
Pyrene	37.1	PAH
Arochlor 1,016	75	PCB
Arochlor 1,254	1.96	PCB
Benzene(monochloro)	846	Chlorobenzenes
Benzene(dichloro)	1,650	Chlorobenzenes
Benzene(trichloro)	184	Chlorobenzenes
Benzene(tetrachloro)	184	Chlorobenzenes
2,3,7,8-TCDD	1.7	Dioxin/furan
Dibenzofuran	59.3	Dioxin/furan
BDE-209	1.1	PBDE
Polybrominated biphenyls	431	PBB
Perfluorooctane sulfonate	7.3	PFOS
Perfluorooctanoic acid	0.241	PFOA
Perfluorinated carboxylic acid	3.21	PFCA
Aldrin	16.2	Pesticide
Chlordane	16.04	Pesticide
DDT	564	Pesticide
Dieldrin	64.7	Pesticide
Endosulfans	0.28	Pesticide
Endrin	1.17	Pesticide
Heptachlor	16	Pesticide
Heptachlor epoxide	0.78	Pesticide
Isophorone	0.083	Pesticide
Toxaphene	51	Pesticide
Analine	40.2	Other
Benzene	11.3	Other
Bisphenol-A	32,100	Other
Bis(2-chloroethoxy)methane	0.24	Other

(continued)

Table 2 (continued)

Priority pollutants	Concentration (mg/kg DM)	Compounds family
Bis(2-chloroethyl)ether	0.13	Other
Bis(2-chloroisopropyl)ether	5.7	Other
Chloroethane	24	Other
Chlorophenol congeners	8,490	Other
Chloroform	60	Other
Dichloroethylene	865	Other
Dichloropropane	1,230	Other
Di(2-ethylhexyl)phthalate	130	Other
Ethylbenzene	65.5	Other
Hexachlorobutadiene	8	Other
Hexachloroethane	61.5	Other
N-nitrosodimethylamine	0.053	Other
N-nitrosodiphenylamine	19.7	Other
Nitrophenol congeners	500	Other
Phenol	920	Other
Tetrachloroethane	5	Other
Tetrachloroethylene	50	Other
Trichloroethane	33	Other
Trichloroethylene	125	Other
Toluene	1,180	Other
Toluene(2,4-)dinitro	10	Other

The concentration values are the maximum values reported for these compounds found in the sewage sludge, although lower amounts may be found and do not represent a specific or general pattern.

Acknowledgments This work has been supported by the *Comunidad de Madrid* (Spain) through the Industrial PhD projects (*IND2017/AMB-7720* and *IND2019/AMB-17114*), REMTAVARES Network (*S2018/EMT-4341*) and the European Social Fund. In addition, the authors thank the financial support from the Spanish MICINN through the project CATAD3.0 *PID2020-116478RB-I00.*

References

1. Loizia P, Neofytou N, Zorpas AA (2019) The concept of circular economy strategy in food waste management for the optimization of energy production through anaerobic digestion. Environ Sci Pollut Res 26:14766–14773. https://doi.org/10.1007/s11356-018-3519-4
2. Turovskiy IS, Mathai PK (2005) Sludge quantities and characteristics. In: Wastewater sludge processing. Wiley, pp 30–59
3. Canziani R, Spinosa L (2019) Industrial and municipal sludge: emerging concerns and scope for resource recovery. In: Sludge from wastewater treatment plants. Elsevier

4. Tang K et al (2020) Removal of pharmaceuticals, toxicity and natural fluorescence by ozonation in biologically pre-treated municipal wastewater, in comparison to subsequent polishing biofilm reactors. Water 12:1059–1074. https://doi.org/10.3390/w12041059
5. Metcalf and Eddy (2014) Wastewater engineering treatment and resource recovery.5th edn. McGrawHill
6. Andreoli CV, von Sperling M, Fernandes F (2007) Chapter 2. Sludge characteristics and production. In: Sludge treatment and disposal. IWA Publishing, London, pp 18–44
7. European Commision (2012) Glossary of terms related to urban waste water. https://ec.europa.eu/environment/water/water-urbanwaste/info/glossary_en.htm. Accessed 20 Jan 2021
8. Kacprzak M et al (2017) Sewage sludge disposal strategies for sustainable development. Environ Res 156:39–46
9. European Commission (2001) European Commission DG Environment–B/2. Disposal and recycling routes for sewage sludge – part 3: scientific and technical sub-component report
10. Werther J, Ogada T (1999) Sewage sludge combustion. Prog Energy Combust Sci 25(1):55–116
11. Johnson OA, Affam AC (2019) Petroleum sludge treatment and disposal: a review. Environ Eng Res 24(2):191–201
12. Ali AM, Abu-Hassan MA, Ibrahim RRK, Zaini MAA, Abdulkarim BI, Hussein AS, Su SM, Mohd Halim MAI (2017) Characterization of petroleum sludge from refinery industry biological wastewater treatment unit. Preprints 2017:2017080033. https://doi.org/10.20944/preprints201708.0033.v1
13. Islam B (2015) Petroleum sludge, its treatment and disposal. A review. Int J Chem Sci 13 (4):1584–1602
14. Gadipelly C et al (2014) Pharmaceutical industry wastewater: review of the technologies for water treatment and reuse. Ind Eng Chem Res 53(29):11571–11592
15. Iqbal SA, Mahmud I, Quader AK (2014) Textile sludge management by incineration technique. Procedia Eng 90:686–691. https://doi.org/10.1016/j.proeng.2014.11.795
16. Mani S, Bharagava RN (2018) 3 Textile industry wastewater environmental and health hazards and treatment approaches. CRC Press
17. Lotito AM, de Sanctis M, Di Iaconi C, Bergna G (2014) Textile wastewater treatment: aerobic granular sludge vs activated sludge systems. Water Res 54:337–346. https://doi.org/10.1016/j.watres.2014.01.055
18. Simão L et al (2018) Wastes from pulp and paper mills – a review of generation and recycling alternatives. Cerâmica 64(371):443–453
19. de Azevedo ARG et al (2019) Characterizing the paper industry sludge for environmentally-safe disposal. Waste Manag 95:43–52. https://doi.org/10.1016/j.wasman.2019.06.001
20. Morales AB et al (2016) Agri-food sludge management using different co-composting strategies: study of the added value of the composts obtained. J Clean Prod 121:186–197
21. Pascual JA et al (2018) Characterisation of sludge produced by the agri-food industry and recycling options for its agricultural uses in a typical Mediterranean area, the Segura River basin (Spain). Waste Manag 82:118–128. https://doi.org/10.1016/j.wasman.2018.10.020
22. Seviour R, Nielsen PH (2010) Microbial ecology of activated sludge.1st edn. IWA Publishing, London
23. Hatamoto M et al (2017) Microbial community structure and enumeration of bacillus species in activated sludge. J Water Environ Technol 15(6):233–240
24. Hu M, Wang X, Wen X, Xia Y (2012) Microbial community structures in different wastewater treatment plants as revealed by 454-pyrosequencing analysis. Bioresour Technol 117:72–79. https://doi.org/10.1016/j.biortech.2012.04.061
25. Nascimento AL et al (2018) Sewage sludge microbial structures and relations to their sources, treatments, and chemical attributes. Front Microbiol 9:1–11
26. Ibarbalz FM, Figuerola ELM, Erijman L (2013) Industrial activated sludge exhibit unique bacterial community composition at high taxonomic ranks. Water Res 47(11):3854–3864. https://doi.org/10.1016/j.watres.2013.04.010

27. Meerbergen K et al (2017) Assessing the composition of microbial communities in textile wastewater treatment plants in comparison with municipal wastewater treatment plants. MicrobiologyOpen 6(1):1–13
28. Rosińska A (2019) Traditional contaminants in sludge. In: Industrial and municipal sludge: emerging concerns and scope for resource recovery. Elsevier, pp 425–453
29. I C Consultants Ltd London, Technical University Munich, IRSA Rome, ECA Barcelona (2001) Pollutants in urban waste water and sewage sludge. European Commission. http://scholar.google.com/scholar?hl=en&btnG=Search&q=intitle:Pollutants+in+urban+waste+water+and+sewage+sludge#0
30. EPA (2020b) Polychlorinated biphenyls (PCBs). https://www.epa.gov/pcbs/learn-about-polychlorinated-biphenyls-pcbs#:~:text=PCBs have been demonstrated to system and other health effects. Accessed 6 Nov 2020
31. EPA (2017) Technical fact sheet – polybrominated biphenyls (PBBs). pp 1–5
32. Javier F, Zavala-Díaz de la Serna FJ (2015) Revisión Global de Los Contaminantes Emergentes PBDE y El Caso Particular de México. Rev Int Contam Ambient 31(3):311–320
33. Öberg K, Warman K, Öberg T (2002) Distribution and levels of brominated flame. Chemosphere 48:805–809
34. EPA (2020a) Basic information of PFAS. https://www.epa.gov/pfas/basic-information-pfas#health. Accessed 23 Nov 2020
35. Gómez-Canela C, Barth JA, Lacorte S (2012) Occurrence and fate of perfluorinated compounds in sewage sludge from Spain and Germany. Environ Sci Pollut Res 19(9):4109–4119
36. Sidhu GK et al (2019) Toxicity, monitoring and biodegradation of organophosphate pesticides: a review. Crit Rev Env Sci Tec 49(13):1135–1187. https://doi.org/10.1080/10643389.2019.1565554
37. Hernández-Abreu AB et al (2021) Effective adsorption of the endocrine disruptor compound bisphenol a from water on surface-modified carbon materials. Appl Surf Sci 552:149513
38. Enyoh CE et al (2020) Microplastics exposure routes and toxicity studies to ecosystems : an overview. Environ Anal Health Toxicol 35(1):1–10
39. Zhang Z, Chen Y (2020) Effects of microplastics on wastewater and sewage sludge treatment and their removal : a review. Chem Eng J 382:122955. https://doi.org/10.1016/j.cej.2019.122955
40. Rolsky C, Kelkar V, Driver E, Halden RU (2020) Municipal sewage sludge as a source of microplastics in the environment. Environ Sci Health 14:16–22. https://doi.org/10.1016/j.coesh.2019.12.001
41. Reyes-Contreras C et al (2020) Organic micropollutants in sewage sludge: influence of thermal and ultrasound hydrolysis processes prior to anaerobic stabilization. Environ Technol 41 (11):1358–1365. https://doi.org/10.1080/09593330.2018.1534892
42. Taboada-Santos A et al (2019) Thermal hydrolysis of sewage sludge partially removes organic micropollutants but does not enhance their anaerobic biotransformation. Sci Total Environ 690:534–542. https://doi.org/10.1016/j.scitotenv.2019.06.492
43. USEPA (2020) CFR 40, 2020. https://www.govinfo.gov/content/pkg/CFR-2020-title40-vol31/pdf/CFR-2020-title40-vol31.pdf. Accessed 5 Jul 2021
44. Harrison EZ, Oakes SR, Hysell M, Hay A (2006) Organic chemicals in sewage sludges. Sci Total Environ 367(2–3):481–497
45. Wild SR, Jones KC (1991) Organic contaminants in wastewaters and sewage sludges: transfer to the environment following disposal. Environ Manag Ser (EMISS) 19:133–158

Identifying Emerging Pollutants Using Non-target or Wide-Screening Liquid Chromatography-Mass Spectrometry

Dyana Vitale, Vicente Andreu, and Yolanda Picó

Contents

Abstract In this chapter we review the available data on the application of non-targeted high-resolution mass spectrometry (HRMS) analysis to identify a wide range of substances (mainly emerging pollutants) present in sludge. We focus on non-targeted workflows developed with these instruments: (1) the screening of suspects or broad screening and (2) the analysis of unknown compounds providing a fingerprint of almost all chemical contaminants present in sludge. The different workflows as well as their achievements and pitfalls within the identification of chemicals in sludge are discussed in depth. The pre-treatment of sludge (decisive for the detection of compounds) is also discussed since it cannot be avoided due to the complexity of the sample.

Keywords Emerging pollutants, High-resolution mass spectrometry, Non-target analysis, Suspected screening

D. Vitale, V. Andreu, and Y. Picó (✉)
Food and Environmental Research Group of the University of Valencia (SAMA-UV), Desertification Research Centre, CIDE, CSIC-GV-UV, Valencia, Spain
e-mail: yolanda.pico@uv.es

Avelino Núñez-Delgado and Manuel Arias-Estévez (eds.),
Emerging Pollutants in Sewage Sludge and Soils,
Hdb Env Chem (2023) 114: 209–224, DOI 10.1007/698_2022_856,

Published online: 16 April 2022

1 Introduction

The presence of pollutants in sludge is of concern, mostly because they are increasingly used as agricultural soil amendments to enhance crop yields [1–3]. According to data from the Spanish National Sludge Registry [4], and as in the other European countries, agricultural use has prevailed as the final destination for sludge (approximately 80% of the amount generated). Landfilling has been greatly reduced (approximately 8% at present), and incineration is growing (around 4%). Another destination of lesser quantitative importance is the use of sludge on non-agricultural land [5]. From a technical point of view, sludge can accumulate many organic pollutants that pass from sludge to soil and from soil to crop and can ultimately affect humans, and therefore it is important to know their pollution profile [6–9].

Recently, the introduction of high-resolution mass spectrometry (HRMS) has forced to evolve the analytical schemes from target analysis of "a priori" selected compounds to the detection of any known or unknown compound that may be present in the samples [10]. Although still in its infancy, the few existing examples show great potential for characterizing sludge contamination.

2 Non-target Approaches to Determine Contaminants and Other Chemicals Retained in Sludge

Table 1 shows a summary of the most representative methods published based on non-target analysis of sludge or soil amended sludge. The methods follow the three basic steps of any analytical procedure: (1) extraction, (2) clean-up, and (3) determination. Pérez-Lemus et al. [19] reviewed sample preparation and instrumental methods proposed in the last few years (2012–2018) to assess PPCPs in sewage sludge mostly by classical target approaches. Martín-Pozo et al. [8] also presented an overview focused on the research published for the determination of emerging pollutants (Eps) in sewage sludge. Instrumental and sample preparation methods for detection and quantification of pollutants are reviewed, with an emphasis on the sample treatment techniques. Methods summarized in Table 1 are within the general schemes reported for target determinations in both reviews.

2.1 *Extraction and Clean-Up*

The determination of EPs requires extraction and clean-up. The extraction is commonly based on solid-liquid extraction (SLE) and the clean-up in solid phase extraction (SPE). Most of these methods are developed and optimized to extract as many contaminants as possible. In one of the most widely used, the sludge is extracted with mixtures of equal parts of methanol and acetonitrile after adjusting

Table 1 Analytical methods to identify compounds in sewage sludge using non-target high-resolution mass spectrometry (HR-MS)

Analytes	Extraction	Clean-up	Determination	Reference
EAC	*SLE*: • Neutral (pH 5–7) • Acid (pH 2) MeOH:AcN (1:1) + EDTA-Na_2	Agilent BondElut Plexa SPE and elution with H_2O:MeOH (9:1)	LC-QqTOF ESI ± FWHM 30,000 DIA (MS^E) DDA (list of *m/z*)	[11]
Diverse organic compounds	*SLE*: • Neutral (pH 5–7) • Acid (pH 2) MeOH:AcN (1:1) + EDTA-Na_2	Agilent BondElut Plexa SPE and elution with H_2O:MeOH (9:1)	LC-QqTOF ESI ± + 30,000 FWHM *m/z* 50–1,000 DIA (MS^E)	[12]
EPs and their TPs	*SLE-USE*: MeOH–Milli-Q water (pH 2.5, FA 0.5% and 0.1% EDTA)	No clean-up	LC-QqTOF-MS ESI ± + 30,000 FWHM *m/z* 50–1,000 DIA (MS/MS)	[13]
EPs and their TPs	*SLE-USE*: MeOH:water (50:50) 0.5% FA	No clean-up	LC-QqTOF-MS ESI ± + 30,000 FWHM *m/z* 50–1,000 DIA (MS^E)	[14]
Pharmaceuticals	*QuEChERS*: water +1% AA in AcN + $MgSO_4$ & NaOAc	*d-SPE*: $MgSO_4$, PSA & C_{18}	LC-QTOF-MS ESI + *m/z* 50–1,000 30,000 FWHM DDA (intensity)	[15]
Micropollutants	*MSPD*: homogenized with C_{18} in a mortar, placed in a syringe and eluted with MeOH	In syringe clean-up with diatomaceous earth	LC-QTOF-MS ESI + *m/z* 50–1,000 10,000 FWHM DIA (MS^E) DDA (intensity)	[16]
POPs	*PLE*: Dionex ASE 300 ethyl acetate and acetone 70:30	In-cell either Florisil or neutral alumina	LC-Q orbitrap or GC-MS/MS	[17]

(continued)

Table 1 (continued)

Analytes	Extraction	Clean-up	Determination	Reference
	(% v/v), 3 cycles of 5 min at 100°C with 60% flush		ESI ± m/z 50–1,000 MS: 60000–1,000,000 FWHM MS/MS:7500 FWHM DDA (inclusion list of m/z)	
Non-target organic pollution	*PLE*: Dionex ASE 300 NH_4^+solution (1 mM, pH 9) & MeOH, 5 cycles of 5 min at 80°C with 60% flush	SPE in HLB cartridge and analytes eluted with MeOH	LC-Q-Orbitrap nESI + m/z 75–1,000 MS: 240000 FWHM MS/MS: 15000 FWHM DDA (intensity, inclusion list of m/z)	[18]

AA acetic acid, *AcN* Acetonitrile, *DDA* data-dependent analysis, *DIA* data independent analysis, *dSPE* dispersive solid-phase extraction, *EAC* Endocrine Active Compounds, *EDTA-Na2* ethylene diamino tetra acetic disodium salt, *EPs* emerging pollutants, *ESI* electrospray, *FA* Formic acid, *FWHM* full width at half maximum, *HLB* hydrophilic lipophilic balance, *LC-QqTOF-MS* liquid chromatography-quadrupole time of light mass spectrometry, *MeOH* methanol, *MS* mass spectrometry, *MSPD* matrix solid phase dispersion, *MS/MS* tandem mass spectrometry, *NaOAc* sodium acetate, *nESI* nanoelectrospray, *PLE* pressurized liquid extraction, *POPs* Persistent Organic Pollutants, *PSA* primary-secondary amine, *SLE* solid-liquid extraction, *SLE-USE* SLE-ultrasound assisted extraction, *SPE* solid phase extraction, *TPs* transformation products

the sample to different pH values (acidic and neutral). This allows a wide range of compounds to be extracted. The extracts at different pH values are mixed and purified through an SPE cartridge with state-of-the-art polymeric material [11, 12]. SLE is a process that requires energy, which is normally supplied as mechanical agitation. However, the application of ultrasound (ultrasound-assisted extraction or USA) improves the extraction of contaminants compared to other matrix compounds. Hence, Boix et al. [14] proposed a protocol to extract EPs and transformation products (TPs) with an acidified water + methanol mixture without any purification process. Similarly, Aalizadeh et al. [13] extracted these compounds using a mixture of methanol and water at pH 2.5 with ethylenediamine tetraacetic acid (EDTA) to complex the interfering metals and neither applied any further clean-up process.

As reported in the reviews that cover target analysis of sludge [1, 2], QuEChERS is the most widely extraction scheme. Martínez-Piernas et al. [15] applied it to the non-target analysis of sludge using the AOAC version of the method that provides an acid pH. This is a further indication that the acid pH favors extraction. To apply QuEChERS to sludge, it is necessary to add water as in most cases dehydrated sludge is used. In this method a dispersive SPE purification (dSPE) is applied in which, in addition to a secondary primary amine (PSA), C_{18} is used to retain non-polar compounds. Matrix solid phase dispersion (MSPD) was also proposed [16] using C_{18} as dispersant and soft in-cell clean-up with diatomaceous earth. The efficiency of MSPD extraction, corresponding to three different sludge samples, varied between 69 and 100%. Other alternative extraction method is SLE by pressurized liquid extraction (PLE) (ethyl acetate, dichloromethane, or aqueous mixtures). The sample homogenized with some adsorbents is extracted at high pressure and temperature with an organic solvent. PLE could be followed by an SPE clean-up using hydrophilic/lipophilic balance (HLB) sorbents that eliminate salts and other interferences [18]. However, SPE can be replaced by an in-cell cleanup technique employing either Florisil or neutral alumina as a sorbing phase [17].

Interestingly, and even though the study was not performed by HRMS, García Valverde et al. [20] compared three different extraction methods based on ultrasonic cylindrical probe (UAE), PLE, and QuEChERS for the simultaneous determination of 30 EPs (including pesticides, pharmaceutical products, and some of the main TPs) in agricultural soils. The QuEChERS method showed the best performance (recoveries between 29 and 99%), followed by the UAE (between 12 and 101%) and finally the PLE showed the worst results (between 9 and 89%). In none of the tested methods, 4-amino-antipyrine, 4-methylamino-antipyrine, ciprofloxacin and ofloxacin, compounds with a water solubility higher than 25×10^3 mg/L, moderate acidity ($4 \leq pKa \geq 6$), and very low polarity (Log Kow ≤ 1) could be extracted properly. An operational scheme of the three compared procedures is outlined in Fig. 1, which pointed out how these protocols are not simple, since several steps requiring trained personnel are required.

2.2 High Resolution Mass Spectrometry

HRMS can profile organic compounds in sludge and in soil amended with them without too many limitations, providing much more information that using only target analysis. HRMS can be combined with both, gas and liquid chromatography (GC or LC) to separate compounds. However, the literature search showed that nowadays the focus is on EPs. Hence, reported studies apply LC or ultra-high performance liquid chromatography (UHPLC). This does not have too much sense because sludges are considered able to accumulate non-polar compounds (log Know > 3), which sometimes are more detectable by GC. In any case, there are many contaminants that were identified using LC. HRMS offers three different working modes: target, suspected screening, and non-target (or unknown identification), which are schematized in Fig. 2 (complexity increase from the right to the left). In any case, one important aspect of HRMS is that at least in the MS spectrum the

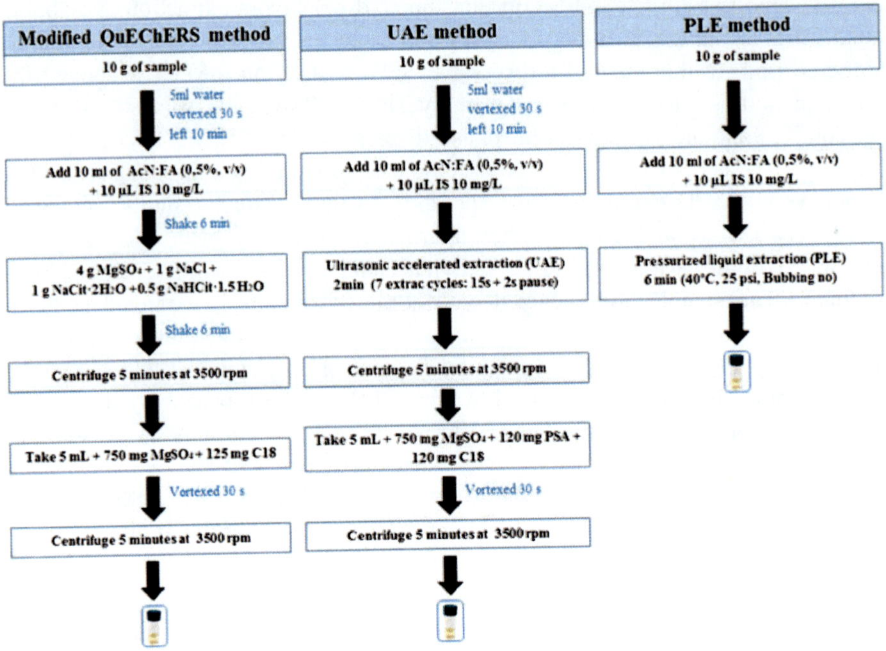

Fig. 1 Diagram of three procedures used for EPs soil sample extraction. Reproduced from García Valverde et al. [20] with permission of Elsevier

whole mass range is recorded and then, suspected and non-target could be performed retrospectively.

The potential of HRMS is such that prior chromatographic separation is considered less important. Reversed-phase liquid chromatography (RPLC) with UHPLC columns (particle size <1.7 μm) is the most common set-up without much further complication. However, for the identification of compounds, target (by suspected screening) or non-target (by HRMS) in addition to mass accuracy, isotopic pattern and MS/MS spectra, the evaluation of retention time (t_R) is also very important. Aalizadeh et al. [13] developed three robust t_R prediction models based on many EPs for the two chromatographic systems most widely reported: reversed-phase liquid chromatography (RPLC), and hydrophilic interaction liquid chromatography (HILIC), in positive and negative electrospray (ESI) ionization. The application of these models with suspected and non-target screening facilitates identification of new EPs by suspect and non-target HRMS screening.

Nowadays, non-target screening analysis requires the use of HRMS either based on the use of time-of-flight (TOF) or Orbitrap mass analyzers. Both are able to provide accurate mass spectra within a wide *m/z* range, throughout the entire chromatographic run [16]. Using low fragmentation energy that provides as main *m/z* the protonated or deprotonated molecule, the accurate mass spectrometers can calculate a list of most probable empirical formulas according to the mass error (must be lower than 5 ppm). Thanks to this information, the search of new pollutants in

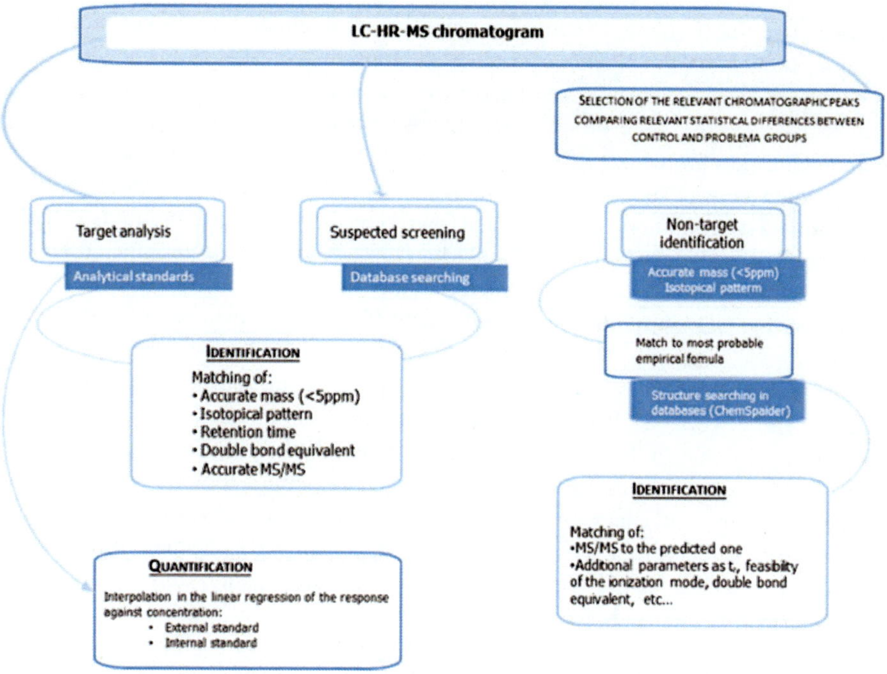

Fig. 2 Outline of the different working modes with HR-MS (Reproduced with permission from Andrés-Costa et al. [21]

complex matrices such as sludge is possible. Furthermore, structural elucidation is also possible because the accurate mass full-spectrum acquisition data are provided, both in MS and MS/MS modes [22]. The further achievement of the MS at higher fragmentation energies (MS^E) or MS/MS fragmentation provides the most important fragments of the molecule facilitating structural elucidation or confirmation.

This information can be obtained by data independent analysis (DIA) or data-dependent analysis (DDA). DIA used in broadband collision induced dissociation (bb-CID), which gives two mass spectra at low and high collision energies (CEs), so that the low CE provides the molecular ion, and the high CE the most characteristic fragments [13]. Other instruments perform DIA applying MS/MS due to the introduction in the collision cell of narrow m/z ranges through the entire spectra. In DDA, a preselected property in the MS triggers the acquisition of the MS/MS. One property frequently selected is m/z ions intensity. Once ions intensity overpass a threshold, the instrument provided the MS/MS of a limited number of the highest intensity ions (commonly, between 10 and 20 can be isolated and fragmented for each cycle) without preselection by the analyst [15]. Another way to perform DDA is using an inclusion (preselected) mass list containing the exact masses of the precursor ion of the parent compound and plausible TPs with an intensity threshold of 1,000 counts [13]. A combination of intensity trigger plus an inclusion list is also feasible [18]. Furthermore, to avoid the repetitive fragmentation of background ions dynamic exclusion or exclusion lists can also be used.

Castro et al. [16], during preliminary method development, demonstrated that both DDA and DIA were prone to report false negatives, particularly when compounds elute in chromatographic regions with high number of molecular features and/or when they do not lead to formation of pseudo-molecular ions during ESI ionization. These authors also reported that, for the selected conditions, DDA performs slightly better than DIA. Similarly, Black et al. [11] also compared DDA (using list of exact mass targets and retention times) and DIA (by MS^E) and concluded that DDA provided best quality MS/MS spectra to facilitate identification of non-target compounds.

The data processing is also an important aspect. So far, suspect screening has been the more well-established and successful approach to detect emerging compounds, mainly due to the different suspect lists available nowadays [17]. Table 2 reports a summary of the different database applied to identify compounds in sewage sludge. Although advances in data treatment have been amazing, the identity confirmation is very time consuming and sometimes only a small fraction of compounds screened can be identified. A thorough investigation of TP fragmentation patterns together with a comparison with spectral libraries and literature evidence were decisive for the structural assignment and classification of unknown compounds [15].

A step further is based on non-target analysis approaches (also called non-target screening), in theory starting without any preconception for the chemicals present in a sample, nor suspect lists. Although in practice, the choice of extraction method and analytical platform limits the chemical space covered. The data analysis is based only on chromatographic or spectrometric peaks. With these, the experimental designs play an essential role in the explorative interpretation of the results. Detected compounds (defined by their t_R and m/z ratio) can be sorted and filtered based on intensity, treatments, locations or time-points to assess their abundance in a sample, the effect of a processes, identify markers or highlight chemical classes sharing similarity [18].

2.3 *Compounds and TPs Identified in Sludge by Non-target Techniques*

The studies that apply non-target screening using HRMS in sludge or soil amended with sludge are still scarce. They are mainly linked to the profiling on contaminants and their transformation products in the sludge. In this sense, Aalizadeh et al. [13] performed the suspect and non-target screening of TPs of three EPs (tramadol, furosemide, and niflumic acid). Ten new TPs were tentatively identified by the combination of the newly developed t_R models and in silico fragmentation models and the results proved the value of t_R prediction for newly identified TPs where the

Table 2 Different database and setups to identify compounds in sludge

Type of analysis	Type of compounds	Database	Reference
Target	Pesticides, pharmaceuticals, organophosphorus flame retardants, musk fragrances, and UV stabilizers	Eawag database	[17]
Suspected screening	Chemical compounds	Additional database with information extracted from several lists of chemicals	[17]
Suspected screening	1,684 pesticides and metabolites (770 with MS/MS spectra), 8,998 forensic toxicants (3,497 with MS/MS spectra), and 1,451 water contaminants (1,083 with MS/MS spectra)	Agilent pesticide database Forensic toxicants and water contaminants personal compound database libraries (PCDLs)	[11]
Suspected screening	Intensity threshold of 1,000 cps, an S/N ratio > 10, a tolerance of 5 ppm mass accuracy and an isotope ratio difference (IRD) $<10\%$	Two spectral libraries: The internal library: All-in one HRMS (Sciex) and the open-access database of mass spectra MassBank Europe (MassBank Europe)	[15]
Non-target	For non-identified compounds in library formula finder verifying formula alignment	In silico fragmentation tool ChemSpider database (ChemSpider) (integrated into the Sciex OS software) was checked to enhance spectra interpretation	[15]
Suspected screening	Spectral peaks were picked, retention time aligned and grouped by isotopologues and adduct peaks into features (called compounds inside compound discoverer)	Several datasets accessed through ChemSpider following NORMAN suspected list	[18]
Suspected screening	273 biocides & pesticides alongside their chemical identifiers, predicted tR and three most common and abundant MS/MS fragments from spectra libraries	Screening homemade database with a complete list of biocides and pesticides (active ingredients), was compiled from regulatory databases	[13]
Non-target screening	Use of characteristic fragmentation (i.e., fragmentation pattern) during data-dependent MS/MS fragmentation events	Background subtraction and peak picking were carried out using metabolite detect (metabolite tools 2.0, Bruker Daltonics, Bremen, Germany): Subtraction algorithm eXpose mode, delta time ≤ 0.1 min, delta mass ≤ 0.05 m/z and ratio 5	[13]
Suspected screening	9,000 compounds alongside chemicals identifiers, predicted tR and MS/MS	Forensic toxicants compound database libraries (PCDLs)	[16]
Suspected screening	Database includes compound name and elemental composition of the parent compounds, fragment ions and tR	An in-house database of more than 600 compounds containing many pharmaceuticals, illicit drugs and pesticides	[14]

reference standards were difficult or impossible to obtain. This workflow was also used to identify 28 biocides in sewage sludge. Two new quaternary ammonium compounds were also tentatively identified via non-target screening strategy. Castro et al. [16] identified a group of 68 micropollutants in sludge of different sludge treatment plants following the DDA approach. Some of them are reported in this compartment for the first time. Semi-quantitative concentration data were reported for a group of 37 pollutants in samples obtained from 16 of these treatment plants. Out of them, ten pharmaceuticals showed detection frequencies >50% and median sludge residues >100 ng g^{-1}. Some of these compounds (*O*-desmethylvenlafaxine and the antimycotic drugs miconazole, clotrimazole, and ketoconazole) were identified as concerning species due to its potential toxicity to biota.

Interestingly and following with the combination of toxicity and substance identification, Black et al. [11] combined non-targeted analysis using HRMS with predictive estrogenic activity modeling on sewage sludge samples. Diisobutyl phthalate and dextrorphan were predicted to exhibit estrogenic activity and identified in >75% of sludge samples, signifying their universal presence and persistence. Ten non-target features were identified (metoprolol, fenofibric acid, erythrohydrobupropion, oleic acid, mestranol, 4′-chlorobiphenyl-2,3-diol, medrysone, scillarenin, sudan I, and N,O-didesmethyltramadol) in treatment set samples and are considered to have influenced the in vitro estrogenic activity observed. The same team in other study analyzed a broader spectrum of endocrine active organic contaminants in sewage sludge being able to discover 118 compounds including hormones, pharmaceuticals, phosphate flame retardants, recreational drugs, antimicrobials, and pesticides. Additionally, 22 of these identified compounds are predicted to interfere with estrogen receptors [11].

An important facet of the study of contaminants in sludge, already pointed out, is to identify most abundant TPs in sludge and to establish their prevalence in the sludge along the time. Covering this aspect, Boix et al. [14] investigated degradation of selected EPs in the aqueous and solid phases of sewage sludge after anaerobic digestion using two different digesters: mesophilic and thermophilic. Initially, sludge samples were screened by UHPLC-QqTOF-MS for identification of EPs in the samples. In the sludge samples thirteen compounds, namely 4-aminoantipyrine (4-AA), 4-acetyl aminoantipyrine (4-AAA), 4-formyl aminoantipyrine (4-FAA), Acesulfame, Benzoylecgonine (BE), Carbamazepine, Diclofenac, Fenofibric Acid, Irbesartan, Salicylic Acid, Thiabendazole, Valsartan (see confirmation in Fig. 3) and Venlafaxine were identified. These compounds were subsequently selected as target analytes and analyzed by a quantitative method based on HPLC-MS/MS QqQ to monitor their behavior, and to evaluate their degradation efficiency and distribution in the two phases of sewage sludge [22]. This is an illustrative example of how HRMS can help to identify most suitable targets that were not commonly detected because they were not selected “a priori” in the methods.

Other selected studies focused on the identification of these EPs and TPs in soil treated with the sludge. These studies hypothesize that while soil amendment with sludge can maintain the fertility of arable soil without the use of synthetic fertilizers, it is necessary to investigate the unintended consequences of contaminants in the

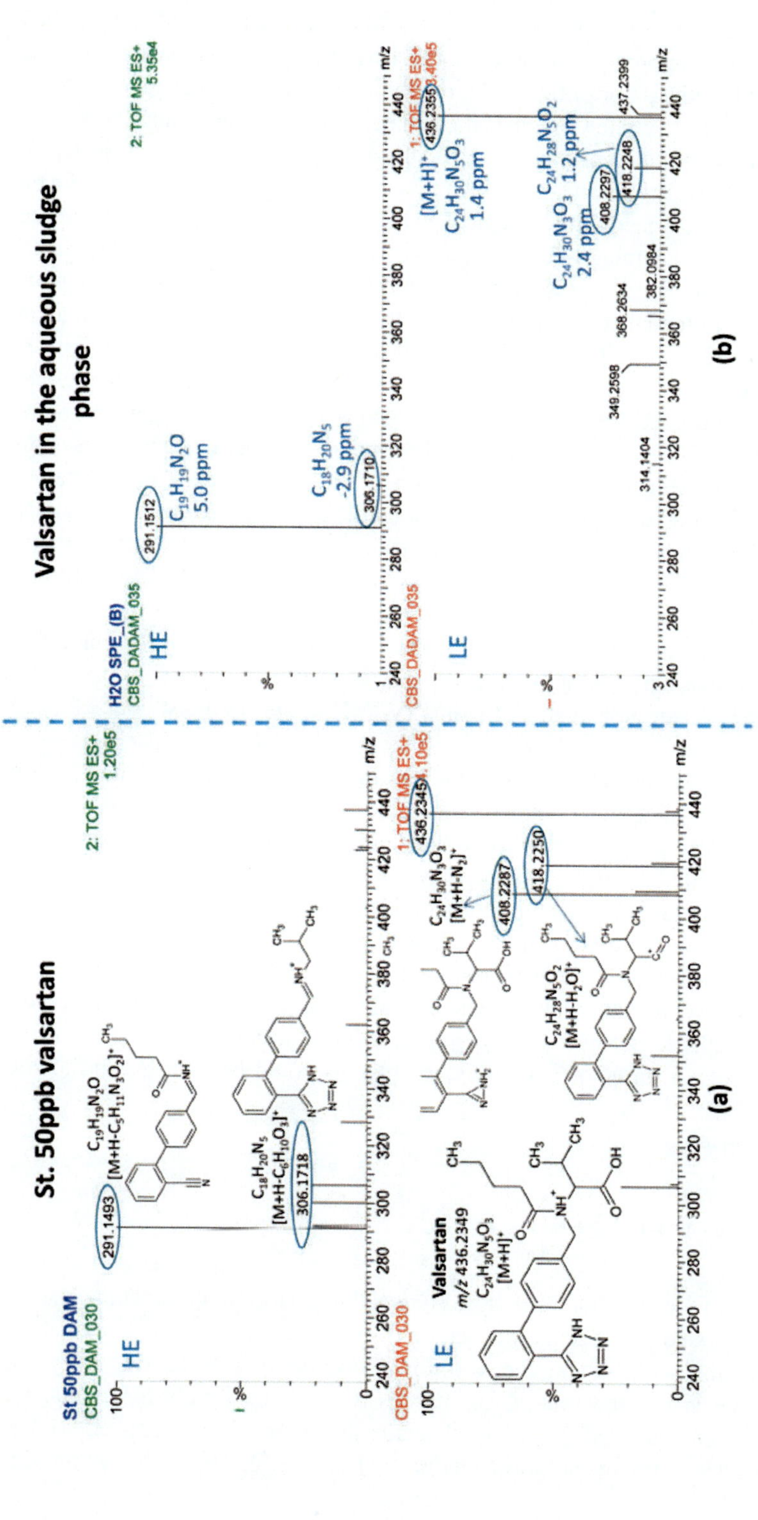

Fig. 3 Valsartan detection and confirmation by UHPLC-QTOF MS in the standard at 50 μg/L (**a**) and aqueous sludge phase (**b**). Low energy (LE) and high energy (HE) spectra showed the protonated molecule and four fragment ions. Possible structures are proposed for each fragment ion detected. Reproduced from Boix et al. [14] with permission from Elsevier

sludge passing into the soil. Within these studies, Martínez-Piernas et al. [15] applied a suspect screening approach for the comprehensive investigation of 262 potential TPs, associated with 20 prioritized pharmaceuticals found in real tomato crops exposed to long-term wastewater irrigation. The occurrence and fate of the TPs was evaluated by the retrospective analysis of (among other matrices) soil. Up to 18 TPs were tentatively identified, of which two were not previously reported. Seven TPs were finally confirmed with analytical standards. These results pointed out the transfer of these contaminants from irrigation water (and also from sludge based amendments) to soil, and remarks the need to control the release of these compounds as well as their potential to be uptake and translocated in crops used for human consumption.

In the same way, Chiaia-Hernández et al. [17] extended a GC-MS/MS target analysis with a suspect screening by LC-HRMS/MS of >500 halogenated compounds obtained from a Swiss database that includes industrial and household chemicals identified. In total, the confirmation of 96 compounds with an overlap of 34 in soil and sediment was achieved. The identified compounds consist generally of esters, tertiary amines, trifluoromethyls, organophosphates, azoles and aromatic azines, with azoles and triazines being the most common groups. Newly identified compounds include transformation products, pharmaceuticals such as the flukicide niclofolan, the antimicrobial cloflucarban, and the fungicide mandipropamid. The results indicate that agricultural and urban soils as well as sediments impacted by agriculture and wastewater treatment plants (WWTPs) are the most contaminated sites.

Interestingly, Gravert et al. [18] studied an experimental agricultural plot, amended along the time with synthetic inorganic fertilizers, human urine, manure, or wastewater treatment sludge at very high rates. The EPs can be separated in groups according to the type of treatment by principal component- and differential-analysis. From cattle manure, natural compounds such as bile acids and steroids were found. Human urine led to pollution with common pharmaceuticals such as metoprolol and propranolol. The highest number was added by wastewater treated sludge, with 25 significant contaminants, spanning blood pressure regulators, anti-depressants, synthetic steroids, and sleep medication. Furthermore, using Kendrick mass defect plots, a series of polypropylene glycols could be revealed in the soil (Fig. 4).

3 Conclusions and Future Trends

The chapter put together data from those studies that perform non-target analysis of contaminants in sludge and soil amended with sludge, considering both, extraction and determination, and showing the workflows successfully applied to identify EPs and TPs by suspected and non-target analysis. One initial conclusion is that the number of studies reported is still very scarce and more studies are required to achieve full knowledge of the pollutants present in sludge. These studies provide

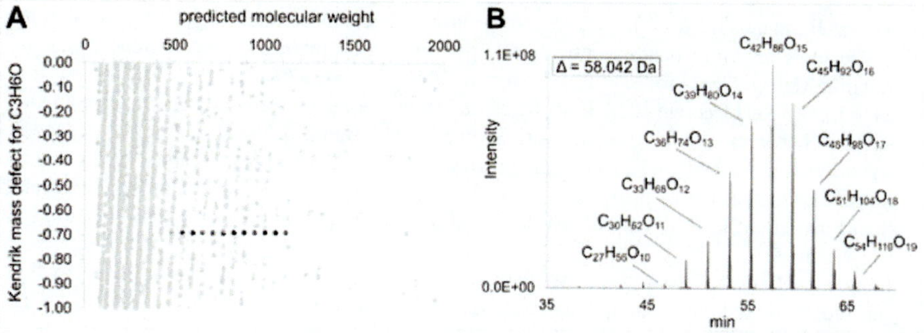

Fig. 4 Kendrick mass defect plot (**a**) used to identify polypropylene glycols in amended soil. PPG share a mass defect of −0.69 marked with black dots. Extracted ion chromatograms (**b**) display equally spaced PPG peaks with the highest intensity at PPG n14 (57.43 min). Reproduced with permission [18]

very important information about the compounds that are present in these sludge materials, their accumulation and their degradation. These dehydrated sludges are used as organic amendments in soils, and these pollutants can be transferred from sludge to soil and crops with the implicit possibility to affect human beings, ultimately. This chapter also pointed out how to apply these techniques and, even though there is still much work to be done, it shows that our understanding of the substances accumulated in sludge has increased exponentially in the last years.

However, most works focused on the identification of relatively polar EPs that are amenable by LC or UHPLC, while there are no studies applying GC. This situation really makes little sense because the non-polar pollutants have the greatest affinity for sludge. The authors believe that in a near future studies on non-target analysis performed by GC will be conducted to complete the profile of contaminants in the sludge.

Acknowledgements Funding: The research received funding from the Spanish Ministry of Science, Innovation and Universities and the European Regional Development Fund through the project CICLIC (RTI2018-097158-B-C31) and from the Generalitat Valenciana through the project ANTROPOCEN@ (PROMETEO/2018/155).

References

1. Biel-Maeso M, Corada-Fernández C, Lara-Martín PA (2019) Removal of personal care products (PCPs) in wastewater and sludge treatment and their occurrence in receiving soils. Water Res 150:129–139. https://doi.org/10.1016/j.watres.2018.11.045
2. Buta M, Hubeny J, Zieliński W, Harnisz M, Korzeniewska E (2021) Sewage sludge in agriculture – the effects of selected chemical pollutants and emerging genetic resistance determinants on the quality of soil and crops – a review. Ecotoxicol Environ Saf 214:112070. https://doi.org/10.1016/j.ecoenv.2021.112070

3. Malvar JL, Santos JL, Martín J, Aparicio I, Alonso E (2021) Occurrence of the main metabolites of the most recurrent pharmaceuticals and personal care products in Mediterranean soils. J Environ Manag 278:111584. https://doi.org/10.1016/j.jenvman.2020.111584
4. BOE (1990) Real Decreto 1310/1990, de 29 de octubre, por el que se regula la utilización de los lodos de depuración en el sector agrario. Span State Off Bull 262:2
5. Vazquez-Roig P, Andreu V, Onghena M, Blasco C, Picó Y (2011) Assessment of the occurrence and distribution of pharmaceuticals in a Mediterranean wetland (L'Albufera, Valencia, Spain) by LC-MS/MS. Anal Bioanal Chem 400:1287–1301. https://doi.org/10.1007/s00216-011-4826-5
6. Campo J, Masiá A, Picó Y, Farré M, Barceló D (2014) Distribution and fate of perfluoroalkyl substances in Mediterranean Spanish sewage treatment plants. Sci Total Environ 472:912–922. https://doi.org/10.1016/j.scitotenv.2013.11.056
7. Llorca M, Farré M, Picó Y, Barceló D (2011) Analysis of perfluorinated compounds in sewage sludge by pressurized solvent extraction followed by liquid chromatography-mass spectrometry. J Chromatogr A 1218:4840–4846. https://doi.org/10.1016/j.chroma.2011.01.085
8. Martín-Pozo L, de Alarcón-Gómez B, Rodríguez-Gómez R, García-Córcoles MT, Çipa M, Zafra-Gómez A (2019) Analytical methods for the determination of emerging contaminants in sewage sludge samples. A review. Talanta 192:508–533. https://doi.org/10.1016/j.talanta.2018.09.056
9. Silva S, Rodrigues JA, Coelho MR, Martins A, Cardoso E, Cardoso VV et al (2021) Occurrence of pharmaceutical active compounds in sewage sludge from two urban wastewater treatment plants and their potential behaviour in agricultural soils. Environ Sci Water Res Technol 7:969–982. https://doi.org/10.1039/d1ew00132a
10. Picó Y, Barceló D (2015) Transformation products of emerging contaminants in the environment and high-resolution mass spectrometry: a new horizon. Anal Bioanal Chem 407:6257–6273. https://doi.org/10.1007/s00216-015-8739-6
11. Black GP, Anumol T, Young T, M. (2019) Analyzing a broader spectrum of endocrine active organic contaminants in sewage sludge with high resolution LC-QTOF-MS suspect screening and QSAR toxicity prediction. Environ Sci: Processes Impacts 21:1099–1114. https://doi.org/10.1039/C9EM00144A
12. Black GP, He G, Denison MS, Young TM (2021) Using estrogenic activity and nontargeted chemical analysis to identify contaminants in sewage sludge. Environ Sci Technol 55:6729–6739. https://doi.org/10.1021/acs.est.0c07846
13. Aalizadeh R, Nika MC, Thomaidis NS (2019) Development and application of retention time prediction models in the suspect and non-target screening of emerging contaminants. J Hazard Mater 363:277–285. https://doi.org/10.1016/j.jhazmat.2018.09.047
14. Boix C, Ibáñez M, Fabregat-Safont D, Morales E, Pastor L, Sancho JV et al (2016) Behaviour of emerging contaminants in sewage sludge after anaerobic digestion. Chemosphere 163:296–304. https://doi.org/10.1016/j.chemosphere.2016.07.098
15. Martínez-Piernas AB, Plaza-Bolaños P, Agüera A (2021) Assessment of the presence of transformation products of pharmaceuticals in agricultural environments irrigated with reclaimed water by wide-scope LC-QTOF-MS suspect screening. J Hazard Mater 412:125080. https://doi.org/10.1016/j.jhazmat.2021.125080
16. Castro G, Ramil M, Cela R, Rodríguez I (2021) Identification and determination of emerging pollutants in sewage sludge driven by UPLC-QTOF-MS data mining. Sci Total Environ 778: 146256. https://doi.org/10.1016/j.scitotenv.2021.146256
17. Chiaia-Hernández AC, Scheringer M, Müller A, Stieger G, Wächter D, Keller A et al (2020) Target and suspect screening analysis reveals persistent emerging organic contaminants in soils and sediments. Sci Total Environ 740:140181. https://doi.org/10.1016/j.scitotenv.2020.140181
18. Gravert TKO, Vuaille J, Magid J, Hansen M (2021) Non-target analysis of organic waste amended agricultural soils: characterization of added organic pollution. Chemosphere 280:130582. https://doi.org/10.1016/j.chemosphere.2021.130582

19. Pérez-Lemus N, López-Serna R, Pérez-Elvira SI, Barrado E (2019) Analytical methodologies for the determination of pharmaceuticals and personal care products (PPCPs) in sewage sludge: a critical review. Anal Chim Acta 1083:19–40. https://doi.org/10.1016/j.aca.2019.06.044
20. García Valverde M, Martínez Bueno MJ, Gómez-Ramos MM, Aguilera A, Gil García MD, Fernández-Alba AR (2021) Determination study of contaminants of emerging concern at trace levels in agricultural soil. A pilot study. Sci Total Environ 782:146759. https://doi.org/10.1016/j.scitotenv.2021.146759
21. Andrés-Costa MJ, Andreu V, Picó Y (2017) Liquid chromatography–mass spectrometry as a tool for wastewater-based epidemiology: assessing new psychoactive substances and other human biomarkers. TrAC Trends Anal Chem 94:21–38. https://doi.org/10.1016/j.trac.2017.06.012
22. Boix C, Ibáñez M, Fabregat-Safont D, Morales E, Pastor L, Sancho JV et al (2016) Analytical methodologies based on LC-MS/MS for monitoring selected emerging compounds in liquid and solid phases of the sewage sludge. MethodsX 3:333–342. https://doi.org/10.1016/j.mex.2016.04.010

Fate of Emerging Pollutants During Anaerobic Digestion of Sewage Sludge

Lorena Gonzalez-Gil, Marta Carballa, and Juan M. Lema

Contents

Abstract Organic micropollutants (OMPs) enter municipal wastewater treatment plants (WWTP) and tend to accumulate in sewage sludge, regardless of their sorption properties. Using sewage sludge as a soil fertilizer is a milestone within a circular economy view, although the presence of these OMPs entails potential risks that should be minimized, even if there are no legal restrictions regarding the presence of OMPs in sludge yet. OMPs biotransformation efficiency during Anaerobic Digestion (AD) depends upon the compound characteristics, but it appears quite different on each WWTP. The reasons behind these variabilities are not clear since typical operational parameters of AD seem to exert a minor effect on biotransformation. To overcome the limitations of this process, it is crucial to firstly understand the occurrence of OMPs in sewage sludge, secondly, evaluate the factors and

L. Gonzalez-Gil
Defence University Centre, Spanish Naval Academy, Marín, Spain

M. Carballa and J. M. Lema (✉)
CRETUS, Department of Chemical Engineering, Universidade de Santiago de Compostela, Santiago de Compostela, Spain
e-mail: juan.lema@usc.es

Avelino Núñez-Delgado and Manuel Arias-Estévez (eds.),
Emerging Pollutants in Sewage Sludge and Soils,
Hdb Env Chem (2023) 114: 225–242, DOI 10.1007/698_2022_857,

Published online: 16 April 2022

mechanisms involved in the fate of OMPs, and finally find an accurate model to predict the behaviour of OMPs during AD. These are the main issues examined in this chapter.

Keywords Anaerobic digestion, Biotransformation, Modelling, Occurrence, Pharmaceuticals, Sorption

1 Introduction

The term organic micropollutants (OMPs) encompasses a wide variety of chemical compounds present in the environment at very low concentrations (from ng/L to μg/L). Among these compounds, we can find pharmaceuticals, illicit drugs, personal care products, polycyclic aromatic hydrocarbons (PAHs), industrial chemicals and pesticides. The increasing concern for emerging pollutants lies in their potential toxicity, oestrogenicity, mutagenicity, antibiotic resistance and oxidative stress [1, 2] even at low concentrations.

Currently, the removal of OMPs from wastewater within the European Union is not legally required (European Water Framework Directive 2013/39/EU, [3]), even though two Watch Lists of substances were implemented (Decision 2015/495/EU [4]; Decision 2018/840/EU) to temporary monitor and gather information regarding their potential risks in aquatic environments. Moreover, in some EU countries reference compounds in wastewater at a national level are monitored [5]. Despite the fact that sewage sludge is used in many countries as biosolid for agricultural purposes, less legal attention was drawn to the presence of OMPs in this source of pollution. Although an updated regulation for land application of biosolids is expected [6, 7], the European regulation in force (Directive 86/278/EEC, [8]) ignores the potential harm to the environment and public health derived from the accumulation and persistence of OMPs in soils [9, 10] and only a few EU countries have limited the concentration of some OMPs in sludge in their national legislation [11].

Anaerobic digestion (AD), the standard sludge stabilization process used in wastewater treatment plants (WWTP), was designed to effectively biodegrade organic matter into biogas. Since many OMPs are as well biodegradable, several studies have demonstrated the capacity of AD to substantially reduce their concentration [12–14]. Understanding the fate of OMPs during AD is essential to broaden the capabilities of this process, but also to know its limitations, evaluate strategies to overcome them and weigh the need for additional technologies that ensure safe application of sewage sludge in agriculture.

2 Why Do OMPs Accumulate in Sewage Sludge?

Sewage sludge is usually considered a sink where many OMPs tend to accumulate. To explain this fact and to determine the occurrence of OMPs in sewage sludge, three factors should be considered: (1) the concentration of the compound entering the WWTP (C_{ww}); (2) the hydrophobicity, quantified by the partition coefficient (K_d) between the solid and liquid phases; and (3) the biological removal in the conventional activated sludge (CAS) unit, that can be estimated through the biotransformation constant, k_{biol}. It is widely considered that only OMPs with a high K_d appear at relevant concentrations in sludge, though we will demonstrate in this section that the presence of hydrophilic OMPs is also significant.

2.1 *Occurrence and Removal of OMPs in Wastewater*

As previously mentioned, three parameters determine the concentration of OMPs in sewage sludge, being their values shown in Table 1 for a selection of OMPs typically present in WWTPs. As can be observed, there is an important variability in the inlet concentration range of some compounds (e.g. ibuprofen and naproxen), which could be due to their different usage rates among countries, seasons and even WWTPs [18]. Within the OMPs entering WWTP there are compounds, such as musk fragrances, with a high affinity for the solids (high K_d) and others, such as ibuprofen and sulfamethoxazole, which are hydrophilic (low K_d). Finally, since most WWTP remove organic matter via CAS units, it is also important to consider the biotransformation rate of OMPs in this system. If they are hardly biotransformed, as happens with carbamazepine (low k_{biol}), they will be present in both the primary and the secondary sludge; on the contrary, if they are easily biotransformed, like the hormones oestradiol and estrone, their concentration will be only significative in the primary sludge.

2.2 *Mass Balances to Explain the Fate of OMPs in WWTPs*

In this section, we will show how sludge is the endpoint of not only hydrophobic but also hydrophilic pollutants. As an example, we have selected four OMPs with different sorption and biodegradability characteristics: 17β-oestradiol (E2, lipophilic and highly biodegradable), triclosan (TCS, lipophilic and recalcitrant), ibuprofen (IBP, hydrophilic and highly biodegradable) and carbamazepine (CBZ, hydrophilic and recalcitrant) (Table 1).

To perform a simple mass balance, the concentrations of OMPs in the liquid and solid phases along the WWTP were estimated considering the average inlet concentrations (C_{ww}) and average values of K_d and k_{biol} (Table 1). The mass balance was

Table 1 Application, occurrence range in wastewater entering WWTPs (C_{ww}) and in (primary/secondary) sewage sludge (C_{ss}), partition coefficient (K_d) and biotransformation rate in conventional activated sludge units (k_{biol}) of several OMPs. Mean values of C_{ww} are in brackets, when available

OMP	Abbr.	Application	C_{ww} (μg/L)[a]	log K_d[b]	k_{biol} (L/g_{VSS} d)	C_{ss} (μg/g)[c,b,d]
17β-oestradiol	E2	Oestrogen	0.01–3 (0.17)	2.5–3.2	170–350[c]	LOQ–0.08
Estrone	E1	Oestrogen	0.002–0.67 (0.08)	2.4–2.9	2–200[c]	LOQ–0.60
Celestolide	ADBI	Fragrance	LOQ–0.02[e]	3.0–3.5[f]	6–63[f]	LOQ–0.04[c,e]
Galaxolide	HHCB	Fragrance	0.04–13[e]	3.3–4.3	7–41[f]	4.2–31[c,e]
Tonalide	AHTN	Fragrance	0.11–5.4[e]	3.4–4.3	2–38[f]	0.4–9[c,e]
17α-ethinylestradiol	EE2	Contraceptive/oestrogen	0.002–0.07 (0.02)	2.4–2.8	0.02–20[c]	0.004–0.42
Bisphenol A	BPA	Fungicide/plasticizer	0.02–5.8[c]	2.3–3.0[c]	0.2–16[c]	LOQ–4.7
Fluoxetine	FLX	Antidepressant	0.01–2.3 (0.54)	3.7–4.0	5–9[a]	0.07–0.43[b,g]
Triclosan	TCS	Antiseptic	0.38–4.2 (1.9)	3.0–3.9	0.02–0.8[c]	0.04–17
4-nonylphenol	NP	Surfactant	–	3.8–4.3	–	6.0–21
Ibuprofen	IBP	Anti-inflammatory	0.8–373 (37)	0.3–2.7	1.5–35	LOQ–4.1
Roxithromycin	ROX	Antibiotic	0.01–0.21	1.6–2.4	0.2–9	<0.002–0.18
Erythromycin	ERY	Antibiotic	0.06–10 (1.8)	1.4–2.4	0.2–6	LOQ–0.20
Naproxen	NPX	Anti-inflammatory	0.04–52 (6.0)	1.1–1.5	0.2–2.9	LOQ–1.0
Diclofenac	DCF	Anti-inflammatory	0.06–11 (1.0)	1.2–2.7	0.04–1.2	LOQ–7.0
Sulfamethoxazole	SMX	Antibiotic	LOQ–7.9 (0.92)	0.5–2.5	0.3	LOQ–68
Trimethoprim	TMP	Antibiotic	LOQ–4.7 (0.76)	1.9–2.5	0.2	LOQ–41
Carbamazepine	CBZ	Anticonvulsant	LOQ–21 (1.2)	0.1–2.5	0.1	LOQ–1.7
Diazepam	DZP	Anxiolytic	21	1.3–2.1	0.1	LOQ–0.58

LOQ value below limit of quantification
[a] Verlicchi et al. [15]
[b] Verlicchi and Zambello [16]
[c] Tran et al. [17]
[d] Mejías et al. [18]
[e] Clara et al. [19]
[f] Alvarino et al. [20]
[g] Gonzalez-Gil et al. [12]

performed in the WWTP of Santiago de Compostela (Spain), whose operating parameters were previously described by Carballa et al. [21]. The main characteristics of the WWTP (hydraulic retention time, flow rate, total and volatile suspended solids concentration) are summarized in Fig. 1.

From the results shown in Fig. 1 it can be concluded that, regardless of their characteristics, OMPs tend to accumulate in the sewage sludge (stream 7, mixture of primary and secondary sludge). Thus, when comparing the total concentration (liquid + solid phase) of the four OMPs in this stream with stream 1 (inlet to the WWTP), an increase in the concentrations is observed for the four compounds. This can be explained because the suspended solids concentration is much higher in sludge than in wastewater, leading to a greater contribution of the solid phase to the total OMP concentration. Actually, for lipophilic substances, like E2 and TCS, a 10-times increment could be attained; on the contrary, the contribution of the liquid phase to the total concentration is very high in the case of hydrophilic substances (e.g. IBP and CBZ). Finally, it should be pointed out that even compounds that biotransform efficiently in the secondary treatment, such as IBP and E2, would be present at relevant concentrations in the sewage sludge due to the contribution of the primary sludge, where biotransformation is negligible [22] and whose flow rate is usually higher than that of the secondary sludge.

In summary, when regarding the fate of OMPs in WWTPs, it should be noted that even when OMPs are removed from the wastewater line (the concentration in the treated water – stream 5 – is lower than the raw wastewater – stream 1), they tend to accumulate in sewage sludge reaching concentrations potentially harmful for the environment [16, 23], and for an agricultural usage [24].

2.3 *Occurrence of OMPs in Sewage Sludge*

While the occurrence of OMPs in wastewater has been extensively investigated during the last decades, their analysis in the sludge line has been disregarded in most studies [17]. The reviews of Mejías et al. [18], Tran et al. [17] and Verlicchi and Zambello [16] offer a good and updated summary of the concentrations of more than 150 OMPs found in different kinds of sludge. The concentration range of the compounds selected in this chapter in a mixture of primary and secondary sludge (C_{SS}) are gathered in Table 1. As can be observed, hydrophobic substances such as TCS and HHCB are detected at relatively high concentrations in sewage sludge (>10 μg/g), but also concentrations of hydrophilic pharmaceuticals such as SMX and TMP are in the same order of magnitude. For these four OMPs, the concentration at the WWTP entrance is similar; thus, hydrophilic compounds in sludge could be expected at lower concentrations. However, as demonstrated via mass balance, the high concentration of hydrophilic compounds in sludge might be explained by the contribution of the liquid phase of sludge to the total concentration and the lack of biotransformation in primary and secondary sludge (low k_{biol}, Table 1).

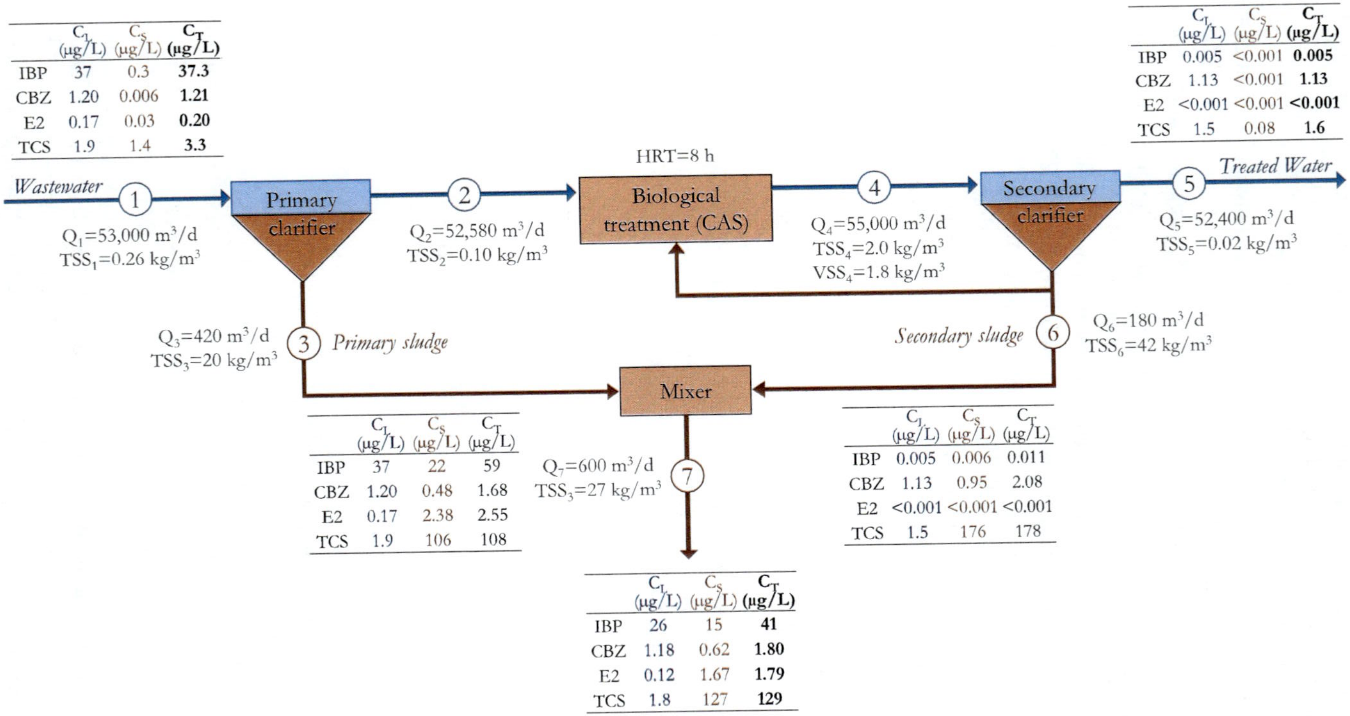

Fig. 1 Fate of four OMPs in the water and sludge line of a typical WWTP. The concentrations in the liquid and solid phases of each stream are represented by C_L and C_S, respectively. The sum of both represents the total concentration (C_T). The solid phase concentration in µg/g can be obtained by dividing C_s (µg/L) by the corresponding TSS (g/L). Operational parameters of the WWTP (flow rate (Q), total suspended solids (TSS), volatile suspended solids (VSS) and hydraulic retention time (HRT)) were calculated from Carballa et al. [21]

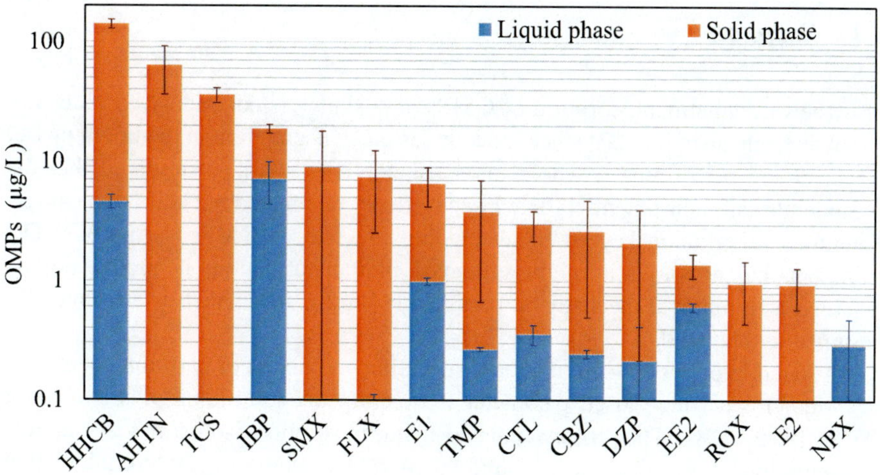

Fig. 2 Occurrence of OMPs in the liquid and solid phases of sewage sludge. The data were obtained from a municipal WWTP located in Spain [12]

It should be noted that the concentration of OMPs in sludge is frequently expressed in ng or μg per dry weight of solids, which in some cases might refer to the total concentration (if the water phase of sludge was evaporated) or only to the solid phase (if water is removed after centrifugation of samples). In spite of the fact that measuring OMPs in both sludge phases provides more information about their fate than just reporting the total or the solid phase concentration, it is very unusual to find studies splitting the concentration into the two phases, maybe due to the complexity of the liquid matrix [12]. Figure 2 shows the concentration in both phases of mixed sludge for the OMPs found by Gonzalez-Gil et al. [12], where can be observed that compounds such as TCS and E2 were only detected in the solid phase, while in the case of CBZ and especially IBP the contribution of the liquid phase is significant.

3 Fate of OMPs During Sewage Sludge Anaerobic Digestion

It is generally assumed that the main mechanisms governing the fate of OMPs in anaerobic systems are sorption and biotransformation, while volatilization is negligible since there is no aeration [20, 25, 26]. These two mechanisms and the factors affecting the removal during AD will be discussed in this section.

3.1 Sorption

In wastewater treatment, sorption of OMPs onto sludge is commonly considered a removal mechanism [20, 27] since it decreases the concentration in the liquid phase. However, by performing a whole mass balance (Fig. 1), it is observed that the compounds and their potential toxic effects are not eliminated but shifted to sludge. Therefore, sorption is key to understanding the fate of OMPs during WWTPs treatment and although it might influence biotransformation, it should not be regarded as a removal mechanism, particularly during AD, which is a technology designed to treat sludge.

As stated in previous sections, the distribution coefficient or partition coefficient (K_d, Eq. 1) is a widely used parameter to quantify the solid-liquid distribution of OMPs [9, 27, 28]. This corresponds to a linear sorption isotherm valid at low concentrations [29], like those found in WWTPs. Moreover, it considers that equilibrium between phases is reached almost instantaneously [27, 30].

$$K_d = \frac{C_s}{C_L \cdot \mathrm{TSS}} \tag{1}$$

where C_s is the concentration in the solid phase (μg/L), C_L is the concentration in the liquid phase and TSS is the concentration of total suspended solids (kg/L).

According to Eq. (1), the concentration ratio of the compounds in solution and sorbed on the solids remains constant. However, this constant value (K_d) varies among treatment units, because besides the physicochemical properties of the OMPs (i.e. the octanol-water coefficient), sorption depends on the composition of the solid matrix and the ambient conditions (temperature, pH, ion strength, complexing agents), which affect electrostatic interactions, cationic exchanges, cationic bridges, surface complexation and hydrogen bonding ([9, 28, 31–34]). Furthermore, sludge suffers a shift in the pH during AD (from pH of 5–6 to 7–8) [32] and its composition is altered [35]. Accordingly, the sorption equilibrium in primary/secondary sludge and digested sludge could be different. Nevertheless, few studies compare both data, and fewer K_d values are reported for digested sludge [16, 36, 37].

Carballa et al. [9] and Gonzalez-Gil et al. [12] suggest that in most compounds K_d values are not affected by the type of sludge (primary/secondary or digested) nor by AD operational conditions, yet some variations are reported for specific OMPs. Narumiya et al. [32] found that OMPs with a p*K*a around nine increase their hydrophobicity because a significant increment in the fraction of neutral species during AD takes place. In contrast, in compounds with p*K*a around four deprotonated species become dominant during AD, decreasing its partition coefficient, as might be the case of DZP (p*K*a = 3.4) [12]. In summary, Narumiya et al. [32] hypothesize that the hydrophobicity of OMPs depends on the concentration of neutral species, which might be altered for some OMPs due to the pH shifts that occur during AD (pH of 5–6 in sewage sludge and 7–8 in the digestates).

3.2 *Biotransformation Mechanisms and Efficiencies*

Once OMPs enter an anaerobic digester they might undergo a biological removal, providing their functional moieties are susceptible to being attacked by the catalytic activities of microorganisms. Several authors have demonstrated that the predominant mechanism for OMPs removal in AD relies on biological activities [36, 38, 39], that can transform OMPs via metabolism when they allow bacterial growth and/or cometabolism [17, 40]. Therefore, cometabolism occurs in the presence (for growing cells) or absence (for resting cells) of a growth substrate, usually at low OMPs concentration due to the unspecific activity of some enzymes [41, 42].

Biotransformation is a general expression that includes moderate and unknown modifications in of OMPs chemical structure, whereas biodegradation is usually employed when OMPs' chemical structure is deeply changed, and mineralization is the term used if the final product is CO_2. In many cases, the structure of the transformation products (TPs) is quite similar to the parent compound.

In most cases, biotransformation is performed by cometabolic processes [25, 43, 44] and ultimately depends on the enzymatic activities present in the anaerobic digester. In the last years, many attempts were made to link OMPs biotransformation to specific enzymes [45]. This task is quite difficult since a huge number of enzymes of different types are involved (e.g. kinases, transferases, hydrolases and lyases) [44, 46, 47], and many of them still remain unknown [48–50].

As can be observed in Fig. 3, some compounds are removed to a high extent (SMX, ROX, NPX and TMP) while others, such as CBZ, are quite recalcitrant, with significant discrepancies among studies in the biotransformation efficiencies of some compounds such as hormones (EE2, E1, E1+E2), musk fragrances (AHTN, HHCB) and other OMPs (DCF, NP, BPA), which result in interquartile boxes bigger than 20 percentual points. Although the use of raw or spiked with OMPs sludge could affect results [57], this fact hardly explains such important divergences, and future studies focused on understanding the biotransformation routes of these OMPs are required [45].

Overall, in comparison with other sludge treatment technologies (i.e. composting and aerobic digestion), AD seems to be more efficient in removing a wide spectrum of OMPs [16, 62]. However, despite several OMPs being biotransformed during AD, many TPs and persistent OMPs (Fig. 3) remain at significant concentrations in the digested sludge. In order to assure a safe application in soils, bio-analytical tools that evaluate the potential toxic effects of sludge are needed [12, 14].

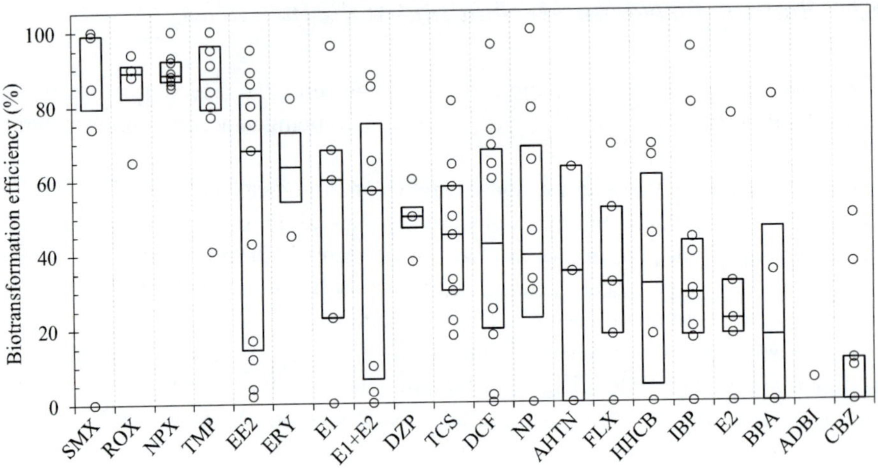

Fig. 3 Summary of reported biotransformation efficiencies in sewage sludge AD. Each circle represents the published data of one of the source studies [12, 19, 32, 38, 47, 51–61]. The box shows the interquartile range obtained as the difference between the upper (75th percentile) and lower quartile (25th percentile). The bar inside the box represents the median value (50th percentile). Sometimes, the total removal of E1+E2 is reported because E1 is reduced to E2 under anaerobic conditions

3.3 *Can Operational Strategies Promote the Biotransformation of OMPs During AD?*

In this section, we will analyze the effect of the operating parameters such as sludge retention times (SRT), organic loading rate (OLR), temperatures and feeding compositions on OMPs biotransformation.

Operation at higher SRT enables longer times for the OMPs to be biotransformed and theoretically a broader microbial diversity in the digester. The first effect would not have an impact on most OMPs, since biotransformation is not kinetically limited when working at these very high retention time values [63]. Actually, several studies confirm that biotransformation of most OMPs is not affected by the SRT in the typical range of sewage sludge anaerobic digesters (10–30 days) [12, 52, 60, 61]. Similarly, possible variations in the microbial composition due to SRT neither seem to have a relevant influence on OMP removal.

A possible strategy to promote microbial/enzymatic activities, and thus the cometabolic biotransformation of OMPs during AD, could consist in increasing OLR, always under the limits assuring a stable sewage sludge digestion. The typical OLR range applied to AD is between 1 and 2 g COD/L d and, according to reported results [12, 52, 59, 60, 64], biotransformation of OMPs under this range is not significantly affected. These findings suggest that even when working at the lower limit, OLR seems to be sufficiently high to promote cometabolism.

Sewage sludge AD can be performed under mesophilic (30–37°C) or thermophilic conditions (55°C). Temperature is a key parameter in anaerobic digestion processes since it determines the microbial composition of the reactor and thus it could influence OMPs bioavailability and biotransformation kinetics. However, the differences in the biotransformation efficiency at mesophilic or thermophilic conditions were shown to be negligible for many OMPs [12, 38, 52, 59, 61]. Only in some cases (e.g. EE2, NP, polycyclic aromatic hydrocarbons), higher temperatures seem to favour the biotransformation rate of OMPs [12, 56, 65]. Yet, thermophilic conditions offer some advantages, such as higher sludge hygienization and detoxication (i.e. reducing oestrogenic activity and total antibiotic resistance genes) [12, 13].

It was already demonstrated that the main substrate characteristics could influence the removal of certain OMPs (e.g. polycyclic aromatic hydrocarbons, NP and polychlorinated biphenyls), although few studies have evaluated this relationship so far [56, 66] and they were only testing mono-digestion of different sludge types (primary, secondary, mixed or thermally pretreated). Changing the feeding composition via codigesting sewage sludge with other substrates might be a feasible and efficient strategy to optimize, not just biogas production, but also digested sludge characteristics in terms of both conventional parameters and OMPs. If the microbial/enzymatic activities involved in OMPs anaerobic biotransformation are known, appropriate cosubstrates could be selected to encourage specific cometabolic activities towards OMPs or even to further transform TPs. Another promising alternative could be based on supplementing specific nutrients to the digesters. For instance, adding electron acceptors (iron or sulphate) increased the removal of quite persistent OMPs (i.e. codeine, climbazole and benzotriazole) [67].

In summary, the effect of SRT, OLR and temperature seem to be slight and compound dependent, thus the selection of the best strategy will depend on the target OMPs to be removed. In general, it could be stated that thermophilic conditions are more effective than mesophilic ones to decrease toxic activities and that codigestion with specific cosubstrates or nutrients may have the potential to overcome biotransformation limitations. Additional strategies based on sludge pre- and post-treatments (e.g. ultrasonic, thermal, chemical and enzymatic methods) should be evaluated to increase OMPs removal if necessary [36, 61].

4 Modelling the Fate of OMPs During Sewage Sludge Anaerobic Digestion

Modelling the fate of OMPs in AD is essential to accurately predict the release of OMPs into the environment, but it is also a useful tool to understand the mechanisms behind their biotransformation and to include OMP removal criteria in process design. Unfortunately, there are still few models describing the behaviour of OMPs in AD [27].

The pseudo-first-order model is by far the most commonly used in biological wastewater treatments [27]. It considers that the biotransformation rate (r, $g_{OMP}/L{\cdot}d$)

depends on the concentration of OMPs (C_{OMP}, g_{OMP}/L), the biotransformation kinetic constant (k_{biol}, L/g_{VSS}·d) and the concentration of volatile suspended solids (VSS, g_{VSS}/L) as an indirect measure of the biomass (Eq. 2). Nevertheless, this model does not consider the activity of the biomass and it does not seem to be appropriate for AD [25, 31, 63].

Pseudo-first order kinetics:

$$r = k_{\text{biol}} \cdot \text{VSS} \cdot C_{\text{OMP}} \tag{2}$$

Barret et al. [31] and Delgadillo-Mirquez et al. [25] proposed models based on cometabolic kinetics, since they found a link between polycyclic aromatic hydrocarbons and the main substrate removal rates during AD. Cometabolic models (Eq. 3) consider this interaction between the primary substrate (i.e. COD) and the OMPs.

Cometabolic kinetics:

$$r = \left(T_{\text{c}} \cdot \frac{\mu}{Y} + k_{\text{c}}\right) \cdot \left(\frac{C_{\text{OMP}}}{K_{\text{sc}} + C_{\text{OMP}}}\right) \cdot \text{VSS} \tag{3}$$

In Eq. (3), T_c means the OMPs biotransformation capacity (g_{OMP}/g_{COD}), namely the maximum amount of OMPs that can be biotransformed per unit of weight of primary substrate consumed; K_{sc} (g_{OMP}/L) is the half-saturation constant and indicates the OMP concentration where half the maximum biotransformation rate is reached; k_c is the maximum specific rate of OMPs biodegradation in absence of primary substrate ($mg_{OMP}/g_{VSS} \cdot d$); μ is the biomass growth rate ($1/d$); and Y is the biomass growth yield (g_{VSS}/g_{COD}). Both k_c and K_{sc} depend on the response of the OMP biotransformation route, the OMP bioavailability and the affinity of the enzymes. Finally, μ and Y are linked to the primary substrate uptake.

Not all the OMPs that are cometabolically biotransformed show a direct relationship with the primary substrate uptake rate, as is the case of pharmaceuticals and personal care products [64]. Moreover, both the pseudo-first-order and the cometabolic model are not able to mimic kinetics where the biotransformation rate of OMPs drastically slows down or comes to a halt (i.e. appearance of concentration plateaus). This phenomenon has been observed for different kinds of OMPs and in several biological systems: activated sludge units [68, 69], nitrifying reactors [70], anaerobic/anoxic/aerobic-membrane bioreactors [71] and anaerobic digesters [63, 64]. Therefore, it seems to be independent of the biological system, the phase partitioning of the OMP or even the biotransformation mechanism (cometabolism/metabolism). Gonzalez-Gil et al. [63] proposed different mechanistic models to explain this incomplete biotransformation in anaerobic digesters. According to this study, the two most plausible hypotheses are: (1) the reversibility of the biological reactions could lead to a chemical equilibrium between the parent compound and the TP, or (2) the sequestration of OMPs into the solids might limit the bioavailability of the compounds to microorganisms and enzymes. The reversible model (Eq. 4) is based on pseudo-first-order kinetics but includes a reversible biotransformation constant (k_{rev}, L/g_{VSS}·d).

Reversible kinetics:

$$r = k_{biol} \cdot \mathrm{VSS} \cdot C_{OMP} - k_{rev} \cdot \mathrm{VSS} \cdot C_{TP} \tag{4}$$

The sequestration model is more complex and it assumes pseudo-first-order biotransformation kinetics in the liquid phase of sludge and only in the fraction of the solid phase where OMPs are not sequestered. Both mechanisms are feasible, and the selection might depend on the OMP. However, since the reversible model is simpler and it is valid for a wider range of OMPs, it is considered an appropriate and accurate model to generally predict the fate of OMPs in AD [63]. Either if reversibility of reactions and/or OMPs sequestration are limiting the biotransformation of OMPs, no kinetic constraints seem to be responsible for the biotransformation halt [63], thus explaining why the increase of HRT/SRT does not promote the OMP removal.

Besides the selection of the kinetic equation, modelling the fate of OMPs also requires including the sorption–desorption equilibrium (Eqs. 5 and 6) and defining the compartment where biotransformation takes place. Most biological models assume that only the dissolved fraction of OMPs is biotransformed [27], while the sorbed fraction is not accessible for microbial degradation [25]. However, recent studies demonstrate that OMPs are biotransformed by intracellular enzymes when sorbed into biomass [44, 47]. Hence, to model the fate of OMPs during AD in a realistic way, biotransformation in both the liquid and solid phases should be included [63].

$$j_{sor} = k_{sor} \cdot X_{TSS} \cdot C_w \tag{5}$$

$$j_{des} = k_{des} \cdot C_s \tag{6}$$

where j_{sor} is the OMP sorption rate (μg/L d), k_{sor} is the sorption kinetic constant (L/gTSS d), j_{des} is the desorption rate (μg/L d), k_{des} is the desorption kinetic constant (1/d), X_{TSS} is the total suspended solids concentration (g/L), C_w is the concentration of OMP in the liquid phase (μg/L) and C_s is the concentration of OMP in the solid phase (μg/L).

5 Conclusions

OMPs tend to accumulate in the sludge line of WWTP achieving higher concentrations than those measured in the inlet wastewater. This is the case of hydrophobic pollutants, but also of hydrophilic OMPs. Even those OMPs biologically removed in activated sludge systems remain almost unaltered in primary sludge and thus are present at relevant concentrations in the final mixture of sludge.

Biotransformation, in both the liquid and solid phases of sludge, should be considered the only removal mechanism of OMPs during AD, while sorption should be regarded as a factor affecting the OMP fate. Although biotransformation likely occurs through cometabolism, cometabolic models are not appropriate to simulate the fate of most OMPs during AD. On the contrary, reversible kinetics can predict the biotransformation halt that leads to incomplete removal of many OMPs. Operational strategies based on increasing the HRT/SRT or the OLR, do not seem to promote the removal of OMPs, but thermophilic conditions and codigestion strategies might be key to achieving these upcoming goals.

Overall, AD has the capacity to biotransform several OMPs and reduce specific toxicities from sewage sludge, although many compounds and transformation products are still present at significant concentrations in digested sludge. Since legal limits of OMPs in sludge are still missing, it might be too early to ensure that this technology is enough for a safe application of sludge in agricultural fields and perhaps new (pre- or post-) treatment strategies should be analyzed.

Acknowledgements The authors acknowledge the financial support provided by the Spanish Government (AEI) through Antares project (PID2019-110346RB-C21) and the Defence University Centre at the Defense University Center at the Spanish Naval Academy (CUD-ENM) for all the support provided for this research. Authors from Universidade de Santiago de Compostela belong to the Galician Competitive Research Group (GRC ED431C 2017/29). All these programmes are co-funded by FEDER (EU).

References

1. Escher B, Leusch F (2011) Bioanalytical tools in water quality assessment. IWA Publishing, London
2. Marti E, Variatza E, Balcazar JL (2014) The role of aquatic ecosystems as reservoirs of antibiotic resistance. Trends Microbiol 22:36–41
3. European Commission (2013) Council directive 2013/39/EU amending directives 2000/60/EC and 2008/105/EC as regards priority substances in the field of water policy. Off J Eur Union L 226
4. European Commission (2015) Commission implementing decision (EU) 2015/495 establishing a watch list of substances for union-wide monitoring in the field of water policy pursuant to Directive 2008/105/EC of the European Parliament and of the Council. Off J Eur Union L 78: 40–42
5. Kosek K, Luczkiewicz A, Fudala-Książek S, Jankowska K, Szopińska M, Svahn O, Tränckner J, Kaiser A, Langas V, Björklund E (2020) Implementation of advanced micropollutants removal technologies in wastewater treatment plants (WWTPs) – examples and challenges based on selected EU countries. Environ Sci Pol 112:213–226. https://doi.org/10.1016/j.envsci.2020.06.011
6. European Commission (2010) Working document on sludge and biowaste, pp 1–22
7. Inglezakis V, Zorpas AA, Karagiannidis A, Samaras P, Voukkali I, Sklari S (2014) European Union legislation on sewage sludge management. Fresenius Environ Bull 23:635–639
8. European Commission (1986) Council directive 86/278/EEC on the protection of the environment, and in particular of the soil, when sewage sludge is used in agriculture. Off J Eur Commun. Off J Eur Union L 181. 2004R0726 – v.7 of 05.06.2013

9. Carballa M, Fink G, Omil F, Lema JM, Ternes T (2008) Determination of the solid-water distribution coefficient (Kd) for pharmaceuticals, estrogens and musk fragrances in digested sludge. Water Res 42:287–295. https://doi.org/10.1016/j.watres.2007.07.012
10. Mininni G, Blanch AR, Lucena F, Berselli S (2014) EU policy on sewage sludge utilization and perspectives on new approaches of sludge management. Environ Sci Pollut Res 22:7361–7374. https://doi.org/10.1007/s11356-014-3132-0
11. Kelessidis A, Stasinakis AS (2012) Comparative study of the methods used for treatment and final disposal of sewage sludge in European countries. Waste Manag 32:1186–1195. https://doi.org/10.1016/j.wasman.2012.01.012
12. Gonzalez-Gil L, Papa M, Feretti D, Ceretti E, Mazzoleni G, Steimberg N, Pedrazzani R, Bertanza G, Lema JM, Carballa M (2016) Is anaerobic digestion effective for the removal of organic micropollutants and biological activities from sewage sludge? Water Res 102:211–220. https://doi.org/10.1016/j.watres.2016.06.025
13. Jang HM, Shin J, Choi S, Shin SG, Park KY, Cho J, Kim YM (2017) Fate of antibiotic resistance genes in mesophilic and thermophilic anaerobic digestion of chemically enhanced primary treatment (CEPT) sludge. Bioresour Technol 244:433–444. https://doi.org/10.1016/j.biortech.2017.07.153
14. Papa M, Dogruer G, Bailey D, Leusch F (2022) Anaerobic digestion of sewage sludge has no effect on glucocorticoid and anti-progestagenic activity but increases estrogenicity three-fold. Chemosphere 286(2):131753. https://doi.org/10.1016/j.chemosphere.2021.131753
15. Verlicchi P, Al Aukidy M, Zambello E (2012) Occurrence of pharmaceutical compounds in urban wastewater: removal, mass load and environmental risk after a secondary treatment-a review. Sci Total Environ 429:123–155. https://doi.org/10.1016/j.scitotenv.2012.04.028
16. Verlicchi P, Zambello E (2015) Pharmaceuticals and personal care products in untreated and treated sewage sludge: occurrence and environmental risk in the case of application on soil – a critical review. Sci Total Environ 538:750–767. https://doi.org/10.1016/j.scitotenv.2015.08.108
17. Tran NH, Reinhard M, Yew-Hoong Gin K (2018) Occurrence and fate of emerging contaminants in municipal wastewater treatment plants from different geographical regions-a review. Water Res 133:182–207. https://doi.org/10.1016/j.watres.2017.12.029
18. Mejías C, Martín J, Santos JL, Aparicio I, Alonso E (2021) Occurrence of pharmaceuticals and their metabolites in sewage sludge and soil: a review on their distribution and environmental risk assessment. Trends Environ Anal Chem 30. https://doi.org/10.1016/j.teac.2021.e00125
19. Clara M, Gans O, Windhofer G, Krenn U, Hartl W, Braun K, Scharf S, Scheffknecht C (2011) Occurrence of polycyclic musks in wastewater and receiving water bodies and fate during wastewater treatment. Chemosphere 82:1116–1123. https://doi.org/10.1016/j.chemosphere.2010.11.041
20. Alvarino T, Suarez S, Lema JM, Omil F (2014) Understanding the removal mechanisms of PPCPs and the influence of main technological parameters in anaerobic UASB and aerobic CAS reactors. J Hazard Mater 278:506–513. https://doi.org/10.1016/j.jhazmat.2014.06.031
21. Carballa M, Omil F, Lema JM, Llompart M, García-Jares C, Rodríguez I, Gómez M, Ternes T (2004) Behavior of pharmaceuticals, cosmetics and hormones in a sewage treatment plant. Water Res 38:2918–2926. https://doi.org/10.1016/j.watres.2004.03.029
22. Carballa M, Omil F, Lema JM (2005) Removal of cosmetic ingredients and pharmaceuticals in sewage primary treatment. Water Res 39:4790–4796. https://doi.org/10.1016/j.watres.2005.09.018
23. Chen F, Ying G-G, Ma Y-B, Chen Z-F, Lai H-J, Peng F-J (2014) Field dissipation and risk assessment of typical personal care products TCC, TCS, AHTN and HHCB in biosolid-amended soils. Sci Total Environ 470–471:1078–1086. https://doi.org/10.1016/j.scitotenv.2013.10.080
24. Ivanová L, Mackuľak T, Grabic R, Golovko O, Koba O, Staňová AV, Szabová P, Grenčíková A, Bodík I (2018) Pharmaceuticals and illicit drugs – a new threat to the application

of sewage sludge in agriculture. Sci Total Environ 634:606–615. https://doi.org/10.1016/j.scitotenv.2018.04.001

25. Delgadillo-Mirquez L, Lardon L, Steyer J-P, Patureau D (2011) A new dynamic model for bioavailability and cometabolism of micropollutants during anaerobic digestion. Water Res 45: 4511–4521. https://doi.org/10.1016/j.watres.2011.05.047
26. Semblante GU, Hai FI, Huang X, Ball AS, Price WE, Nghiem LD (2015) Trace organic contaminants in biosolids: impact of conventional wastewater and sludge processing technologies and emerging alternatives. J Hazard Mater 300:1–17. https://doi.org/10.1016/j.jhazmat.2015.06.037
27. Pomiès M, Choubert J-MJM, Wisniewski C, Coquery M (2013) Modelling of micropollutant removal in biological wastewater treatments: a review. Sci Total Environ 443:733–748. https://doi.org/10.1016/j.scitotenv.2012.11.037
28. Ternes TA, Herrmann N, Bonerz M, Knacker T, Siegrist H, Joss A (2004) A rapid method to measure the solid-water distribution coefficient (Kd) for pharmaceuticals and musk fragrances in sewage sludge. Water Res 38:4075–4084. https://doi.org/10.1016/j.watres.2004.07.015
29. Limousin G, Gaudet JP, Charlet L, Szenknect S, Barthès V, Krimissa M (2007) Sorption isotherms: a review on physical bases, modeling and measurement. Appl Geochem 22:249–275. https://doi.org/10.1016/j.apgeochem.2006.09.010
30. Dionisi D, Bertin L, Bornoroni L, Capodicasa S, Papini MP, Fava F (2006) Removal of organic xenobiotics in activated sludges under aerobic conditions and anaerobic digestion of the adsorbed species. J Chem Technol Biotechnol 81:1496–1505. https://doi.org/10.1002/jctb.1561
31. Barret M, Patureau D, Latrille E, Carrère H (2010) A three-compartment model for micropollutants sorption in sludge: methodological approach and insights. Water Res 44: 616–624. https://doi.org/10.1016/j.watres.2009.08.029
32. Narumiya M, Nakada N, Yamashita N, Tanaka H (2013) Phase distribution and removal of pharmaceuticals and personal care products during anaerobic sludge digestion. J Hazard Mater 260:305–312. https://doi.org/10.1016/j.jhazmat.2013.05.032
33. Spark KM, Swift RS (2002) Effect of soil composition and dissolved organic matter on pesticide sorption. Sci Total Environ 298:147–161
34. Tolls J (2001) Sorption of veterinary pharmaceuticals in soils: a review. Environ Sci Technol 35:3397–3406
35. Barret M, Carrère H, Latrille E, Wisniewski C, Patureau D (2010) Micropollutant and sludge characterization for modeling sorption equilibria. Environ Sci Technol 44:1100–1106
36. Stasinakis AS (2012) Review on the fate of emerging contaminants during sludge anaerobic digestion. Bioresour Technol 121:432–440. https://doi.org/10.1016/j.biortech.2012.06.074
37. Suárez S, Carballa M, Omil F, Lema JM (2008) How are pharmaceutical and personal care products (PPCPs) removed from urban wastewaters? Rev Environ Sci Biotechnol 7:125–138. https://doi.org/10.1007/s11157-008-9130-2
38. Malmborg J, Magnér J (2015) Pharmaceutical residues in sewage sludge: effect of sanitization and anaerobic digestion. J Environ Manag 153:1–10. https://doi.org/10.1016/j.jenvman.2015.01.041
39. Tiwari B, Sellamuthu B, Ouarda Y, Drogui P, Tyagi RD, Buelna G (2017) Review on fate and mechanism of removal of pharmaceutical pollutants from wastewater using biological approach. Bioresour Technol 224:1–12. https://doi.org/10.1016/j.biortech.2016.11.042
40. Tran NH, Urase T, Ngo HH, Hu J, Ong SL (2013) Insight into metabolic and cometabolic activities of autotrophic and heterotrophic microorganisms in the biodegradation of emerging trace organic contaminants. Bioresour Technol 146:721–731
41. Criddle CS (1993) The kinetics of cometabolism. Biotechnol Bioeng 41:1048–1056
42. Horvath RS (1972) Microbial co-metabolism and the degradation of organic compounds in nature. Bacteriol Rev 36(2):146–155
43. Barret M, Delgadillo-Mirquez L, Trably E, Delgenes N, Braun F, Cea-Barcia G, Steyer JP, Patureau D (2012) Anaerobic removal of trace organic contaminants in sewage sludge: 15 years of experience. Pedosphere 22:508–517. https://doi.org/10.1016/S1002-0160(12)60035-6

44. Gonzalez-Gil L, Carballa M, Lema JM (2017) Cometabolic enzymatic transformation of organic micropollutants under methanogenic conditions. Environ Sci Technol 51:2963–2971. https://doi.org/10.1021/acs.est.6b05549
45. Gonzalez-Gil L, Carballa M, Lema JM (2020) Removal of organic micro-pollutants by anaerobic microbes and enzymes. In: Varjani S, Pandey A, Tyagi RD, Ngo HH, Larroche C (eds) Current developments in biotechnology and bioengineering. Elsevier, pp 397–426. https://doi.org/10.1016/B978-0-12-819594-9.00016-4
46. Gonzalez-Gil L, Carballa M, Corvini PF, Lema JM (2019) Reversibility of enzymatic reactions might limit biotransformation of organic micropollutants. Sci Total Environ 665:574–578. https://doi.org/10.1016/j.scitotenv.2019.02.143
47. Gonzalez-Gil L, Krah D, Ghattas A, Carballa M, Wick A, Helmholz L, Lema JM, Ternes TA (2019) Biotransformation of organic micropollutants by anaerobic sludge enzymes. Water Res 152:202–214. https://doi.org/10.1016/j.watres.2018.12.064
48. Abendroth C, Simeonov C, Peretó J, Antúnez O, Gavidia R, Luschnig O, Porcar M (2017) From grass to gas: microbiome dynamics of grass biomass acidification under mesophilic and thermophilic temperatures. Biotechnol Biofuels 10:1–12. https://doi.org/10.1186/s13068-017-0859-0
49. Hanreich A, Schimpf U, Zakrzewski M, Schlüter A, Benndorf D, Heyer R, Rapp E, Pühler A, Reichl U, Klocke M (2013) Metagenome and metaproteome analyses of microbial communities in mesophilic biogas-producing anaerobic batch fermentations indicate concerted plant carbohydrate degradation. Syst Appl Microbiol 36:330–338. https://doi.org/10.1016/j.syapm.2013.03.006
50. Whiteley CG, Lee DJ (2006) Enzyme technology and biological remediation. Enzym Microb Technol 38:291–316. https://doi.org/10.1016/j.enzmictec.2005.10.010
51. Carballa M, Omil F, Alder AC, Lema JM (2006) Comparison between the conventional anaerobic digestion of sewage sludge and its combination with a chemical or thermal pre-treatment concerning the removal of pharmaceuticals and personal care products. Water Sci Technol 53:109–117. https://doi.org/10.2166/wst.2006.241
52. Carballa M, Omil F, Ternes T, Lema JM (2007) Fate of pharmaceutical and personal care products (PPCPs) during anaerobic digestion of sewage sludge. Water Res 41:2139–2150. https://doi.org/10.1016/j.watres.2007.02.012
53. Esperanza M, Suidan MT, Marfil-Vega R, Gonzalez C, Sorial GA, McCauley P, Brenner R (2007) Fate of sex hormones in two pilot-scale municipal wastewater treatment plants: conventional treatment. Chemosphere 66:1535–1544. https://doi.org/10.1016/j.chemosphere.2006.08.020
54. Ifelebuegu AO (2011) The fate and behavior of selected endocrine disrupting chemicals in full scale wastewater and sludge treatment unit processes. Int J Environ Sci Technol 8:245–254. https://doi.org/10.1007/BF03326213
55. Mailler R, Gasperi J, Chebbo G, Rocher V (2014) Priority and emerging pollutants in sewage sludge and fate during sludge treatment. Waste Manag 34:1217–1226. https://doi.org/10.1016/j.wasman.2014.03.028
56. Paterakis N, Chiu TY, Koh YKK, Lester JN, McAdam EJ, Scrimshaw MD, Soares A, Cartmell E (2012) The effectiveness of anaerobic digestion in removing estrogens and nonylphenol ethoxylates. J Hazard Mater 199–200:88–95. https://doi.org/10.1016/j.jhazmat.2011.10.075
57. Phan HV, Wickham R, Xie S, McDonald JA, Khan SJ, Ngo HH, Guo W, Nghiem LD (2018) The fate of trace organic contaminants during anaerobic digestion of primary sludge: a pilot scale study. Bioresour Technol 256:384–390. https://doi.org/10.1016/j.biortech.2018.02.040
58. Samaras VG, Stasinakis AS, Mamais D, Thomaidis NS, Lekkas TD (2013) Fate of selected pharmaceuticals and synthetic endocrine disrupting compounds during wastewater treatment and sludge anaerobic digestion. J Hazard Mater 244–245:259–267. https://doi.org/10.1016/j.jhazmat.2012.11.039
59. Samaras VG, Stasinakis AS, Thomaidis NS, Mamais D, Lekkas TD (2014) Fate of selected emerging micropollutants during mesophilic, thermophilic and temperature co-phased

anaerobic digestion of sewage sludge. Bioresour Technol 162:365–372. https://doi.org/10.1016/j.biortech.2014.03.154
60. Yang S, Hai FI, Price WE, McDonald J, Khan SJ, Nghiem LD (2016) Occurrence of trace organic contaminants in wastewater sludge and their removals by anaerobic digestion. Bioresour Technol 210:153–159. https://doi.org/10.1016/j.biortech.2015.12.080
61. Zhou H, Zhang Z, Wang M, Hu T, Wang Z (2017) Enhancement with physicochemical and biological treatments in the removal of pharmaceutically active compounds during sewage sludge anaerobic digestion processes. Chem Eng J 316:361–369. https://doi.org/10.1016/j.cej.2017.01.104
62. Martín J, Santos JL, Aparicio I, Alonso E (2015) Pharmaceutically active compounds in sludge stabilization treatments: anaerobic and aerobic digestion, wastewater stabilization ponds and composting. Sci Total Environ 503–504:97–104. https://doi.org/10.1016/j.scitotenv.2014.05.089
63. Gonzalez-Gil L, Mauricio-Iglesias M, Carballa M, Lema JM (2018) Why are organic micropollutants not fully biotransformed? A mechanistic modelling approach to anaerobic systems. Water Res 142:115–128. https://doi.org/10.1016/j.watres.2018.05.032
64. Gonzalez-Gil L, Mauricio-Iglesias M, Serrano D, Lema JM, Carballa M (2018) Role of methanogenesis on the biotransformation of organic micropollutants during anaerobic digestion. Sci Total Environ 622–623:459–466. https://doi.org/10.1016/j.scitotenv.2017.12.004
65. Trably E, Patureau D (2006) Successful treatment of low PAH-contaminated sewage sludge in aerobic bioreactors. Environ Sci Pollut Res 13:170–176
66. Barret M, Barcia GC, Guillon A, Carrère H, Patureau D (2010) Influence of feed characteristics on the removal of micropollutants during the anaerobic digestion of contaminated sludge. J Hazard Mater 181:241–247. https://doi.org/10.1016/j.jhazmat.2010.05.003
67. Falås P, Wick A, Castronovo S, Habermacher J, Ternes TA, Joss A (2016) Tracing the limits of organic micropollutant removal in biological wastewater treatment. Water Res 95:240–249. https://doi.org/10.1016/j.watres.2016.03.009
68. Blair B, Nikolaus A, Hedman C, Klaper R, Grundl T (2015) Evaluating the degradation, sorption, and negative mass balances of pharmaceuticals and personal care products during wastewater treatment. Chemosphere 134:395–401. https://doi.org/10.1016/j.chemosphere.2015.04.078
69. Gulde R, Anliker S, Kohler H-P, Fenner K (2018) Ion trapping of amines in protozoa – a novel removal mechanism for micropollutants in activated sludge. Environ Sci Technol 52:52–60. https://doi.org/10.1021/acs.est.7b03556
70. Fernandez-Fontaina E, Carballa M, Omil F, Lema JM (2014) Modelling cometabolic biotransformation of organic micropollutants in nitrifying reactors. Water Res 65:371–383. https://doi.org/10.1016/j.watres.2014.07.048
71. Xue W, Wu C, Xiao K, Huang X, Zhou H, Tsuno H, Tanaka H (2010) Elimination and fate of selected micro-organic pollutants in a full-scale anaerobic/anoxic/aerobic process combined with membrane bioreactor for municipal wastewater reclamation. Water Res 44:5999–6010. https://doi.org/10.1016/j.watres.2010.07.052

Pharmaceuticals and Personal Care Products as Contaminants of Emerging Concern in Sewage Sludge and Soils and the Role of Transformation Products in Their Fate and Environmental Impact

James S. Carsella, Jacob Mutz, Travis Zuniga, and Chad A. Kinney

Contents

J. S. Carsella, J. Mutz, T. Zuniga, and C. A. Kinney (✉)
Chemistry Department; College of Science, Technology, Engineering, and Mathematics, Colorado State University Pueblo, Pueblo, CO, USA
e-mail: chad.kinney@csupueblo.edu

Avelino Núñez-Delgado and Manuel Arias-Estévez (eds.),
Emerging Pollutants in Sewage Sludge and Soils,
Hdb Env Chem (2023) 114: 243–266, DOI 10.1007/698_2022_858,

Published online: 16 April 2022

Abstract Human pharmaceuticals and personal care products (PPCPs) are a class of chemicals that range from medications to cleaning agents, showing diverse chemical properties. These chemicals have been placed in a variety of products, some of which are available over the counter and are commonly used. Since many PPPCPs enter the environment they have been commonly identified as contaminants of emerging concern (CECs). Many PPCPs are released form households where they enter the municipal waste stream and ultimately enter a wastewater treatment facility (WWTF). Researchers have characterized the presence of PPCPs through the waste stream and treatment facilities and ultimately in aquatic and terrestrial environments impacted by wastewater treatment end products. In comparison with parent PPCPs, transformation products (TPs) of these compounds are poorly characterized in the same environments. However, mounting evidence suggests that understanding TPs is as critical to understanding the fate and effects of PPCPs as the parent compound. Human metabolites of many PPCPs are known, but TPs of PPCPs can be produced in WWTFs and after be released in the natural environment. PPCPs that are in WWTFs are commonly transformed by both biotic and abiotic processes and not removed from the water system but released along with the TPs into surface waters, that may be used for municipal drinking water downstream or increasingly used as a source of non-potable water for agricultural uses. Some PPCPs and TPs can partition into the sludge phase during wastewater treatment, with a large portion of sludge produced being land-applied as a soil amendment (biosolids). The effects on the environment at times are difficult to quantify because the repercussions can be far reaching and there is a great need to elucidate the pathways and TPs formed as well as the effects on the biome.

Keywords Bioreactors, Biosolids, Pharmaceuticals and personal care products, Wastewater treatment

1 Background

1.1 PPCPs in Municipal Waste Stream

Pharmaceuticals and personal care products (PPCPs) include chemical classes that are used to treat and prevent diseases as well as enhance the quality of daily life. These classes include everything from prescribed medications like metformin, over the counter medications such as ibuprofen, and personal care products, which for example include chemicals such as triclosan (TCS) and triclocarban (TCC). The last two TCS and TCC are compounds that have been included in toothpaste, soaps, shampoos, and cosmetics [1]. The fate of these chemicals was not on the forefront of concern until researchers began to study them in a variety of environment matrices about 20–25 years ago [1]. The reality was PPCPs went down the drain of many households and ended up in the sewage system. It was later discovered that WWTFs did not completely remove these chemicals from the water effluent and in fact many PPCPs that were identified as "removed" partitioned into the sludge collected at the WWTFs [2–4]. The portioning of these compounds follows a Log Kow model, with the effluent from the WWTF carrying the water-soluble portion of the polar PPCPs out of the WWTF and the sludge becoming the repository of the more nonpolar compounds [1, 3]. This chapter looks at the PPCPs and gives an illustration to the transformation products that can result from the PPCPs. It then examines the fates of these compounds and their potential effects on the environment.

2 Waste Stream and Wastewater Treatment

2.1 Presence, Composition, and Concentration of PPCPs and Transformation Products in the Municipal Waste Stream and Wastewater Influent of TPs in the Municipal Waste Stream

PPCPs can undergo decomposition in the human body as well as by microbes in the environment and by microbes and abiotic chemical processes during wastewater treatment. The resulting products are referred to as transformation products. Although not as widely studied as the output from wastewater treatment, the composition of PPCPs and some transformation products in the influent entering WWTFs around the globe have been reported. Waste streams will reflect both the population and the industries served and represent a complex mixture of trace chemical inputs. When considering typical inputs of PPCPs into the waste stream and treatment facilities, challenges exist when considering existing scientific literature. This arises from the fact that individual studies generally include only one or a handful of treatment facilities and the list of target analytes varies from one study to another making direct comparisons challenging. Consider the comprehensive review of pharmaceutical compounds in urban wastewater by Verlicchi et al. [5], in which the authors work with data pertaining to 118 pharmaceutical compounds in the influent and effluent of some 264 wastewater treatment facilities based on 78 peer-reviewed publications. Among the 118 pharmaceuticals considered, more than 60% are included in three or fewer of the publications reviewed and many are only included in a single publication included in the review.

Of PPCPs measured in WWTF influent their concentration is typically in the low ng/L to mid μg/L concentration range in the dissolved phase [5]. However, this does not account for PPCPs that enter WWTFs as conjugated metabolites or adsorbed to or encapsulated in particles [6–8]. As a result, strictly focusing on the dissolved phase may underestimate total PPCPs entering WWTFs. For instance, the concentration of human metabolites of the pharmaceutical carbamazepine and nicotine routinely exceed the concentration of the parent compounds in wastewater influent reaching treatment facilities [9, 10]. Further, the distribution of parent carbamazepine and nicotine and the metabolites of these biologically active compounds are not substantially changed during wastewater treatment [10]. The behavior of all biologically active compounds and their metabolites is complex. Natural and synthetic estrogens are commonly excreted and enter the waste stream as sulfate and glucuronide conjugates, which are typically deconjugated during wastewater treatment returning these to their parent, active compounds [10, 11]. Further, PPCP's present in wastewater influent can be transformed during wastewater treatment, including production of halogenated products during disinfection using chlorination [12]. Additional details about the behavior and transformation of PPCPs during wastewater treatment will be addressed in a later section.

2.2 *Wastewater Treatment Technologies and Influence on Presence of PPCPs and TPs*

It is important to recognize that in general drinking water treatment plants (DWTPs) and WWTFs were never designed for and do not completely remove pharmaceuticals. Remediation efficiencies can be <10% in the case of such pharmaceuticals as carbamazepine, atenolol, acetylsalicylic acid, diclofenac, mefenamic acid, propranolol, atenolol, clofibric acid, and lincomycin. For WWTFs, treatment systems are unable to fully degrade PPCPs because they are generally designed to handle easily and moderately degradable organics in the mg/L range. However, the solubilities of PPCP, sorption characteristics, volatilities, biodegradability, polarities, and stabilities vary over a wide range, allowing many PPCPs to pass through the system unaffected or only partially transformed. Also, since PPCPs can be active at very low (ng/L−μg/L) concentrations, even if a system can affect the PPCP, it may remain at potentially harmful levels post treatment [13]. Traditionally, WWTFs consist of a primary treatment and a secondary treatment process, with some facilities utilizing a tertiary treatment for additional purification needs. Generally, in the primary treatment solid wastes such as plastics, oils, fats, sand, and grit are isolated from the influent water through mechanical filtration and sedimentation. The secondary treatment step varies widely from facility-to-facility; common techniques used are fixed bed reactors, membrane bioreactors, moving bed biofilm reactors, and the most common technique of conventional activated sludge (CAS). These secondary processes are utilized to remove biological waste from the influent. WWTFs that utilize a tertiary step commonly use it for the purpose of further disinfection. The most common types of tertiary disinfection are accomplished by chlorination or ultraviolet irradiation [14].

Conventional treatment and sand filtration are rarely efficient at removing PPCPs. Biological wastewater treatment is also often not sufficient to mineralize PPCPs, which at most are only converted into stable TPs. It is therefore necessary to consider other processes, such as membrane filtration, adsorption on activated carbon, and ozonation, as options for the removal of PPCPs [15, 16]. Understanding the inefficiencies of WWTFs is critical for addressing improvements to specifically address the presence of PPCPs. This can be accomplished by assessing the water quality to determine presence of PPCPs before and after treatment, which poses its own challenges. There are many methods of sampling that can be utilized, the most common being grab sampling. However, long-term methods are also strongly favored as they can help account for seasonal variations as well as increase sample volume to improve detection limits. An example of these sampling techniques being utilized occurred in Erie, Pennsylvania, USA. Raw (untreated) and treated water grab samples were collected from the local water treatment facility over 10 months. Along with grab samples, long-term samples were collected for 30–60 day periods using passive polar organic chemical integrative samplers (POCIS) utilizing OASIS™ HLB sorbent. The specific results from this study will be assessed in proceeding sections, but the difference in the sampling methodology is applicable to the topic at hand. For grab samples, 6 of the 17 tested PPCPs were found at detectable levels in raw or treated water; for long-term POCIS samples, an additional 4 PPCPs (10 total) were found at detectable levels [17]. This demonstrates the importance of sampling to the determination of PPCPs and TPs, and these factors should always be considered when assessing the validity of research conclusions.

2.3 Presence of PPCPs and TPs in Wastewater Effluent

Release of PPCPs by discharge of wastewater effluent is a worldwide problem, demonstrated by the detection of pharmaceuticals in at least 71 countries and 631 different types of pharmaceuticals/transformation products being detected in aqueous environments [18]. The potential threat many of these PPCPs pose to both the environment and inhabitants, specifically humans, is the major driving factor to continue PPCP research. By improving our knowledge of levels of PPCPs in the environment and understanding appropriate effective removal techniques, solutions to eliminating PPCPs in the environment can be achieved. It is often very difficult to assess PPCP levels in aquatic environments. Varying physicochemical properties make it a challenge to design a single treatment strategy for PPCPs. Furthermore, the biologically effective range for PPCPs can be at very low concentrations (ng/L–μg/L) [13], which then can influence what treatment solutions must be made to achieve acceptable levels of contaminants. Despite the challenges, it is critical to continue to protect human and ecological health.

The primary pathway for PPCPs entering aquatic environments is through WWTFs in which PPCPs are move into soils and surface waters by application of treated sludge or discharge of WWTFs effluent. A continuous influx of PPCPs exacerbates this problem as demonstrated by a study of a heavily urbanized river in Guangzhou City, China, in which the daily consumption of PPCPs was directly linked to the variety of species in the river [19]. A pilot study on the trace organic contaminants in wastewater in Skaneateles Lake in New York State compiled data and found the following PPCPs in various drinking water sources: sulfamethoxazole (ND-0.39 ng/L), triclosan (ND-1.93 ng/L), atenolol (ND-19.5 ng/L), ibuprofen (ND-1.16 ng/L), bisphenol A (ND-1.42 ng/L), and caffeine (ND-11.1 ng/L) [20]. In the United States, the hormone 17α-ethinylestradiol (EE2) has been detected in surface water at a concentration >100 ng/L [18]. The maximum contamination tends to be far worse in less affluent nations; as of 2019, the maximum concentrations of the PPCPs sulfamethoxazole, estriol, ibuprofen, and 17β-estradiol (E2) found in African drinking water sources were 53,828 ng/L, 45,550 ng/L, 17,600 ng/L, and 15,700 ng/L, respectively [21].

2.4 *Transformation During Wastewater Treatment*

Wastewater treatment facilities will typically include disinfection as a final treatment step usually employing chlorination or an advanced oxidation process (ozonation or UV/peroxide) to remove bacteria and oxidize potential contaminates. This can be an important source of TPs for some of the PPCPs (Fig. 1).

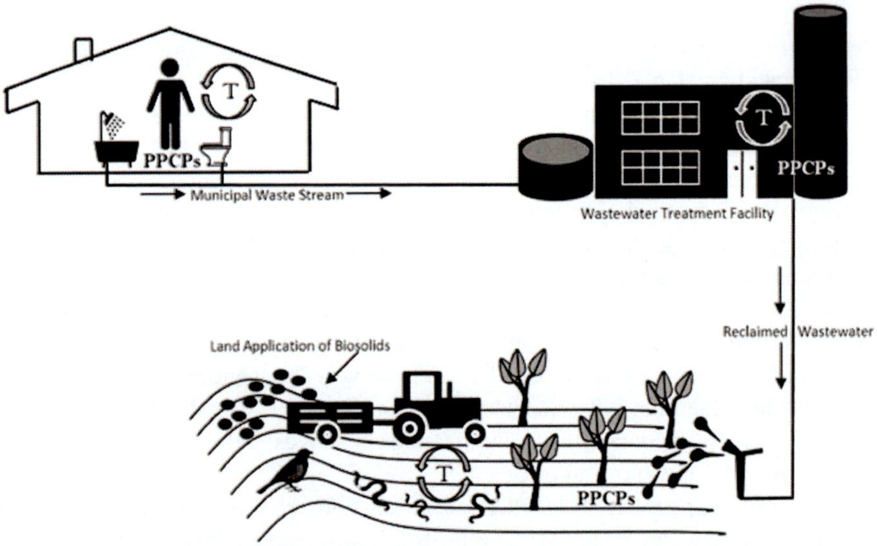

Fig. 1 Release of human pharmaceuticals and personal care products (PPCPs) into the municipal waste stream results in the introduction of these compounds into the terrestrial environment via land application of biosolids (treated sewage sludge) and irrigation with reclaimed wastewater. At numerous points, including the point of use, during wastewater treatment, or after release, PPCPs can be transformed into new products (capital T surrounded by arrows) by biotic and abiotic process. Many transformation products are biologically active or can be converted into biologically active compounds, and therefore, largely represents unknown risks

2.4.1 Chlorination

Chlorination is a very common treatment technique used in DWTPs and WWTFs primarily for biological treatment. However, due to the operating conditions and the mixture of reactive species that can exist, chlorination TPs can be abundant and potentially harmful. It is also possible to see no effects from chlorination on PPCPs as a result of competing reactions. For example, chlorine has a higher reactivity with oxazepam than with diazepam, indicating that diazepam could largely remain intact during chlorine disinfection of wastewater (Y.-Y [22].). Chlorination can introduce unintended side reactions. This is an area of study still requiring attention by researchers. A 2019 study investigated the relations of different reactive species that are generated in chlorination and the effectiveness of removing PPCPs. It was found that the in situ generation of $HO^{\bullet}$, reactive chlorine species ($Cl^{\bullet}$, Cl_2, and $ClO^{\bullet}$) and O_3 during solar irradiation of free available chlorine greatly accelerated the PPCP degradation in simulated drinking water and real water samples. PPCPs containing electron-donating moieties and PPCPs having high reaction rate constants with O_3 can be quickly degraded with half-life <7.3 min. PPCPs short of electron-donating moieties were degraded mainly by $HO^{\bullet}$ with half-life ranging from 7.39 to 14.96 min. Increased oxygen concentrations were found to increase PPCP elimination rates. Enhancing free available chlorine has a minor effect on PPCP degradation. The study concluded that the cytotoxicity did not increase significantly in solar/free available chlorine (FAC) treated water and that the process could be a potential candidate for removing various PPCPs [23]. However, if only the parent compound is being assessed as the measure of removal, but TPs are not accounted for, this may overlook an important source of CECs and biologically active compounds stemming from the presence of PPCPs in the waste stream, which may contribute to the overall toxicity of municipal wastewater.

Due to the wide variety of conditions used during chlorination, it is possible for a single PPCP to undergo many different transformations. A 2021 study by Chen et al. [24] examined the transformation pathways of levofloxacin under chlorination treatment. Using UHPLC-TOF-M/S, GC-MS, and NMR, 15 different transformation products were identified from the chlorination of levofloxacin. Through the structural identification, a plausible pathway for each transformation was proposed. Proposed pathways included chlorination, dealkylation, decarbonation, hydrolysis, hydroxylation, and oxidation. By controlling the pH and the FAC of the treatment, it was demonstrated that the transformations could be regulated to predictable TPs. The group then used quantitative structure-activity relationship (QSAR) techniques to predict the genotoxicity of 11 of the TPs, of which it was determined that 10 of the TPs were more genotoxic than the parent levofloxacin [24]. Herein lies the primary issue with not only chlorination but also ozonation; these treatment procedures are utilized in DWTPs or WWTFs, but little regard is given to the effects they can have on trace levels PPCPs. If the TPs produced are in fact more toxic and/or more environmentally prevalent, then the consumption from DWTPs can have severely adverse reactions for consumers whereas the emission of these species into the environment through WWTFs can have long-term effects on wildlife and biota from either surface water emissions or use of sewage and sludge.

2.4.2 Ozonation

Ozonation, while an excellent disinfectant process, can produce unknown degradation products when the process is not optimized for specific PPCPs [25]. Often the concern with the TPs is the unknown reactivity and toxicity the new TPs poses. The partial oxidation of the antiviral acyclovir is a good example of this. The ozonation process of acyclovir yields the TP N-(4-carbamoyl-2-imino-5-oxoimidazolidin)-formamido-N-methoxyacetic acid which was shown to be a more harmful toxin than the parent compound; 14.1 mg/L (EC_{10}) was able to inhibit green algae growth while 100 mg/L of acyclovir had no effect on algae growth [26]. Further demonstrating the complexity of ozonation due to the large volume of potential TPs is a study by Gulde et al. [27], in which 87 parent micropollutants were exposed to ozonation conditions and ozonation transformation products (OTPs) were investigated. A total of 1,749 potential OTPs were identified, being derived from 70 of the 87 micropollutants. Conditions for treating the 1,749 OTPs were also examined. For OTPs it was determined that granular activated carbon (GAC) yielded abatement for 40% to 53% of the OTPs whereas 87% were abated by powder activated carbon (PAC) [27].

3 PPCPs and TPs Release in Terrestrial Environments

3.1 Reclaimed Wastewater: Irrigation Source

Effluent from WWTFs is often released into rivers and streams where it finds its way into municipal drinking water plants or is used in agriculture for irrigation. Further, direct use of treated wastewater (reclaimed wastewater) containing trace PPCPs may potentially result in accumulation of these compounds in the agricultural crops, grazing animals, and consequently lead to human dietary exposure [28]. The use of reclaimed wastewater is often necessary, especially in arid or semi-arid regions. However, in these locations PPCPs may have a greater tendency to accumulate into edible leaves and fruits due to the generally high plant transpiration rates [29]. Some studies suggest that the levels of accumulation in irrigated crops are at minimal concern. A two-year study on leafy greens revealed limited accumulation of contaminants in soil and plant leaves, their concentration being in the range of 1–30 ng/g and 1–660 ng/g in soil and leaves, respectively [30]. However, as stated previously, for some PPCPs, low ng/g or ng/L levels can have adverse effects on wildlife and humans. Overall, leafy vegetables are more likely to have higher uptake of PPCPs from reclaimed wastewater than fruit produced by crops [31].

Methods by which reclaimed wastewater is applied to plants can affect the concentration of the PPCPs in the environment. For example, plant uptake and bioconcentration is significantly lower in soils amended with biosolids compared to soils irrigated with treated wastewater. In soils amended with biosolids and irrigated with reclaimed wastewater, the bioavailability of PPCPs for plant uptake was moderately decreased as compared to plants grown in soils irrigated with reclaimed wastewater alone. While treated wastewater acts as a continuous source of PPCPs, biosolids act both as a source and a sink for these compounds [32]. With all these factors considered, the continuous introduction of contaminated reclaimed wastewater into the environment through irrigation may be the more important source of PPCP contamination of agricultural areas.

Despite the contamination of agricultural regions, studies demonstrate that PPCPs present in the reclaimed water when used for irrigation do not cause significant contamination of groundwater and accumulation in the receiving soils [33]. In a separate study, average concentrations of PPCPs detected in groundwater were typically two orders of magnitude lower (0.4–37.7 μg/L) than those measured in WWTP effluents (2.0–3,765 μg/L) [34], demonstrating attenuation of PPCPs prior to entering groundwater.

3.2 *Sewage Sludge/Biosolids: Presence and Composition of PPCPs and TPs*

Previously identified PPCPs, which are essential components in modern life [35], enter the municipal waste stream and are incompletely removed during wastewater treatment [2, 3, 36]. It is known that PPCPs are introduced into agricultural systems through land application of biosolids and irrigation with treated wastewater [37]. An environmentally relevant concentration for several frequently detected PPCPs is 50–120 ng/L [9]. Though designed to be nonbioaccumulative and eliminated from humans and animals shortly after being administered, transformation of pharmaceuticals has been observed as part of the metabolic process to facilitate elimination in urine and or feces [38].

Sewage sludge, a by-product of wastewater treatment, is applied to agricultural fields as biosolids for various beneficial reasons. The USEPA [39] defines biosolids as nutrient rich organic residual that may be applied as fertilizer once it has been treated and processed. Some of the advantages of applying biosolids to agricultural fields comprise the relatively low cost as well as the valuable nutrients and organic matter, which can enhance soil physical properties and ultimately crop yield [40]. With respect to just the United States McClellan and Halden [36] indicate that approximately 250 metric tons of biosolids are added to the soil annually. Numerous PPCPs in the concentration ranges of ng/L–μg/L and even mg/L have been reported in several countries including, Mexico, South Africa, Brazil, Tunisia, Vietnam, and other tropical Asian countries [41]. Land application of biosolids is a global practice, and therefore, it is pivotal that researchers continue expand our understanding of the role of biosolids application as it pertains to PPCPs and TPs introduction into the natural environment.

Multiple studies have reported concentration ranges of PPCPs in biosolids, however there remains a lack of understanding and accounting for the potential transformation of the respective parent compound of interest. For instance, Herklots et al. [42] mention that the pharmaceutical compounds carbamazepine, salbutamol, sulfamethoxazole, and trimethoprim have been measured in reclaimed wastewater at concentrations of 93.6 ng/L, 11.4 ng/L, 59.2 ng/L, and 46 ng/L, respectively, while the same compounds were also detected in biosolids destined for land application with average concentrations of: 271.2 μg/kg for salbutamol, 66.4 μg/kg for carbamazepine, 92.95 μg/kg for sulfamethoxazole, and 11.43 μg/kg for trimethoprim. Brezina et al. [43] also confirmed the presence of carbamazepine (CBZ), an antiepileptic drug commonly detected in wastewater end products, along with 10,11-dihydro-10,11-dihydroxycarbamazepine (DiOHCBZ), one of the main metabolites of CBZ, in treated sewage sludge. The reported concentrations for CBZ and DiOHCBZ in treated sewage sludge were 1.29 ± 0.03, and 2.8 ± 0.2 μg/L, respectively. Expanding upon the current understanding of fate and transformation of PPCPs will assist with future studies with respect to concentration.

4 Pharmaceuticals and Personal Care Products and TPs in Soil Environments

4.1 Fate and Behavior of PPCPs in Soil Environments

A variety of studies have considered the persistence and behavior of PPCPs in soils. Researchers have found that many PPCPs will persist in soils for an extended period of time following land applications of biosolids or irrigations with reclaimed wastewater [3, 44]. Mobility of PPCPs is, not surprisingly highly compound dependent with some compounds retained in the upper layers of soils while others more readily migrate within the soil column, and this behavior will be influenced by soil characteristics as well, particularly the quantity of soil organic matter present [44, 45]. Differences in behavior in soils of parent PPCPs and TPs might be expected and have been reported. Two common metabolites of carbamazepine, 10,11-epoxycarbamazepine and 10,11-dihydro-10,11-trans-dihydroxycarbamazepine, present in reclaimed wastewater showed lower sorption and higher mobility in soils compared to the parent carbamazepine in soils [46]. These findings support a common theme highlighted in this chapter that TPs must be considered when assessing the overall exposure and risks associated with PPCPs.

In a 2009 study Xu et al. looked at adsorption in agricultural soils of six selected PPCPs: clofibric acid, ibuprofen, naproxen, triclosan, diclofenac, and BPA [47]. The movement is well described with the Freundlich equation (Eq. 1) [48].

$$Q_e = K_F C_e^{1/n} \tag{1}$$

In this equation Q_e is the amount of phenolic compounds absorbed per unit mass of resin [48]. K_F is the Freundlich constant which represents the absorptive capacity of the substrate. C_e is the equilibrium concentration of phenolic compounds in the system, and $(1/n)$ is the empirical constant of the adsorption intensity of the system [48]. The results showed that the soil had an affinity for the compounds. The movement of the compounds followed the order triclosan > BPA > clofibric acid > naproxen > diclofenac > ibuprofen. The results suggest that a retardation factor exists in layers of soils and allows for the bacteria at each level to have a longer residence time with the compound [47]. It also suggests that the compounds that readily move with the mobile phase, water in this case, have a greater potential of reaching water tables or being eluted out of the soil [47].

Degradation of the PCPPs followed a first order exponential decay which was influenced by microorganisms and the clay content of the soil. When the experiment was repeated with sterile soils the degradation was decreased [47]. High concentrations also decreased the rate of degradation indicating toxicological effects on the microbiome in the soil [47].

4.2 *Transformation in Soils*

Transformation of PPCPs in soil environments typically occur by three mechanisms: The exposure to UV light, abiotic chemical degradation, and by microbially mediated processes. Apart from the surface layer of the soil, chemical and microbial action is the primary means of degrading PPCPs and forming TPs as well as the degradation of TPs. PPCPs enter the soil from exposure to contaminated water sources and by the application of biosolids as fertilizers and soil amendments. Past studies show that mineralization of the PPCPs can occur in the aqueous environment. This was demonstrated with ibuprofen and the herbicide 2,4-D [49]. This work showed that 85% of the 2,4-D and 68% of the ibuprofen was mineralized in the water before it was applied to the soil. This biodegradation was vastly different in soils with 57% and 45%, respectively [49]. This evidence shows that a competition occurs between the abiotic and biotic processes. Mineralization by abiotic processes can sequester compounds while biotic processes tend to create transformation products which in turn may also be sequestered via mineralization or released based on the affinity of the compound between the soil and the water. It is possible to construct models for these environments. For example, in a study by Girardi et al. the mineralization of the aqueous phase followed a logistic model while the soil followed a two-pool first order degradation model [49]. Presumably the latter is due to the greater presence of microbes in the soil.

4.3 Biological Activity and Impacts of PPCPs and TPs in Non-target Organisms

By their very design most PPCPs are biologically active and as noted many of their TPs also have biological activity or can be readily transformed back into the active parent compounds. While a greater number of projects assess the impacts of PPCPs and TPs in aquatic environments [50–52], there is some evidence of potential impact in terrestrial environments. Bioaccumulation of PPCPs and a limited number of TPs in lower trophic terrestrial organisms and plants have been widely reported [35, 53, 54]. While relatively little data is available about the activity of TPs in terrestrial environments there are some data available that indicates TPs can be important when considering the overall fate and impacts of PPCPs [35, 53, 54].

Most available data related to exposure of terrestrial organisms to PPCPs is derived from controlled laboratory experiments and frequently involve acute toxicity (lethality) as the primary endpoint. Generally, these experiments involve individual or select mixtures of parent PPCPs frequently at concentrations that exceed those observed in the natural environment. In studies using the earthworm *Eisenia fetida* the LC_{50} for many PPCPs are in the mg/kg concentration range [55, 56]. Earthworms represent a popular test organism for contaminants in soils given that they are well distributed globally, are in contact with the soil and soil solution, and are primary consumers of organic matter in soils [57–59]. Earthworms are exposed to organic chemical contaminants in soils through diet as well as dermal exposure [60].

Reclaimed wastewater and biosolids represent complex mixtures of PPCPs that can be introduced into soil environments. In addition to organic chemical contaminants reclaimed wastewater and biosolids can contain a variety of organic chemical contaminants other than PPCPs as well as inorganic constituents. Further, the bioavailability of PPCPs and TPs can be limited in soil environments because of the high organic matter content of biosolids [35, 61, 62]. While controlled laboratory experiments allow researchers to better elucidate accumulation and effects of specific contaminants or mixtures of contaminants, controlled laboratory exposure experiments may not adequately represent exposure to complex mixtures of contaminants or address factors that influence the bioavailability of such contaminants. Bioaccumulation of PPCPs in earthworms in agricultural soils amended with biosolids has been observed [53, 63]. Exposure to environmentally relevant quantities of biosolids has been demonstrated to increase earthworm mortality and reduce reproductive success (i.e., cocoon production and numbers of juveniles) [64]. Earthworms represent a low trophic level in terrestrial food webs. As such, bioaccumulation of PPCPs as well as possible transformation of PPCPs represents a potential route of exposure to higher trophic organisms. Further disruption of earthworm populations has the potential to influence higher trophic organisms that rely on earthworms for their diet. Evidence suggests that earthworms can be an important route of exposure to PPCPs for avian species and may exert an influence on birds [65, 66]. Markman et al. [65, 67] demonstrated endocrine disrupting compounds present in earthworms collected from trickling filter beds at a WWTF can cause a significant enlargement of the high vocal center of the brain in male European starlings (*Sturnus vulgaris*) resulting in a change in song length and complexity. Exposure to the endocrine disrupting compounds further decreased immune function in the exposed birds.

Reports of exposure and effects of PPCPs in non-photosynthetic terrestrial organisms are relatively limited. This likely reflects challenges monitoring higher trophic organisms that may be exposed to PPCPs and TPs in their native habitats. Traditional sampling methods for wildlife are commonly invasive, stressful, or lethal. However, alternative approaches relying on samples of convenience such as hair, eggshells, feathers, or fecal matter offer noninvasive approaches that may be employed [68]. Fecal samples from North American otters and mink have been employed as a diagnostic sample to document exposure to organic and inorganic chemical contaminants, including some PPCPs, in wild populations of these organisms [69].

The accumulation of PPCPs in reclaimed wastewater and biosolids in plants, primarily crop plants, has been widely reported [70–72]. Generally, such studies have focused on potential human exposure and risk as a primary concern when considering accumulation of PPCPs in plants. Human risk of exposure to PPCPs via consumption of contaminated crop has been deemed low due to the low concentrations of PPCPs reported in edible portions of plants [73–75]. However, exposure and uptake of PPCPs in plants can impact growth and development of these plants. Exposure to biosolids and PPCPs can negatively impact germination and plant development [64, 76, 77]. Accumulation of PPCPs in plants also represents some of the rare examples where researchers have demonstrated extensive transformation of parent PPCPs [61, 78–81]. Macherius et al. [80] reported a majority of triclosan accumulated in carrots as TPs and Riemenschneider et al. [81] reported about 45% of carbamazepine in tomato plants present as TPs with the ratio of TPs to parent carbamazepine up to 2.5 in the fruits. Substantial transformation of PPCPs in plants demonstrates the importance of PPCP TPs in terrestrial environments and importance of TPs when considering the fate and impacts of PPCPs in terrestrial systems.

The toxicity of some TPs, especially well characterized human metabolites has been assessed. For example, carbamazepine 10,11-epoxide and acradine, both metabolites of the antiepileptic pharmaceutical carbamazepine, have been identified as potential genotoxic as well as widely produced in crops exposed to carbamazepine contaminated soil [81, 82]. Carbamazepine is among the most widely studied PPCP given its low removal efficiencies during wastewater treatment and high frequency of detection in terrestrial environments. Where researchers have recently assessed the toxicity of the most abundant TPs of select PPCPs in select test organism they have generally found lower toxicity among the TPs compared to the parent compounds, yet it is important to note that the TPs still posed biological activity and could potentially be transferred further [83, 84]. Notable exceptions to this observation include greater toxicity of the R enantiomers of ibuprofen and naproxen [84]. While limited data about the widespread presence of PPCP TPs in wastewater end products and environments impacted by these products is available, where data exists there are examples of toxicological concern as well as concerns about the overall biological activity that TPs may contribute in exposed environments and organisms.

5 Measuring PPCPs and Transformation Products

5.1 Analytical Tools and Techniques

Where analytical standards exist, which is the case for many important and well-characterized human metabolites or TPs, analytical techniques for the quantitative analysis of PPCP transformation products are like that of the parent PPCPs. This will include an extraction or isolation method, which are abundantly referenced in the scientific literature. This may be by liquid extraction that includes process like QuEChERs or Solid Phase Extraction (SPE), which includes the use of a sorbent in the extraction process to aid in analyte isolation, concentration, and sample clean-up [3, 85, 86]. More aggressive extraction processes may be necessary for some matrices such as biosolids or biological samples. In such instances, ultrasonic assisted extraction (UAE) [31, 87], microwave assisted extract (MAE) [88], or pressurized liquid extraction (PLE) [66, 71] may be more appropriate.

Given the complex nature of many samples (e.g., soils, biosolids, and biological samples) chromatographic methods are generally coupled with mass spectrometry for qualitative and quantitative analysis. Advances in analytical tools, namely liquid chromatography-mass spectrometry (LC/MS) and to a lesser extent gas chromatography-mass spectrometry (GC/MS), as it pertains to PPCP analysis, readily available analytical standards of PPCPs, and a growing availability of isotope labeled PPCPs has made the analysis of parent PPCPs in a variety of matrices accessible to researchers. While early research on the presence and fate of PPCPs relied primarily on single quadrupole mass spectrometers, more recently liquid chromatography coupled to a tandem mass spectrometer (MS/MS) has become the standard, especially triple quadrupole mass spectrometers, which allows for greater sensitivity and selectivity (USEPA Method 1694). Triple quadrupole MS for PPCP analysis of complex samples employ multiple-reaction-monitoring (MRM). When operated in the MRM mode the first quadrupole serves as a mass filter that allows only ions of the m/z ratio of target analytes to pass through to the second quadrupole commonly referred to as a collision cell. The collision cell contains a collision gas, typically argon or nitrogen. Collision of the target ions with the collision gas results in the reproducible production of fragment ions. The third quadrupole, like the first quadrupole, is operated as a mass filter that allows the characteristic fragment ions of the target analytes to pass to the detector. For most analytes a minimum of two MRM transitions will be monitored: one for quantitative purposes and the other for qualitative application.

5.2 *Analytical Standards: Measuring Transformation Products and Matrix Effects*

The use of LC/MS/MS is a powerful tool for the quantitative analysis of PPCPs in complex matrices for analytes for which analytical standards exist. This includes parent PPCPs and some metabolites/transformation products, especially for important human metabolites. However, due to the selective nature of LC/MS/MS operation it is not well suited for monitoring potential transformation products for which analytical standards are not available or for previously unidentified transformation products. Increasingly researchers are reporting the presence of a variety of PPCP transformation products in a variety of matrices including plants exposed to PPCPs [61, 80, 81]. More advanced (and more expensive) analytical tools/techniques (e.g., time-of-flight mass spectrometer or Orbitrap mass spectrometer) are frequently required for non-target analysis need to identify some TPs for which standards are not available or that were previously unknown. Application of accurate mass techniques can be used to identify the molecular formula of TPs, and, when coupled with fragmentation information and the natural abundance of stable isotopes, tentative structures can be proposed [80].

Another attribute of analyzing PPCPs in complex matrices is the potential for "matrix effects," which commonly results in ionization suppression of target PPCPs compared to analysis of the compounds in calibration standards. Researchers can take steps to "clean" extracts prior to instrumental analysis, such as employing semi-selective solid phase extraction. Employing, multi-step extraction and sample preparation steps generally results in analyte loss and can contribute to low recoveries. A common approach to overcome matrix effects and/or low analyte recoveries is the use of "isotope dilution" where stable isotope labeled standards of the target analytes are added to the samples (USEPA Method 1694) [39]. The labeled version of the target PPCPs are expected to experience the same losses and matrix effects as the unlabeled PPCPs effectively making the labeled compounds the ideal internal standard. While the use of isotope labeled standards are effective analytical tools to address matrix effects and most factors contributing poor analyte recovery, the drawback is the increased analysis costs associated with adding such standards, which typically cost substantially more than their non-labeled counterparts.

5.3 Quantitative Measurements and Interpretation

When researchers can make direct comparison to analytical standards, especially when employing the use of an internal standard(s) or isotope dilution, quantitative measurement of transformation products is straightforward. However, in the absence of analytical standards for transformation products identification, or qualitative analysis, does not readily translate to quantitative determinations. Approaches to estimate the concentration of transformation products without available analytical standards can be performed by estimating a response factor from chromatographic peaks. Macherius et al. [80] employed this method to estimate the formation and the presence of conjugated transformation products of the disinfectant triclosan in carrots to be five times the amount of parent triclosan in the vegetable. However, this method carries with it several assumptions and does not benefit from techniques to address matrix effects or transformation product recovery. Alternatively, mass balance of the parent compound and transformation products with available analytical standards can be used to determine the estimated contribution for newly identified transformation products.

Modern mass spectrometry techniques are capable of measuring low ng/mL concentration of many PPCPs and TPs in extracts/samples. When preconcentration steps are employed as part of sample preparation the ability to detect even lower concentrations of such compounds in the original sample or extract is possible. A consistent area of inquiry to the presence of low concentrations of PPCPs and their TPs in environmental or biological matrices relates to "what does it mean" or "what are the risks" to human and ecological health. These are challenging questions to adequately answer. Acute exposure can be monitored for a variety of model organisms. In such instances, it is common that the LC_{50} or EC_{50} exceeds typical environmental concentrations, prompting the conclusion that the presence of these compounds in the environment is of little risk. A couple of notable examples where risk has been clearly identified at environmentally relevant concentrations includes the collapse of the vulture population on the Indian subcontinent due to the veterinary use of nonsteroidal anti-inflammatory diclofenac in the region [89], and the reproductive disruption of male fish as a result of exposure to natural and synthetic estrogens that is commonly found in wastewater effluent and receiving waters [50, 90]. It should be noted that much of the toxicological research on PPCPs has focused on aquatic organisms with limited work on terrestrial organisms in general with even less information available for TPs, contributing to a general paucity of information a paucity of information in the scientific literature related to environmental toxicology of PPCP TPs.

Research addressing the impacts of chronic exposure to PPCPs and their TPs is limited, especially as it relates to terrestrial biota. Traditional approaches employed to assess acute exposure, such as survival and reproduction, are most likely inadequate to assess the impacts of long-term exposure to low environmental concentrations of PPCPs and their TPs. More subtle effects to chronic exposure may include behavior changes, which could include alteration in predatory response, feeding patterns and behavior, or activity. Such changes in behavior can have long-term consequences for survival, reproductive success, and development, which can lead to long-term changes at the population level. Examples of identified behavioral changes include behavioral defects in *Caenorhabditis elegans* (soil nematode) and their offspring following exposure to the antibiotic sulfamethoxazole [91]. While this work included exposure at environmentally relevant concentrations, it does represent a short-term exposure experiment. Another example of behavior changes in response to short-term exposure to PPCPs involves *Pimephales promelas* (fathead minnow) exposed to antidepressants. Exposure to the antidepressants resulted in reduction in measures of predator avoidance, which may impact survival and reproductive success [52]. While these short-term exposure experiments demonstrate some concerns related to exposure to PPCPs, they do not address the potential impacts of chronic exposure and generational consequences of exposure to low concentrations of PPCPs. Further, impacts of chronic exposure to transformation products of PPCPs remain unknown.

6 Conclusions and Future Needs

In the decade before the start of the twenty-first century researchers began a concerted effort to understand the presence and consequences of PPCPs being released into the natural environment. While a few examples of endocrine disruption in aquatic and terrestrial organisms have been identified, few clear connections to human health concerns have been identified. Accumulation of parent PPCPs in crops does not appear to be a serious human health concern. However, the presence of trace quantities of antibiotics coupled with the observation of antibiotic-resistant genes and bacteria raises concerns of indirect health concerns for humans. Experts predict a higher mortality rate due to infectious diseases by antibiotic-resistant pathogens than by cancer in 2050 [92]. Thus far ecological risk has largely been based on measures of acute toxicology. More subtle chronic effects at the population level resulting from long-term exposure to sub-acute concentrations of PPCPs remains an area of concern. To date, toxicological work has predominantly focused on parent PPCPs with an absence of consideration of TPs for many of the reasons previously cited. The overall importance of TPs on overall exposure and effects of PPCPs remains unknown and an area of research need. Further work needs to be done on the origin and effects of TPs on the often-ignored environmental compartments such as insects and microorganisms that stabilize the biota and allow for the diversity of life to grow and exist. This possibility for TPs to adversely affect this microbiome and therefore impact the larger biota and potentially wildlife and agriculture exists.

References

1. Boxall AB, Rudd MA, Brooks BW, Caldwell DJ, Choi K, Hickmann S et al (2012) Pharmaceuticals and personal care products in the environment: what are the big questions? Environ Health Perspect 120(9):1221–1229
2. Kinney CA, Furlong ET, Werner SL, Cahill JD (2006) Presence and distribution of wastewater-derived pharmaceuticals in soil irrigated with reclaimed water. Environ Toxicol Chem Int J 25(2):317–326
3. Kinney CA, Furlong ET, Zaugg SD, Burkhardt MR, Werner SL, Cahill JD, Jorgensen GR (2006) Survey of organic wastewater contaminants in biosolids destined for land application. Environ Sci Technol 40(23):7207–7215
4. Ternes TA, Joss A, Siegrist H (2004) Peer reviewed: scrutinizing pharmaceuticals and personal care products in wastewater treatment. Environ Sci Technol 38(20):392A–399A
5. Verlicchi P, Al Aukidy M, Zambello E (2012) Occurrence of pharmaceutical compounds in urban wastewater: removal, mass load and environmental risk after a secondary treatment—a review. Sci Total Environ 429:123–155. https://doi.org/10.1016/j.scitotenv.2012.04.028
6. Göbel A, McArdell CS, Joss A, Siegrist H, Giger W (2007) Fate of sulfonamides, macrolides, and trimethoprim in different wastewater treatment technologies. Sci Total Environ 372(2): 361–371. https://doi.org/10.1016/j.scitotenv.2006.07.039
7. Kimura K, Hara H, Watanabe Y (2005) Removal of pharmaceutical compounds by submerged membrane bioreactors (MBRs). Desalination 178(1):135–140. https://doi.org/10.1016/j.desal.2004.11.033
8. Zorita S, Mårtensson L, Mathiasson L (2009) Occurrence and removal of pharmaceuticals in a municipal sewage treatment system in the south of Sweden. Sci Total Environ 407(8): 2760–2770. https://doi.org/10.1016/j.scitotenv.2008.12.030
9. Bahlmann A, Brack W, Schneider RJ, Krauss M (2014) Carbamazepine and its metabolites in wastewater: analytical pitfalls and occurrence in Germany and Portugal. Water Res 57:104–114. https://doi.org/10.1016/j.watres.2014.03.022
10. Ekpeghere KI, Sim W-J, Lee H-J, Oh J-E (2018) Occurrence and distribution of carbamazepine, nicotine, estrogenic compounds, and their transformation products in wastewater from various treatment plants and the aquatic environment. Sci Total Environ 640–641:1015–1023. https://doi.org/10.1016/j.scitotenv.2018.05.218
11. Kumar V, Johnson AC, Nakada N, Yamashita N, Tanaka H (2012) De-conjugation behavior of conjugated estrogens in the raw sewage, activated sludge and river water. J Hazard Mater 227–228:49–54. https://doi.org/10.1016/j.jhazmat.2012.04.078
12. Bulloch DN, Nelson ED, Carr SA, Wissman CR, Armstrong JL, Schlenk D, Larive CK (2015) Occurrence of halogenated transformation products of selected pharmaceuticals and personal care products in secondary and tertiary treated wastewaters from Southern California. Environ Sci Technol 49(4):2044–2051. https://doi.org/10.1021/es504565n
13. Patel M, Kumar R, Kishor K, Mlsna T, Pittman Jr CU, Mohan D (2019) Pharmaceuticals of emerging concern in aquatic systems: chemistry, occurrence, effects, and removal methods. Chem Rev 119(6):3510–3673
14. Batt AL, Kim S, Aga DS (2007) Comparison of the occurrence of antibiotics in four full-scale wastewater treatment plants with varying designs and operations. Chemosphere 68(3):428–435
15. Petrie B, Barden R, Kasprzyk-Hordern B (2015) A review on emerging contaminants in wastewaters and the environment: current knowledge, understudied areas and recommendations for future monitoring. Water Res 72:3–27
16. Yang Y, Ok YS, Kim K-H, Kwon EE, Tsang YF (2017) Occurrences and removal of pharmaceuticals and personal care products (PPCPs) in drinking water and water/sewage treatment plants: a review. Sci Total Environ 596:303–320
17. Kim H, Homan M (2020) Evaluation of pharmaceuticals and personal care products (PPCPs) in drinking water originating from Lake Erie. J Great Lakes Res 46(5):1321–1330. https://doi.org/10.1016/j.jglr.2020.06.019

18. aus der Beek T, Weber F-A, Bergmann A, Hickmann S, Ebert I, Hein A, Küster A (2016) Pharmaceuticals in the environment—global occurrences and perspectives. Environ Toxicol Chem 35(4):823–835. https://doi.org/10.1002/etc.3339
19. Yuan X, Hu J, Li S, Yu M (2020) Occurrence, fate, and mass balance of selected pharmaceutical and personal care products (PPCPs) in an urbanized river. Environ Pollut 266:115340. https://doi.org/10.1016/j.envpol.2020.115340
20. Subedi B, Codru N, Dziewulski DM, Wilson LR, Xue J, Yun S et al (2015) A pilot study on the assessment of trace organic contaminants including pharmaceuticals and personal care products from on-site wastewater treatment systems along Skaneateles Lake in New York state, USA. Water Res 72:28–39. https://doi.org/10.1016/j.watres.2014.10.049
21. Fekadu S, Alemayehu E, Dewil R, Van der Bruggen B (2019) Pharmaceuticals in freshwater aquatic environments: a comparison of the African and European challenge. Sci Total Environ 654:324–337
22. Yang Y-Y, Zhao J-L, Liu Y-S, Liu W-R, Zhang Q-Q, Yao L et al (2018) Pharmaceuticals and personal care products (PPCPs) and artificial sweeteners (ASs) in surface and ground waters and their application as indication of wastewater contamination. Sci Total Environ 616:816–823
23. Cheng S, Zhang X, Song W, Pan Y, Lambropoulou D, Zhong Y et al (2019) Photochemical oxidation of PPCPs using a combination of solar irradiation and free available chlorine. Sci Total Environ 682:629–638. https://doi.org/10.1016/j.scitotenv.2019.05.184
24. Chen M, Wei D, Wang F, Yin J, Li M, Du Y (2021) Bioassay- and QSAR-based screening of toxic transformation products and their formation under chlorination treatment on levofloxacin. J Hazard Mater 414:125495. https://doi.org/10.1016/j.jhazmat.2021.125495
25. Westerhoff P, Yoon Y, Snyder S, Wert E (2005) Fate of endocrine-disruptor, pharmaceutical, and personal care product chemicals during simulated drinking water treatment processes. Environ Sci Technol 39(17):6649–6663
26. Schlüter-Vorberg L, Prasse C, Ternes TA, Mückter H, Coors A (2015) Toxification by transformation in conventional and advanced wastewater treatment: the antiviral drug acyclovir. Environ Sci Technol Lett 2(12):342–346
27. Gulde R, Rutsch M, Clerc B, Schollée JE, von Gunten U, McArdell CS (2021) Formation of transformation products during ozonation of secondary wastewater effluent and their fate in post-treatment: from laboratory-to full-scale. Water Res 200:117200
28. Shahriar A, Tan J, Sharma P, Hanigan D, Verburg P, Pagilla K, Yang Y (2021) Modeling the fate and human health impacts of pharmaceuticals and personal care products in reclaimed wastewater irrigation for agriculture. Environ Pollut 276:116532. https://doi.org/10.1016/j.envpol.2021.116532
29. Dodgen LK, Ueda A, Wu X, Parker DR, Gan J (2015) Effect of transpiration on plant accumulation and translocation of PPCP/EDCs. Environ Pollut 198:144–153. https://doi.org/10.1016/j.envpol.2015.01.002
30. Manasfi R, Brienza M, Ait-Mouheb N, Montemurro N, Perez S, Chiron S (2021) Impact of long-term irrigation with municipal reclaimed wastewater on the uptake and degradation of organic contaminants in lettuce and leek. Sci Total Environ 765:142742. doi:https://doi.org/10.1016/j.scitotenv.2020.142742
31. Mordechay EB, Mordehay V, Tarchitzky J, Chefetz B (2021) Pharmaceuticals in edible crops irrigated with reclaimed wastewater: evidence from a large survey in Israel. J Hazard Mater 416:126184
32. Mordechay EB, Tarchitzky J, Chen Y, Shenker M, Chefetz B (2018) Composted biosolids and treated wastewater as sources of pharmaceuticals and personal care products for plant uptake: a case study with carbamazepine. Environ Pollut 232:164–172
33. Chen W, Xu J, Lu S, Jiao W, Wu L, Chang AC (2013) Fates and transport of PPCPs in soil receiving reclaimed water irrigation. Chemosphere 93(10):2621–2630
34. Kibuye FA, Gall HE, Elkin KR, Ayers B, Veith TL, Miller M et al (2019) Fate of pharmaceuticals in a spray-irrigation system: from wastewater to groundwater. Sci Total Envir 654:197–208. https://doi.org/10.1016/j.scitotenv.2018.10.442

35. Mordechay EB, Tarchitzky J, Chen Y, Shenker M, Chefetz B (2018) Composted biosolids and treated wastewater as sources of pharmaceuticals and personal care products for plant uptake: a case study with carbamazepine. Environ Pollut 232:164–172
36. McClellan K, Halden RU (2010) Pharmaceuticals and personal care products in archived U.S. biosolids from the 2001 EPA national sewage sludge survey. Water Res 44(2):658–668. https://doi.org/10.1016/j.watres.2009.12.032
37. Wu X, Fu Q, Gan J (2016) Metabolism of pharmaceutical and personal care products by carrot cell cultures. Environ Pollut 211:141–147
38. Katzung BG, Masters SB, Trevor AJ (2004) Basic & clinical pharmacology. McGraw-Hill
39. USEPA (2007) EPA method 1694: pharmaceuticals and personal care products in water, soil, sediment, and biosolids by HPLC/MS/MS. EPA, Washington. https://www.epa.gov/sites/default/files/2015-10/documents/method_1694_2007.pdf
40. Walters E, McClellan K, Halden RU (2010) Occurrence and loss over three years of 72 pharmaceuticals and personal care products from biosolids–soil mixtures in outdoor mesocosms. Water Res 44(20):6011–6020
41. Duran-Alvarez JC, Prado B, Gonzalez S, Y., Jimenez-Cisneros, B. (2015) Environmental fate of naproxen, carbamazepine, and triclosan in wastewater, surface water, and wastewater irragated soil – results of laboratory experiments. Sci Total Environ 538:350–362
42. Herklotz PA, Gurung P, Heuvel BV, Kinney CA (2010) Uptake of human pharmaceuticals by plants grown under hydroponic conditions. Chemosphere 78(11):1416–1421
43. Brezina E, Prasse C, Meyer J, Muckter H, Ternes TA (2017) Investigation and risk evaluation of the occurrence of carbamazepine, oxcarbazepine, their human metabolites and transformation products in the urban water cycle. Environ Pollut 225:261–269
44. Yager TB, Furlong ET, Kolpin DW, Kinney CA, Zaugg SD, Burkhardt MR (2014) Dissipation of contaminants of emerging concern in biosolids applied to non-irrigated farmland in eastern Colorado. J Am Water Resour Assoc 50:343–357
45. Chefetz B, Mualem T, Ben-Ari J (2008) Sorption and mobility of pharmaceutical compounds in soil irrigated with reclaimed wastewater. Chemosphere 73(8):1335–1343
46. Paz A, Gadmor G, Malchi T, Blotevogel J, Borch T, Polubesova T, Chefetz B (2016) Fate of carbamazepine, its metabolites, and lamotrigine in osils irrigated with reclaimed wastewater: sorption, leaching and plant uptake. Chemosphere 160:22–29
47. Xu J, Wu L, Chang AC (2009) Degradation and adsorption of selected pharmaceuticals and personal care products (PPCPs) in agricultural soils. Chemosphere 77(10):1299–1305
48. Mu T-H, Sun H-N (2019) Chapter 22 – sweet potato leaf polyphenols: preparation, individual phenolic compound composition and antioxidant activity. In: Watson RR (ed) Polyphenols in plants2nd edn. Academic Press, pp 365–380
49. Girardi C, Nowak KM, Carranza-Diaz O, Lewkow B, Miltner A, Gehre M et al (2013) Microbial degradation of the pharmaceutical ibuprofen and the herbicide 2, 4-D in water and soil—use and limits of data obtained from aqueous systems for predicting their fate in soil. Sci Total Environ 444:32–42
50. Kidd KA, Blanchfield PJ, Mills KH, Palace VP, Evans RE, Lazorchak JM, Flick RW (2007) Collapse of a fish population after exposure to a synthetic estrogen. Proc Natl Acad Sci 104 (21):8897–8901
51. Muir D, Simmons D, Wang X, Peart T, Villella M, Miller J, Sherry J (2017) Bioaccumulation of pharmaceuticals and personal care product chemicals in fish exposed to wastewater effluent in an urban wetland. Sci Rep 7(1):16999. https://doi.org/10.1038/s41598-017-15462-x
52. Painter MM, Buerkley MA, Julius ML, Vajda AM, Norris DO, Barber LB et al (2009) Antidepressants at environmentally relevant concentrations affect predator avoidance behavior of larval fathead minnows (Pimephales promelas). Environ Toxicol Chem 28(12):2677–2684. https://doi.org/10.1897/08-556.1
53. Kinney CA, Furlong ET, Kolpin DW, Burkhardt MR, Zaugg SD, Werner SL et al (2008) Bioaccumulation of pharmaceuticals and other anthropogenic waste indicators in earthworms

from agricultural soil amended with biosolid or swine manure. Environ Sci Technol 42 (6):1863–1870
54. Kinney CA, Heuvel BV (2020) Translocation of pharmaceuticals and personal care products after land application of biosolids. Curr Opin Environ Sci Health 14:23–30
55. Pino MR, Val J, Mainar AM, Zuriaga E, Español C, Langa E (2015) Acute toxicological effects on the earthworm Eisenia fetida of 18 common pharmaceuticals in artificial soil. Sci Total Environ 518:225–237
56. Snyder EH, O'Connor GA, McAvoy DC (2011) Toxicity and bioaccumulation of biosolids-borne triclocarban (TCC) in terrestrial organisms. Chemosphere 82(3):460–467
57. Bouché M (1983) The establishment of earthworm communities. In: Earthworm ecology. Springer, pp 431–448
58. Bouché M (1992) Earthworm species and ecotoxicological studies. Ecotoxicology of earthworms 1:269
59. Hopkin SP (1989) Ecophysiology of metals in terrestrial invertebrates. Elsevier Applied Science Publishers
60. Jager T, Baerselman R, Dijkman E, de Groot AC, Hogendoorn EA, de Jong A et al (2003) Availability of polycyclic aromatic hydrocarbons to earthworms (Eisenia andrei, Oligochaeta) in field-polluted soils and soil-sediment mixtures. Environ Toxicol Chem 22(4):767–775
61. Fu Q, Wu X, Ye Q, Ernst F, Gan J (2016) Biosolids inhibit bioavailability and plant uptake of triclosan and triclocarban. Water Res 102:117–124
62. Shao Y, Yang K, Jia R, Tian C, Zhu Y (2018) Degradation of triclosan and carbamazepine in two agricultural and garden soils with different textures amended with composted sewage sludge. Int J Environ Res Public Health 15(11):2557
63. Kinney CA, Furlong ET, Kolpin DW, Zaugg SD, Burkhardt MR, Bossio JP, Werner SL (2010) Earthworms: diagnostic indicators of wastewater derived anthropogenic organic contaminants in terrestrial environments. In: Contaminants of emerging concern in the environment: ecological and human health considerations. ACS Publications, pp 297–317
64. Kinney CA, Campbell BR, Thompson R, Furlong ET, Kolpin DW, Burkhardt MR et al (2012) Earthworm bioassays and seedling emergence for monitoring toxicity, aging and bioaccumulation of anthropogenic waste indicator compounds in biosolids–amended soil. Sci Total Environ 433:507–515
65. Markman S, Leitner S, Catchpole C, Barnsley S, Müller CT, Pascoe D, Buchanan KL (2008) Pollutants increase song complexity and the volume of the brain area HVC in a songbird. PLoS One 3(2):e1674
66. Sherburne JJ, Anaya AM, Fernie KJ, Forbey JS, Furlong ET, Kolpin DW et al (2016) Occurrence of triclocarban and triclosan in an agro-ecosystem following application of biosolids. Environ Sci Technol 50(24):13206–13214
67. Markman S, Guschina IA, Barnsley S, Buchanan KL, Pascoe D, Müller CT (2007) Endocrine disrupting chemicals accumulate in earthworms exposed to sewage effluent. Chemosphere 70 (1):119–125
68. Taggart MA, Richards N, Kinney CA (2015) Impacts of pharmaceuticals on terrestrial wildlife. Issues Environ Sci Technol 41:216–254
69. Richards NL, Tomy G, Kinney CA, Nwanguma FC, Godwin B, Woollett DAS (2018) Using scat detection dogs to monitor environmental contaminants in sentinel species and freshwater ecosystems. In: Using detection dogs to monitor aquatic ecosystem health and protect aquatic resources. Springer, pp 193–262
70. Carter LJ, Harris E, Williams M, Ryan JJ, Kookana RS, Boxall AB (2014) Fate and uptake of pharmaceuticals in soil–plant systems. J Agric Food Chem 62(4):816–825
71. Holling CS, Bailey JL, Heuvel BV, Kinney CA (2012) Uptake of human pharmaceuticals and personal care products by cabbage (Brassica campestris) from fortified and biosolids-amended soils. J Environ Monit 14(11):3029–3036
72. Wu X, Dodgen LK, Conkle JL, Gan J (2015) Plant uptake of pharmaceutical and personal care products from recycled water and biosolids: a review. Sci Total Environ 536:655–666

73. Christou A, Agüera A, Bayona JM, Cytryn E, Fotopoulos V, Lambropoulou D et al (2017) The potential implications of reclaimed wastewater reuse for irrigation on the agricultural environment: the knowns and unknowns of the fate of antibiotics and antibiotic resistant bacteria and resistance genes–a review. Water Res 123:448–467
74. Christou A, Antoniou C, Christodoulou C, Hapeshi E, Stavrou I, Michael C et al (2016) Stress-related phenomena and detoxification mechanisms induced by common pharmaceuticals in alfalfa (Medicago sativa L.) plants. Sci Total Environ 557:652–664
75. Christou A, Karaolia P, Hapeshi E, Michael C, Fatta-Kassinos D (2017) Long-term wastewater irrigation of vegetables in real agricultural systems: concentration of pharmaceuticals in soil, uptake and bioaccumulation in tomato fruits and human health risk assessment. Water Res 109:24–34
76. Carter LJ, Williams M, Böttcher C, Kookana RS (2015) Uptake of pharmaceuticals influences plant development and affects nutrient and hormone homeostases. Environ Sci Technol 49 (20):12509–12518
77. Schmidt W, Redshaw CH (2015) Evaluation of biological endpoints in crop plants after exposure to non-steroidal anti-inflammatory drugs (NSAIDs): implications for phytotoxicological assessment of novel contaminants. Ecotoxicol Environ Saf 112:212–222
78. Huber C, Bartha B, Harpaintner R, Schröder P (2009) Metabolism of acetaminophen (paracetamol) in plants—two independent pathways result in the formation of a glutathione and a glucose conjugate. Environ Sci Pollut Res 16(2):206–213
79. Huber C, Bartha B, Schröder P (2012) Metabolism of diclofenac in plants–hydroxylation is followed by glucose conjugation. J Hazard Mater 243:250–256
80. Macherius A, Eggen T, Lorenz W, Moeder M, Ondruschka J, Reemtsma T (2012) Metabolization of the bacteriostatic agent triclosan in edible plants and its consequences for plant uptake assessment. Environ Sci Technol 46(19):10797–10804. https://doi.org/10.1021/es3028378
81. Riemenschneider C, Seiwert B, Moeder M, Schwarz D, Reemtsma T (2017) Extensive transformation of the pharmaceutical carbamazepine following uptake into intact tomato plants. Environ Sci Technol 51(11):6100–6109. https://doi.org/10.1021/acs.est.6b06485
82. Malchi T, Maor Y, Tadmor G, Shenker M, Chefetz B (2014) Irrigation of root vegetables with treated wastewater: evaluating uptake of pharmaceuticals and the associated human health risks. Environ Sci Technol 48(16):9325–9333. https://doi.org/10.1021/es5017894
83. Maldonado-Torres S, Gurung R, Rijal H, Chan A, Acharya S, Rogelj S et al (2018) Fate, transformation, and toxicological impacts of pharmaceutical and personal care products in surface waters. Environ Health Insights 12:1178630218795836
84. Grabarczyk Ł, Mulkiewicz E, Stolte S, Puckowski A, Pazda M, Stepnowski P, Białk-Bielińska A (2020) Ecotoxicity screening evaluation of selected pharmaceuticals and their transformation products towards various organisms. Environ Sci Pollut Res 27(21):26103–26114
85. Cerqueira MB, Guilherme JR, Caldas SS, Martins ML, Zanella R, Primel EG (2014) Evaluation of the QuEChERS method for the extraction of pharmaceuticals and personal care products from drinking-water treatment sludge with determination by UPLC-ESI-MS/MS. Chemosphere 107:74–82
86. Sunyer-Caldú A, Diaz-Cruz MS (2021) Development of a QuEChERS-based method for the analysis of pharmaceuticals and personal care products in lettuces grown in field-scale agricultural plots irrigated with reclaimed water. Talanta 230:122302
87. Bossio JP, Harry J, Kinney CA (2008) Application of ultrasonic assisted extraction of chemically diverse organic compounds from soils and sediments. Chemosphere 70(5):858–864
88. Brown TJ, Kinney CA (2011) Rapid lab-scale microwave-assisted extraction and analysis of anthropogenic organic chemicals in river sediments. Int J Geosci 2(03):267

89. Oaks JL, Gilbert M, Virani MZ, Watson RT, Meteyer CU, Rideout BA et al (2004) Diclofenac residues as the cause of vulture population decline in Pakistan. Nature 427(6975):630–633. https://doi.org/10.1038/nature02317
90. Vajda AM, Barber LB, Gray JL, Lopez EM, Woodling JD, Norris DO (2008) Reproductive disruption in fish downstream from an estrogenic wastewater effluent. Environ Sci Technol 42 (9):3407–3414. https://doi.org/10.1021/es0720661
91. Yu Z, Jiang L, Yin D (2011) Behavior toxicity to Caenorhabditis elegans transferred to the progeny after exposure to sulfamethoxazole at environmentally relevant concentrations. J Environ Sci 23(2):294–300. https://doi.org/10.1016/S1001-0742(10)60436-6
92. Klatte S, Schaefer H-C, Hempel M (2017) Pharmaceuticals in the environment–a short review on options to minimize the exposure of humans, animals and ecosystems. Sustain Chem Pharm 5:61–66

Current Methodology for Extraction, Separation, Identification, and Quantification of Microplastics in Terrestrial Systems

D. Arenas-Lago, V. Santás-Miguel, and A. Rodríguez-Seijo

Contents

Abstract Microplastic (MP) contamination is a current global problem in the environment and specifically in soils. Plastics enter the soils through multiple pathways, such as agricultural application, organic amendments from wastewater treatment plants, landfills, or industrial sources. This chapter provides a description of the current state of the analytical techniques used for the extraction, quantification, and identification of MP in soils. In this sense, the methodology for determination and quantification of MP is currently easier and much more developed for aqueous

D. Arenas-Lago (✉) and V. Santás-Miguel
Área de Edafoloxía e Química Agrícola, Facultade de Ciencias, Universidade de Vigo, Ourense, Spain
e-mail: darenas@uvigo.es

A. Rodríguez-Seijo
Centro Interdisciplinar de Investigação Marinha e Ambiental (CIIMAR), University of Porto, Matosinhos, Portugal

Biology Department, Faculty of Sciences, University of Porto, Porto, Portugal

Avelino Núñez-Delgado and Manuel Arias-Estévez (eds.),
Emerging Pollutants in Sewage Sludge and Soils,
Hdb Env Chem (2023) 114: 267–288, DOI 10.1007/698_2022_859,

Published online: 13 February 2022

media than for soils, due to the difficulties affecting analytical methods for extraction and identification of MP in complex matrices such as soils. The procedures and processes needed to isolate MP from soils vary considerably among studies. Currently, there is no single method to isolate MP from soil. In general, the main techniques for MP separation from soil include sieving, density flotation, filtration, and digestion processes. Microplastics extracted from soil samples can be identified and quantified mainly by means of visual classification, scanning electronic microscopy (SEM), spectroscopic techniques (Raman Spectroscopy, Fourier Transform Infrared Spectroscopy (FT-IR), or Time of Flight Secondary Ion Mass Spectrometry (TOF-SIMS)), and thermoanalytical techniques (Thermal Extraction and Desorption Gas Chromatography coupled to Mass Spectrometry (TED-GC/MS), Pyrolysis-Gas Chromatography coupled to Mass Spectrometry (Pyr-GC/MS), and Thermogravimetric Analysis Mass Spectrometry (TGA-MS)), among others. The main advantages and disadvantages of these techniques are also discussed.

Keywords Ecosystems, Microplastics, Spectrometry, Thermoanalytical techniques

1 Introduction

Plastics have become indispensable in our daily lives, and their use and production have increased exponentially over the last 40 years. Plastics and microplastics (MP) (plastic debris smaller than 1 mm) have been identified as one of the most significant environmental issues in the last years due to their spread in all ecosystems, including remote ecosystems such as Antarctica. This represents an ecological issue, as waste management technology has not evolved at the same rate as plastic production/consumption. Although plastic pollution in marine ecosystems has been studied since the 1970 decade, the study of this issue in the context of terrestrial ecosystems is a recent concept (mainly since 2016). This even though plastics reaching the oceans are mainly originating from terrestrial ecosystems and there are 4–23 times more plastics in terrestrial ecosystems than in marine ecosystems [1–3].

After plastic debris discharging and once in the environment, they undergo degradation when exposed to physical, chemical, and/or biological agents, resulting in small particles (usually <1 mm). They may be easily dispersed, and consequently enter into the food chain, and damage organisms, which increases the potential to affect several ecosystem functions at many levels [1, 3–7]. In this sense, plastics can enter the terrestrial ecosystems through multiple (and some still unknown) pathways, such as agricultural applications, organic amendments from wastewater treatment plants, landfills, industrial sources, and others. Recently, the potential relevance of two understudied sources (probably the main sources) of plastics to the terrestrial ecosystems has been indicated: plastics released from the washing of synthetic clothes and car tire abrasion [2, 3, 7–13]. Besides, the emergence of COVID-19 crisis raised significant implications at many levels, including economic, health, and social issues. Nevertheless, this natural disaster has also become one of the major environmental threats due to the wide use of disposable face masks and gloves as one of the most relevant safety measures, which are produced from different types of plastic polymers. Most recent publications claim that the increase in production/consumption of these products is contributing to the accumulation of large amounts of plastic and plastic particles waste in the environment, reaching waterways (freshwater and marine environments) and the surrounding soils [14].

(Micro-)plastics may have a wide range of properties such as source, color (red, white, clear, blue, green, black, purple, yellowish, and brown), erosion grade, size (from 5 to <100 μm as nanoplastics), shape (fibers, triangular, rectangular, circular, polygonal, column, spherical, pellet-shaped, cuboid, cone-shaped, pyramidal, and irregular shapes), and chemical composition (low- and high-density polyethylene (LDPE and HDPE, respectively)), polystyrene (PS), polypropylene (PP), polyethylene terephthalate (PET), polyamides (PA), etc.). Besides, much plastic debris (mainly as fibers or smaller items (<250 μm)) can be confused with soil components such as organic matter, clays, and oxides, which are difficult to distinguish and estimate quantitatively MP into the soils [15, 16].

In this sense, the methodology for the determination and quantification of MP is currently much easier and developed in aqueous media than in soils [17]. Although the number of studies focused on the measurement of MP concentrations in terrestrial environments has increased in the last years, there is still scarce knowledge about their extraction, quantification, identification, and the relationship between MP and adverse effects, fate, and toxicology in soils [1, 18, 19].

For example, plastics used as covers in agricultural activities may be fragmented due to abrasion and erosion over time, giving rise to the formation of MP, which cannot be removed easily and usually remain deposited in the soil [3, 20]. However, the amount, fate, and effects of MP need to be assessed in depth, being an actual challenge due to the lack of sufficient analytical methods. Mainly, difficulties affecting analytical methods for extraction and identification of MP are some of the reasons for the limited data of MP in soils. Unlike aquatic systems, the identification and quantification of MP in complex matrices such as soils are much more problematic. In addition, no validated and standardized methodologies are published for soils and terrestrial environments, unlike aquatic systems and sediments, and these questions are responsible for the scarce advances taking place in the characterization of the problem in this environmental compartment [7]. This chapter delves into the most recent methods related to the separation, identification, determination, and quantification of MP in soils.

2 Collecting Soil Samples for MP Analysis

A correct sampling of soil contaminated by plastics is essential to perform a subsequent appropriate analysis. Among the factors that must be taken into account are the type of soil studied – since it may determine the sampling scheme –, the depth of sampling, the sampling area, and the number of samples to be collected [21]. It is essential that the sampling points are representative of the contamination in the study area. Soil sampling can be punctual or compound. The MP distribution in soils tends to be heterogeneous and their size used to be variable. Thus, composite soil samples – within a determined sampling subarea – are usually more representative of the global content of MP in the entire study area. Besides, a higher number of samples collected in the study area, as well as a greater amount of soil sampled at each point, is more representative than sampling with few samples and little amount of soil, due to the heterogeneous distribution of MP in soils. Also, it is essential to identify possible areas of potential accumulation, such as cavities, where MP can be deposited as a consequence of surface runoff, or areas where the deposition of suspended particles may be greater. The choice of sampling depth is another factor that decisively influences a correct sampling. When sampling is done on undisturbed soil, taking the first few centimeters (10 cm) may be sufficient, since this is the most likely range to find most of the MP. However, in disturbed soils, such as plowed soils, sampling depth must be adapted accordingly to the vertical distribution of MP in the soil [21]. Up to now, samplings have been carried out for MP analysis in agricultural soils, alluvial plains, paddy soils, urban zones, or forest areas [10, 22–28].

Sample collection is done with typical sampling materials used to collect soils, such as stainless steel shovel, stainless steel corer, soil auger, or Lenz sampler [29–32]. Then, samples must be transferred to labeled glass or metal containers with lids (without plastic) for transport and storage. It is necessary to avoid contamination of samples with MP from tools, clothing, or the ambient air of laboratories and storage facilities.

3 Methodology for MP Separation from Soil

The procedures and processes used to isolate MP from soil vary considerably among studies. Currently, there is no single method to isolate MP from soil samples, although many of them share stages of the process. In general, the main techniques for MP separation from soil include sieving, density flotation, filtration, and digestion (Fig. 1).

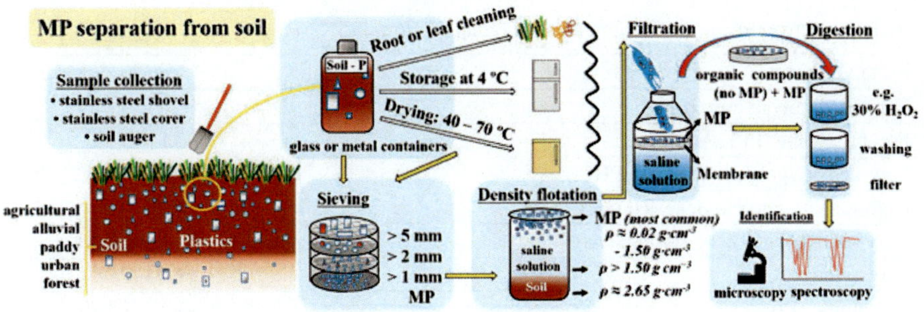

Fig. 1 Most common steps to isolate MP from soils and subsequent identification

3.1 *Soil Sieving*

The initial step to separate MP from soils is dry soil sieving, having the objective of removing oversized debris plastic. This pretreatment – used before density flotation – allows to separate and collect all plastic particles >5 mm, or even smaller (up to 2 mm) through the use of meshes of different sizes [21, 33].

Sieving is usually the simplest method for plastic and MP separation from soil, which allows a mechanical separation and manual classification, through the use of stereoscopic microscopes. Before sieving, soil samples may be subjected to storage at 4°C in the laboratory, and/or air drying [33]. Likewise, soil samples can also be dried in an oven with temperatures between 40°C and 70°C, although it is possible for MP to be deformed and break if heating exceeds the thermal deformation temperature of plastic. When dry soil contains a high content of plant material, such as roots or leaves, it must be previously cleaned manually. After sieving, MP remaining on sieves must be collected and classified for later identification and characterization. Although the use of a 5-mm sieve seems to be more suitable for the classic MP definition, 1-mm and 2-mm sieves should be used, depending on the content of plant material or other “impurities” that soils may contain [33].

3.2 *Extraction*

3.2.1 Density Flotation

Density flotation is one of the most used procedures for the separation of MP from soil [34]. In general, soils present a density between 2.6 and 2.7 g $\cdot$ cm^{-3}. The rationale behind this technique is based on taking advantage of the different densities of soil mineral particles and MP (usually between 0.9 and 1.45 g $\cdot$ cm^{-3}), which is used to separate them by flotation in a high-density saline aqueous solution (of known density) as an extraction medium [35]. Soil particles with high densities

Table 1 Density of most common plastics (adapted from [15, 36])

Plastic polymers	Density ($g \cdot cm^{-3}$)
Expanded polystyrene (EPS)	0.01–0.04
Polypropylene (PP)	0.85–0.92
Low-density polyethylene (LDPE)	0.89–0.93
High-density polyethylene (HDPE)	0.94–0.98
Acrylonitrile-butadiene-styrene (ABS)	1.04–1.06
Polystyrene (PS)	1.04–1.08
Polyamide (PA)	1.13–1.16
Polymethyl methacrylate (PMMA)	1.16–1.20
Polycarbonate (PC)	1.20–1.22
Cellulose acetate (CA)	1.30
Polyethylene terephthalate (PET)	1.38–1.41
Polyvinyl chloride (PVC)	1.38–1.41
Polytetrafluoroethylene (PTFE)	2.10–2.30

settle to the bottom of a container or vial, while MP with low density floats on the surface of the solution. Table 1 shows the densities of the most common plastics.

The densities of the most common MP are in the range of 0.015–1.50 $g \cdot cm^{-3}$; thus, the flotation solution needs to have a density >1.50 $g \cdot cm^{-3}$, except for polytetrafluoroethylene "Teflon" (PTFE). Among the high-density saline aqueous solution most used to separate plastics from soils by flotation are binary salts. Saturated solutions of NaCl are the most used, due to their low cost and risk [37–39], although these are not effective to separate plastics with densities higher than 1.18 $g \cdot cm^{-3}$, such as polyvinyl chloride (PVC) or PET (polyethylene terephthalate) [7, 21]. As an alternative to NaCl, several studies have proposed other saline saturated solutions, and even flotation methods based on more than one extraction using different saline solutions such as: (1) sodium iodide (NaI) (1.8 $g \cdot cm^{-3}$) [40–42], (2) zinc chloride ($ZnCl_2$) (1.6–1.7 $g \cdot cm^{-3}$) [43, 44], (3) sodium bromide (NaBr) (1.55 $g \cdot cm^{-3}$) [45], (4) sodium polytungstate ($Na_6[H_2W_{12}O_{40}]$) (1.4 $g \cdot cm^{-3}$) [46], and (5) potassium formate (HCO_2K) [47].

Besides, the extraction of MP has been proposed to be carried out through the combined use of different saline solutions. For example, extractions with NaCl (1.2 $g \cdot cm^{-3}$) and NaI (1.6 $g \cdot cm^{-3}$) have been used with a continuous flow and a floating separation device [48]. Also, an extraction in three times has been used with distilled water (1 $g \cdot cm^{-3}$), NaCl (1.20 $g \cdot cm^{-3}$) and followed by $ZnCl_2$ (1.55 $g \cdot cm^{-3}$) with shaking and centrifugation [49] or with NaCl (1.2 $g \cdot cm^{-3}$) and $CaCl_2$ (1.5 $g \cdot cm^{-3}$) by using a sedimentation cylinder [50]. Likewise, distilled water can be used for MP with a density lower than 1.0 $g \cdot cm^{-3}$, such as PE (especially light-density PE [LDPE]) and PP, widely used in agriculture as plastic mulch [16].

The use of each of these other solutions shows a series of advantages and disadvantages that must be considered. From the perspective of density, cost, and respect for the environment, $CaCl_2$ seems to be a good option for the separation of

MP, even better than NaCl. However, it has been reported that Ca^{2+} may agglomerate with soil organic matter [50] and adhere to the filter membrane, interfering with the counting and identification of MP [33]. $ZnCl_2$ is considered corrosive and toxic, and NaI may be hazardous, while HCO_2K and $Na_6[H_2W_{12}O_{40}]$ are rather expensive. The density of NaI solutions is high enough ($1.8\ g \cdot cm^{-3}$); however, it is expensive. Thus, the combined use of NaCl and NaI solutions in a 1:1 (v/v) ratio is considered one of the most efficient and relatively economical extractions [34]. Also, NaBr was suggested for the extraction of MP due to its high efficiency for most polymers in soils and its advantages of being a low-cost and environmentally friendly solution [51].

Besides, other compounds have been suggested for plastic extraction from soils, such as canola, castor, and/or olive oil, due to their lipophilic conditions [52–54]. These alternative methods have several advantages such as low cost, easy access, and being environmentally friendly solutions, with good results for extraction of high-density polymers with recovery rates over 90% [21, 54, 55].

In general, all density separation solutions show limitations and affect the separation efficiency of MP. Therefore, the choice of solutions must be made based on the specific characteristics of the MP present in the soil samples. However, when the overall picture of MP distribution is still unclear, a separation solution with a high density is more effective [3].

In general, MP aggregate with soil components, such as organic matter and clays. To separate them from the soil, other procedures have been described – in addition to density flotation – such as agitation, centrifugation, ultrasonication, aeration, and continuous flow techniques [34, 56, 57], usually applied in combination with density extraction [57].

(Micro-)plastics, like polymers, show hydrophobic, lipophilic, and electrostatic properties, which allow the use of alternative methods for separating from soils [58]. Some studies have indicated that, based on the hydrophobicity of MP, foam flotation can be used to extract them from the soil. For that, air bubbles bind selectively to the more hydrophobic particles moving them toward the surface and separating them from the less hydrophobic matrix [59, 60]. Also, new methods using electrostatic separation have been described to separate MP from multiple environmental matrices, including water, sediments, and bleach sands, but it is uncertain that these methods can be suitable for large-scale separations of MP from soils [61, 62]. Recently, a method that uses extraction with pressurized fluids and solvents – such as methanol, hexane, and dichloromethane – has been proposed to extract MP from soils under high-temperature (180–190°C) and high-pressure conditions (103 bar). After extraction, solvents are removed from the extracted residues by evaporation under a stream of nitrogen [56]. Despite the high efficiency of this method, it has some deficiencies such as the need for organic solvents which are highly dangerous for the environment, the fact that natural organic matter may also dissolve and part of MP can be pyrolyzed at high temperatures, destroying their shape and microstructure without being able to be identified and characterized [51, 63].

These techniques, especially those based on Na-salts, should be applied carefully on soils with higher clay contents to avoid clay dispersion by Na, to avoid interference with MP identification. For these reasons, each soil requires a previous study of its characteristics, as well as the adjustment of each soil extractant according to its properties.

3.3 *Filtration*

After the density flotation step, it is necessary to separate the MP from the saline solution. One of the most used methods is solid-liquid filtration through fiberglass, cellulose acetate, cellulose nitrate, polycarbonate, nylon, or alumina membranes that allow the separation of MP, and their subsequent identification by optical or spectroscopic analysis [17]. Filter pore sizes range from 0.2 μm [50] to 20 μm [31].

The filter composition may have consequences in the subsequent identification of MP. For example, the surface of glass fiber filter membranes is very rough, and also has the disadvantage of releasing fibers, so it is not suitable for a later identification by IR spectroscopy. The polycarbonate filter membrane is not hydrophilic, which does not favor the retention of MP. In addition, polycarbonate has a strong infrared signal, which causes disturbance when MPs are identified by FT-IR. In contrast, the membrane surfaces of cellulose or nylon filters are smooth and convenient for extraction [30].

3.4 *Digestion of Organic Matter*

There is no standardized analysis scheme for separating MP in soil, mainly because soil is rich in various complex organic compounds [64, 65]. It is difficult to completely separate (and then distinguish) organic compounds from MP by conventional salt solution density separation, whether focusing on farmland soil, sludge, or other stable soil structures [65]. The treatment of organic matter becomes the key step in detecting MP in soil samples. Hydrogen peroxide (H_2O_2) and Fenton reagent (H_2O_2 combined with $FeSO_4$) are still common reagents for the digestion of organic compounds [66]. Hydrogen peroxide at 70°C has a better effect on the digestion of organic compounds [67]. However, some studies have found that H_2O_2 will reduce the extraction efficiency of MP [68]. Sodium hydroxide (NaOH) and KOH are also used to digest organic matter in soil samples. Although they effectively remove organic matter from the soil, these two reagents can cause MP discoloration and degradation [69]. A recent study used 98% H_2SO_4 to separate MP from farmland soil to obtain pure polyethylene [65].

A large amount of organic matter is also a problem to be solved in the extraction of microplastics in soils, which cannot be completed by density separation [2, 50, 64]. Some of these authors evaluated the efficiency of the most common reagents used to remove the soil organic fraction (30% H_2O_2, 65% HNO_3, 50% NaOH, 96% H_2SO_4 and 13% KClO), and found that 65% HNO_3 reacting with the sample at 90°C showed higher efficiency for organic matter removal. It can remove most organic matter in a reduced time, while HNO_3 treatment can cause partial decomposition of PA, PET, and PMMA polymers. Due to the time-consuming process, the removal efficiency of hydrogen peroxide (10–30% H_2O_2), which is most often used to remove organic matter from environmental matrices, has been also questioned [66]. For these reasons, Fenton's reagent, an advanced oxidation process using H_2O_2 in the presence of Fe^{2+} as a catalyzer, has been indicated as an efficient method with reduced harmful effects in organic material removal [17, 33, 66]. However, Fenton's reagent has been associated with several issues for soils with higher carbonate contents, which can impact the pH and reduce the reaction efficiency. Besides, not all laboratories have appropriate conditions for using these methodologies, since organic matter digestion with Fenton's reagent needs to be carried out with an ice bath to avoid trouble due to high temperatures.

4 Quantification and Identification

The methodology for characterization, quantification, and classification of MP is not standardized, which forces researchers to reach a consensus on the protocol for determining MP. After extraction and separation processes, MP can be analyzed, identified, and classified. Potential plastic particles that are extracted from soil samples can be identified and quantified mainly by visual classification (including microscopy), spectroscopic techniques (Raman Spectroscopy and Fourier Transform Infrared Spectroscopy (FT-IR)), and also techniques such as thermoanalytical procedures, including Thermal Extraction and Desorption Gas Chromatograph coupled to Mass Spectrometry (TED-GC/MS), Pyrolysis-Gas Chromatography coupled to Mass Spectrometry (Pyr-GC/MS), and Thermogravimetric Analysis Mass Spectrometry (TGA-MS), among others.

Currently, visual analysis (with or without the aid of an optical microscope or stereomicroscope) continues to be the simplest method and one of the most widely used to differentiate between MP and other compounds [70], despite its low precision and potential misidentification and confusing with soil components. Nowadays, the most widely used methods for the identification and characterization of MP in soils are Fourier Transform Infrared Spectrometry (FT-IR), Raman spectrometry, and Scanning Electron Microscopy (SEM) [23, 71–74].

In general, visual characterization is used as a preliminary identification or first screening of MP, which makes it possible to reduce the number of particles that must be chemically characterized in subsequent analyses [75, 76]. In this line, Lorenzo-Navarro et al. [77] developed a computational method based on deep learning networks to automatically count and classify MP from 1 to 5 mm from images larger than 16 megapixels. However, this method can be imprecise and requires that the MP to be determined is specified previously.

Optical microscopy makes it possible to quickly determine the shape (e.g., spherical, cylindrical, long and thin fibers, disc, round, irregular ovoid, etc.) and the color (mainly white or similar) of MP [78]. In any case, visual inspection is not totally reliable and has marked limitations, such as its low sensitivity (valid for particle sizes >200 nm) [79], low differentiation among MP and other inorganic compounds, such as clays or organic fibers of cellulose and starch, due to the characteristics of these compounds (similar colors, sizes, and shapes) [80]. Likewise, the visual quantification process of MP is imprecise, which does not facilitate their correct identification and quantification, especially for those MP with a small size [81]. Some studies have reported that false identification of non-plastic particles under a microscope is generally above 20–70%, which was confirmed by spectroscopic analysis [82].

To differentiate MP from other particles, different studies have used the hot needle test. The foundation of this technique is based on the thermoplastic characteristics of MP. With this methodology, microscopic images of the samples are obtained before and after subjecting the MP and other compounds of the sample to thermal heating ($\approx$ 130°C). It is considered that the MPs are particles that in the microscopy images appear fused after thermal heating [83, 84]. In any case, this technique presents limitations in identification, since (for example) some high-density and thermosetting plastics do not melt under these conditions, which does not allow determining the MP polymer composition [85]. Due to the approximate results obtained with the optical microscopic examination, the use of additional analysis techniques [86], spectroscopic techniques or electron microscopy, is recommended to obtain results about the characteristics and chemical composition of the polymers [87].

Fourier Transform Infrared Spectroscopy (FT-IR) is one of the most common and advanced analytical methods for studying MP in soils. This technique is used to accurately identify the chemical composition of MP, determining and distinguishing between polymers and other substances in the sample. The FT-IR and its optimized technologies, such as μ-FT-IR, Total Attenuated Reflectance-FT-IR (ATR-FT-IR), and Focal Plane Array-FT-IR (FPA-FT-IR) have also been the most widely used techniques for the chemical characterization of plastic in soils [23, 72, 88–90].

FT-IR is a simple, efficient, non-destructive, and low-cost MP identification technology, although the size of detectable MP is limited. Its foundation is based on the fact that the spectrometer signal – known as infrared absorption bands of the polymer under study – depends on the change in the permanent dipole moment of a chemical bond, making it sensitive to the polar functional groups in different plastic polymers [91]. The FT-IR technique is also a surface-based technology, which means that for correct identification, plastic particles must be absent from any coating, film, or material [92]. Furthermore, its spectral quality can be influenced by external factors, such as the presence of organic matter and/or water [92, 93]. The FT-IR devices have spatial resolutions up to 5 μm [91]. However, this technique requires minimum sample thicknesses of 150 nm [94] and the deposition of the sample on an IR transparent substrate [95]. Therefore, FT-IR has good reproducibility and it is not affected by sample color, fluorescence, and other conditions, being suitable for identification of MP with a diameter of more than 20 mm or samples that contain a strong polar functional group [96]. The FT-IR signal can be obtained by reflection or transmission of IR signal or also by attenuated total reflection (ATR), which can be used to identify particles >500 μm [97]. It has been reported that FT-IR can be connected to a microscope, allowing measurements (spectral identification) of smaller plastic particles (up to 10 μm) [98]. On the other hand, for black particles, the FT-IR is usually inaccurate due to the high absorption of infrared radiation. Agglomerates or smaller particles can be analyzed using a focal plane matrix (FPA) – μ-FT-IR, with which particle sizes ranging from 10 to 500 μm can be identified [95].

Raman spectroscopy is a suitable technique to identify MP easily, also being fast and non-destructive, with a higher sensitivity than FT-IR, and with the ability to identify particles up to approximately 1 μm [99]. Raman spectroscopy is based on the fact that when MP particles are exposed to monochromatic light, molecular vibrations cause light scattering and allow the identification of the polymer type. The main advantage of Raman spectroscopy is that the identification is based on the full wavelength spectrum and amorphous carbon can be detected. Raman spectroscopy can compensate for some shortcomings of FT-IR. For example, this technique provides a better response to non-polar information and vibration of the molecular structure of the sample; it presents a better spatial resolution (approximately 1 μm), so the shape and thickness of the small particles do not influence the measurement [100]. In addition, this technique presents less interference due to humidity and CO_2 [91]. Surface-Enhanced Raman Spectroscopy (SERS) can identify MP lesser than 500 nm. For example, Xu et al. [101] identified with this methodology MPs with sizes up to 450 nm. These advantages make Raman spectroscopy better than FT-IR for identifying MP, but the process takes longer than FT-IR [102]. However, organic matter, microorganisms [102], background fluorescence [103], and inorganic material [91] can strongly interfere with the spectrum, preventing the correct identification of MP.

FT-IR and Raman are the most used techniques for the identification of MP in soils. However, they do not guarantee that every particle will be detected and counted, due to size limitations [104]. In addition, the analysis of soil samples requires extensive and time-consuming cleaning procedures to degrade organic materials before particles are separated for analysis [92]. Both techniques provide data on types of plastic, number of particles in each plastic, size, shape, and color, but not on mass or concentration of plastic in the soil samples [105]. Many times, FT-IR and Raman spectroscopy are used to analyze a subset of plastic particles isolated from each sample, and then extrapolate the results to estimate their distribution within the entire sample. This makes the reported data prone to errors, omissions, and a lack of representativeness [92].

Scanning Electron Microscopy (SEM) has been indicated with potential use for the identification of smaller plastic items (<0.2 μm), because this technique provides high-resolution images of MP by means of a high-intensity electron beam that is indexed on the surface of the sample [84]. Detailed images of the surface of MP can be obtained, as well as differentiating them from other organic or inorganic compounds of soils, by analyzing the surface morphology of the particles. Among the limitations of the technique are the inability to detect the color of the particles, the lack of information on polymer composition, and the time and cost to prepare and analyze each sample [82].

There are other analytical techniques such as Thermal Extraction and Desorption Gas Chromatograph coupled to Mass Spectrometry (TED-GC/MS), Pyrolysis-Gas Chromatography coupled to Mass Spectrometry (Pyr-GC/MS), Thermogravimetric Analysis Mass Spectrometry (TGA-MS), which are effective for identifying and quantifying plastic particles in soils, by analyzing characteristic thermal degradation products of plastic [106, 107]. The results obtained with these techniques indicate that Pyr-GC/MS is a sensitive and robust method for the qualitative and quantitative analysis of polymers in samples, but it is not suitable for the analysis of samples with PMs of unknown composition. Likewise, thermogravimetric analysis is also limited, since, for example, some MP, such as PE and PP, can only be analyzed when the technique is combined with other analytical methods [21, 106]. Although these techniques allow to analyze MP with size ranges lower than those required by Raman spectrometry and FT-IR, and they can also be used for mass-based concentrations, they are destructive and cannot provide morphological information about particles such as size, color, shape, or number of plastic particles [92] (Table 2).

Other methods such as vis-NIR spectroscopy were used to quickly evaluate the concentrations of plastic (LDPE, PET, and PVC) in soils without a previous extraction method, although the technique showed a low prediction precision [108]. It has been reported as a relatively practical technique compared to conventional FT-IR analysis, being faster, avoiding extraction steps, and quantifying directly the sum of MP in the samples. Also, Du et al. [109] successfully identified plastic particles (PP, PVC, PET, and polyamide) in soils in terms of particle size and abundance using the Time-of-Flight Secondary Ion Mass Spectrometry Technique (TOF-SIMS). The results of this study showed that TOF-SIMS can be used to quickly analyze plastic particles in soils while simultaneously obtaining imaging and mass spectrometry. However, the fragmentation ions of some plastic particles do not allow differentiating from each other in mass spectrometry. This occurs for PP and PE, which could not be distinguished based on their observed ions. Therefore, the relative ionic intensity of the suspected areas of PP was calculated and compared with those obtained from the PE and PP standards to get its identification [109]. Likewise, other methods, such as the Near-infrared (NIR) spectroscopy process, have been used in combination with chemometric for the detection of plastic particles (PET, PP, PVC, PE, and PS; <125 μm) in soils. The Pressurized Liquid Extraction (PLE) with Pyr-GC/MS or FT-IR, or Hyperspectral Imaging Technology with chemometrics, have also been used to detect MP in soils [56, 110–112].

Table 2 Advantages and disadvantages/limitations of the most common techniques for quantification and identification of MP in soils

Group technique	Technique	Advantages	Disadvantages/limitations
Visual techniques	Visual characterization	• Fast • Cheap • Simple identification of shape, size, and colors • Non-destructive	• Low precision • Potential misidentification • Subjective analysis • No small plastic particles (<1 mm) • No chemical composition • No quantification
	Optical microscopy/ stereomicroscope	• Fast • Cheap • Simple identification of shape, size, and colors • Non-destructive	• Low precision • Potential misidentification • Subjective analysis • No particle sizes <200 nm low differentiation between MP and other no MP compounds • No chemical composition • No quantification
	SEM	• High precision (<200 nm) • Shape and size identification • Non-destructive, • Differentiation from no MP compounds	• High cost • Long time to prepare and analyze samples • No color detection • No chemical composition
Spectroscopic techniques	Raman	• Easy sample preparation • Non-destructive, • High-resolution images • Identification of MP composition • Good spatial resolution (1 μm) • SERS (limit detection 500 nm)	• High cost • Long time analysis • Samples require refinements • Small area analysis • Interferences with organic/inorganic materials
	FT-IR	• Fast • Simple analysis • Non-destructive, particles • Limit detection (> 500 μm) • Low-cost identification • Good reproducibility • Automated • μ-FT-IR (detection 10 to 500 μm)	• Strong interference from water and organic matter • Particles without coating, film, or other materials • Sample thicknesses (> 150 nm) • Sample preparation on IR transparent material

(continued)

Table 2 (continued)

Group technique	Technique	Advantages	Disadvantages/limitations
Spectroscopic techniques	TOF-SIMS	• Direct analysis of plastic particles • Imaging and mass spectrometry simultaneously • High spatial resolution • High sensitivity • Easy sample preparation	• Fragmentation of samples in ions • No direct identification
Thermoanalytical techniques	TED-GC/MS	• Fast analysis • Quantification of MP • Identification of MP • High samples masses	• High cost • MP destruction • No morphological information (size, color, shape, or number of plastic particles) • Dry sample needed • Pre-concentration
	Pyr-GC/MS	• Easy and reliable • Sensitive and robust • Quantification of MP • Identification of MP • Identification of additives	• High cost • Small samples masses • MP destruction • No morphological information (size, color, shape, or number of plastic particles) • Dry sample needed • Pre-concentration
	TGA-MS	• Detection of all particle sizes • Analysis of large samples (15–25 mg) • Quantification of MP • Identification of MP	• MP destruction • No morphological information (size, color, shape, or number of plastic particles)

5 Conclusions

As we have just seen, different issues remain to be improved in terms of extraction, identification, and others, regarding MP in soils, mainly due to the absence of a standardized protocol covering their extraction, quantification, and identification. Unlike aquatic ecosystems, terrestrial ecosystems are very heterogeneous. The development of standardized protocols for plastic extraction from soils is very challenging because it is matrix affected by high complexity and diversity among regions or even in soils collected within a few meters from each other. Therefore, future studies should focus on how to improve the extraction of microplastics without affecting the integrity of the soil, especially when digesting organic matter, as well as on the creation of protocols adapted to different soil properties.

Acknowledgments This work has been supported by the LABEX DRIIHM, French program "Investissements d'Avenir" (ANR-11-LABX-0010), which is managed by the Agence Nationale de la Recherche (ANR) within the Observatoire Hommes-Milieux Estarreja (OHM-E/2018/Proj.4); by national funds through FCT – Foundation for Science and Technology within the scope of UIDB/04423/2020 and UIDP/04423/2020 (CIIMAR); and by the project associated to the Xunta de Galicia postdoctoral grants (ED481D 2019/007). ARS wants to acknowledge the COST Action CA20101, and the FCT and CIIMAR for their research contract (CEECIND/03794/2017). VSM thanks for her pre-doctoral fellowship (ED481A-2020/089) financed by Xunta de Galicia. DAL thanks the Xunta de Galicia and the University of Vigo for the postdoc grant (ED481D 2019/007).

References

1. Guo J-J, Huang X-P, Xiang L, Wang Y-Z, Li Y-W, Li H, Cai Q-Y, Mo C-H, Wong M-H (2020) Source, migration and toxicology of microplastics in soil. Environ Int 137. https://doi.org/10.1016/j.envint.2019.105263
2. Kumar M, Xiong X, He M, Tsang DCW, Gupta J, Khan E, Harrad S, Hou D, Ok YS, Bolan NS (2020) Microplastics as pollutants in agricultural soils. Environ Pollut 265. https://doi.org/10.1016/j.envpol.2020.114980
3. Rodríguez-Seijo A, Santos B, Ferreira Da Silva E, Cachada A, Pereira R (2019) Low-density polyethylene microplastics as a source and carriers of agrochemicals to soil and earthworms. Environ Chem 16:8–17. https://doi.org/10.1071/EN18162
4. de Souza Machado AA, Kloas W, Zarfl C, Hempel S, Rillig MC (2018) Microplastics as an emerging threat to terrestrial ecosystems. Glob Chang Biol 24:1405–1416. https://doi.org/10.1111/gcb.14020
5. De Souza Machado AA, Lau CW, Kloas W, Bergmann J, Bachelier JB, Faltin E, Becker R, Görlich AS, Rillig MC (2019) Microplastics can change soil properties and affect plant performance. Environ Sci Technol 53:6044–6052. https://doi.org/10.1021/acs.est.9b01339
6. De Souza MacHado AA, Lau CW, Till J, Kloas W, Lehmann A, Becker R, Rillig MC (2018) Impacts of microplastics on the soil biophysical environment. Environ Sci Technol 52:9656–9665. https://doi.org/10.1021/acs.est.8b02212
7. Rodríguez-Seijo A, Pereira R (2020) Small plastic wastes in soils: what is our real perception of the problem? In: Streit-Bianchi M, Cimadevila M, Trettnak W (eds) Mare plasticum – the Plastic Sea. Springer, Cham. https://doi.org/10.1007/978-3-030-38945-1_9
8. Henry B, Laitala K, Klepp IG (2019) Microfibres from apparel and home textiles: prospects for including microplastics in environmental sustainability assessment. Sci Total Environ 652:483–494. https://doi.org/10.1016/j.scitotenv.2018.10.166
9. Knight LJ, Parker-Jurd FNF, Al-Sid-Cheikh M, ThoMPon RC (2020) Tyre wear particles: an abundant yet widely unreported microplastic? Environ Sci Pollut Res 27:18345–18354. https://doi.org/10.1007/s11356-020-08187-4
10. Ng EL, Lin SY, Dungan AM, Colwell JM, Ede S, Huerta Lwanga E, Meng K, Geissen V, Blackall LL, Chen D (2021) Microplastic pollution alters forest soil microbiome. J Hazard Mater 409. https://doi.org/10.1016/j.jhazmat.2020.124606
11. Qi R, Jones DL, Li Z, Liu Q, Yan C (2020) Behavior of microplastics and plastic film residues in the soil environment: a critical review. Sci Total Environ 703. https://doi.org/10.1016/j.scitotenv.2019.134722

12. Qi Y, Ossowicki A, Yang X, Huerta Lwanga E, Dini-Andreote F, Geissen V, Garbeva P (2020) Effects of plastic mulch film residues on wheat rhizosphere and soil properties. J Hazard Mater 387. https://doi.org/10.1016/j.jhazmat.2019.121711
13. Sanderson P, Qi F, Seshadri B, Wijayawardena A, Naidu R (2018) Contamination, fate and management of metals in shooting range soils—a review. Curr Pollut Rep. https://doi.org/10.1007/s40726-018-0089-5
14. Ammendolia J, Saturno J, Brooks AL, Jacobs S, Jambeck JR (2021) An emerging source of plastic pollution: environmental presence of plastic personal protective equipment (PPE) debris related to COVID-19 in a metropolitan city. Environ Pollut 269. https://doi.org/10.1016/j.envpol.2020.116160
15. Rodríguez-Seijo A, Pereira R (2017) Morphological and physical characterization of microplastics. Compr Anal Chem. https://doi.org/10.1016/bs.coac.2016.10.007
16. Zhang S, Wang J, Liu X, Qu F, Wang X, Wang X, Li Y, Sun Y (2019) Microplastics in the environment: a review of analytical methods, distribution, and biological effects. Trends Anal Chem 111:62–72. https://doi.org/10.1016/j.trac.2018.12.002
17. Zhou Y, Wang J, Zou M, Jia Z, Zhou S, Li Y (2020) Microplastics in soils: a review of methods, occurrence, fate, transport, ecological and environmental risks. Sci Total Environ 748. https://doi.org/10.1016/j.scitotenv.2020.141368
18. Wang J, Liu X, Li Y, Powell T, Wang X, Wang G, Zhang P (2019) Microplastics as contaminants in the soil environment: a mini-review. Sci Total Environ 691:848–857. https://doi.org/10.1016/j.scitotenv.2019.07.209
19. Xu B, Liu F, Cryder Z, Huang D, Lu Z, He Y, Wang H, Lu Z, Brookes PC, Tang C, Gan J, Xu J (2020) Microplastics in the soil environment: occurrence, risks, interactions and fate–a review. Crit Rev Environ Sci Technol 50:2175–2222. https://doi.org/10.1080/10643389.2019.1694822
20. Steinmetz Z, Wollmann C, Schaefer M, Buchmann C, David J, Tröger J, Muñoz K, Frör O, Schaumann GE (2016) Plastic mulching in agriculture. Trading short-term agronomic benefits for long-term soil degradation? Sci Total Environ 550:690–705. https://doi.org/10.1016/j.scitotenv.2016.01.153
21. Möller JN, Löder MGJ, Laforsch C (2020) Finding microplastics in soils: a review of analytical methods. Environ Sci Technol 54:2078–2090. https://doi.org/10.1021/acs.est.9b04618
22. Álvarez-Lopeztello J, Robles C, del Castillo RF (2021) Microplastic pollution in neotropical rainforest, savanna, pine plantations, and pasture soils in lowland areas of Oaxaca, Mexico: preliminary results. Ecol Indic 121. https://doi.org/10.1016/j.ecolind.2020.107084
23. Boughattas I, Hattab S, Zitouni N, Mkhinini M, Missawi O, Bousserrhine N, Banni M (2021) Assessing the presence of microplastic particles in Tunisian agriculture soils and their potential toxicity effects using Eisenia andrei as bioindicator. Sci Total Environ 796. https://doi.org/10.1016/j.scitotenv.2021.148959
24. Choi YR, Kim Y-N, Yoon J-H, Dickinson N, Kim K-H (2021) Plastic contamination of forest, urban, and agricultural soils: a case study of Yeoju City in the Republic of Korea. J Soils Sediments 21:1962–1973. https://doi.org/10.1007/s11368-020-02759-0
25. Dong Y, Bao Q, Gao M, Qiu W, Song Z (2022) A novel mechanism study of microplastic and as co-contamination on indica rice (Oryza sativa L.). J Hazard Mater 421. https://doi.org/10.1016/j.jhazmat.2021.126694
26. Liu M, Lu S, Song Y, Lei L, Hu J, Lv W, Zhou W, Cao C, Shi H, Yang X, Yang X, He D (2018) Microplastic and mesoplastic pollution in farmland soils in suburbs of Shanghai, China. Environ Pollut 242:855–862. https://doi.org/10.1016/j.envpol.2018.07.051
27. Mani T, Hauk A, Walter U, Burkhardt-Holm P (2015) Microplastics profile along the Rhine River. Sci Rep 5. https://doi.org/10.1038/srep17988
28. Ragoobur D, Huerta-Lwanga E, Somaroo GD (2021) Microplastics in agricultural soils, wastewater effluents and sewage sludge in Mauritius. Sci Total Environ 798. https://doi.org/10.1016/j.scitotenv.2021.149326

29. Crossman J, Hurley RR, Futter M, Nizzetto L (2020) Transfer and transport of microplastics from biosolids to agricultural soils and the wider environment. Sci Total Environ 724. https://doi.org/10.1016/j.scitotenv.2020.138334
30. Ding L, Zhang S, Wang X, Yang X, Zhang C, Qi Y, Guo X (2020) The occurrence and distribution characteristics of microplastics in the agricultural soils of Shaanxi Province, in North-Western China. Sci Total Environ 720. https://doi.org/10.1016/j.scitotenv.2020.137525
31. Lv W, Zhou W, Lu S, Huang W, Yuan Q, Tian M, Lv W, He D (2019) Microplastic pollution in rice-fish co-culture system: a report of three farmland stations in Shanghai, China. Sci Total Environ 652:1209–1218. https://doi.org/10.1016/j.scitotenv.2018.10.321
32. van den Berg P, Huerta-Lwanga E, Corradini F, Geissen V (2020) Sewage sludge application as a vehicle for microplastics in eastern Spanish agricultural soils. Environ Pollut 261. https://doi.org/10.1016/j.envpol.2020.114198
33. Yang L, Zhang Y, Kang S, Wang Z, Wu C (2021) Microplastics in soil: a review on methods, occurrence, sources, and potential risk. Sci Total Environ 780. https://doi.org/10.1016/j.scitotenv.2021.146546
34. Han X, Lu X, Vogt RD (2019) An optimized density-based approach for extracting microplastics from soil and sediment samples. Environ Pollut 254. https://doi.org/10.1016/j.envpol.2019.113009
35. He D, Luo Y (2020) Microplastics in terrestrial environments. Emerging contaminants and major challenges. Springer, Cham. https://doi.org/10.1007/978-3-030-56271-7
36. Driedger AGJ, Dürr HH, Mitchell K, Van Cappellen P (2015) Plastic debris in the Laurentian Great Lakes: a review. J Great Lakes Res 41:9–19. https://doi.org/10.1016/j.jglr.2014.12.020
37. Eriksen M, Mason S, Wilson S, Box C, Zellers A, Edwards W, Farley H, Amato S (2013) Microplastic pollution in the surface waters of the Laurentian Great Lakes. Mar Pollut Bull 77: 177–182. https://doi.org/10.1016/j.marpolbul.2013.10.007
38. ThoMPon RC, Olsen Y, Mitchell RP, Davis A, Rowland SJ, John AWG, McGonigle D, Russell AE (2004) Lost at sea: where is all the plastic? Science 304:838
39. Wang W, Ndungu AW, Li Z, Wang J (2017) Microplastics pollution in inland freshwaters of China: a case study in urban surface waters of Wuhan, China. Sci Total Environ 575:1369–1374. https://doi.org/10.1016/j.scitotenv.2016.09.213
40. Claessens M, Van Cauwenberghe L, Vandegehuchte MB, Janssen CR (2013) New techniques for the detection of microplastics in sediments and field collected organisms. Mar Pollut Bull 70:227–233. https://doi.org/10.1016/j.marpolbul.2013.03.009
41. Dekiff JH, Remy D, Klasmeier J, Fries E (2014) Occurrence and spatial distribution of microplastics in sediments from Norderney. Environ Pollut 186:248–256. https://doi.org/10.1016/j.envpol.2013.11.019
42. Kapp KJ, Yeatman E (2018) Microplastic hotspots in the Snake and lower Columbia rivers: a journey from the greater yellowstone ecosystem to the Pacific Ocean. Environ Pollut 241: 1082–1090. https://doi.org/10.1016/j.envpol.2018.06.033
43. Bergmann M, Wirzberger V, Krumpen T, Lorenz C, Primpke S, Tekman MB, Gerdts G (2017) High quantities of microplastic in Arctic Deep-Sea sediments from the HAUSGARTEN observatory. Environ Sci Technol 51:11000–11010. https://doi.org/10.1021/acs.est.7b03331
44. Imhof HK, Schmid J, Niessner R, Ivleva NP, Laforsch C (2012) A novel, highly efficient method for the separation and quantification of plastic particles in sediments of aquatic environments. Limnol Oceanogr Methods 10:524–537. https://doi.org/10.4319/lom.2012.10.524
45. Li C, Cui Q, Zhang M, Vogt RD, Lu X (2021) A commonly available and easily assembled device for extraction of bio/non-degradable microplastics from soil by flotation in NaBr solution. Sci Total Environ 759. https://doi.org/10.1016/j.scitotenv.2020.143482
46. Martin J, Lusher A, ThoMPon RC, Morley A (2017) The deposition and accumulation of microplastics in marine sediments and bottom water from the Irish continental shelf. Sci Rep 7. https://doi.org/10.1038/s41598-017-11079-2

47. Zhang K, Xiong X, Hu H, Wu C, Bi Y, Wu Y, Zhou B, Lam PKS, Liu J (2017) Occurrence and characteristics of microplastic pollution in Xiangxi Bay of three gorges reservoir, China. Environ Sci Technol 51:3794–3801. https://doi.org/10.1021/acs.est.7b00369
48. Zhou Q, Zhang H, Fu C, Zhou Y, Dai Z, Li Y, Tu C, Luo Y (2018) The distribution and morphology of microplastics in coastal soils adjacent to the Bohai Sea and the Yellow Sea. Geoderma 322:201–208. https://doi.org/10.1016/j.geoderma.2018.02.015
49. Corradini F, Meza P, Eguiluz R, Casado F, Huerta-Lwanga E, Geissen V (2019) Evidence of microplastic accumulation in agricultural soils from sewage sludge disposal. Sci Total Environ 671:411–420. https://doi.org/10.1016/j.scitotenv.2019.03.368
50. Scheurer M, Bigalke M (2018) Microplastics in Swiss floodplain soils. Environ Sci Technol 52:3591–3598. https://doi.org/10.1021/acs.est.7b06003
51. Liu M, Song Y, Lu S, Qiu R, Hu J, Li X, Bigalke M, Shi H, He D (2019) A method for extracting soil microplastics through circulation of sodium bromide solutions. Sci Total Environ 691:341–347. https://doi.org/10.1016/j.scitotenv.2019.07.144
52. Crichton EM, Noël M, Gies EA, Ross PS (2017) A novel, density-independent and FTIR-compatible approach for the rapid extraction of microplastics from aquatic sediments. Anal Methods 9:1419–1428. https://doi.org/10.1039/c6ay02733d
53. Mani T, Frehland S, Kalberer A, Burkhardt-Holm P (2019) Using castor oil to separate microplastics from four different environmental matrices. Anal Methods 11:1788–1794. https://doi.org/10.1039/c8ay02559b
54. Scopetani C, Chelazzi D, Mikola J, Leiniö V, Heikkinen R, Cincinelli A, Pellinen J (2020) Olive oil-based method for the extraction, quantification and identification of microplastics in soil and compost samples. Sci Total Environ 733. https://doi.org/10.1016/j.scitotenv.2020.139338
55. Lechthaler S, Hildebrandt L, Stauch G, Schüttrumpf H (2020) Canola oil extraction in conjunction with a plastic free separation unit optimises microplastics monitoring in water and sediment. Anal Methods 12:5128–5139. https://doi.org/10.1039/d0ay01574a
56. Fuller S, Gautam A (2016) A procedure for measuring microplastics using pressurized fluid extraction. Environ Sci Technol 50:5774–5780. https://doi.org/10.1021/acs.est.6b00816
57. Zhang S, Yang X, Gertsen H, Peters P, Salánki T, Geissen V (2018) A simple method for the extraction and identification of light density microplastics from soil. Sci Total Environ 616–617:1056–1065. https://doi.org/10.1016/j.scitotenv.2017.10.213
58. Zhai J, Yan J, Rao Y, Yang T (2021) Microplastic in wetlands: a review on distribution and detection methods. Tumu yu Huanjing Gongcheng Xuebao/Journal Civ Environ Eng 43:158–167. https://doi.org/10.11835/j.issn.2096-6717.2020.137
59. Jian M, Zhou L, Li W, Rao D, Wu X, Zhang C, Liu S (2020) Microplastics in the sediments: from Rivers to lakes investigated in the biggest freshwater Lake (Poyang Lake) basin and Yangtze River in China. Handb Environ Chem. https://doi.org/10.1007/698_2020_464
60. Liu S, Jian M, Zhou L, Li W (2020) Distribution and composition of microplastics in sediments in the estuaries of the Ganjiang River leading into Lake Poyang. Acta Pedol Sin 57:908–916. https://doi.org/10.11766/trxb201901080016
61. Felsing S, Kochleus C, Buchinger S, Brennholt N, Stock F, Reifferscheid G (2018) A new approach in separating microplastics from environmental samples based on their electrostatic behavior. Environ Pollut 234:20–28. https://doi.org/10.1016/j.envpol.2017.11.013
62. He D, Luo Y, Lu S, Liu M, Song Y, Lei L (2018) Microplastics in soils: analytical methods, pollution characteristics and ecological risks. Trends Anal Chem 109:163–172. https://doi.org/10.1016/j.trac.2018.10.006
63. Huerta Lwanga E, Gertsen H, Gooren H, Peters P, Salánki T, Van Der Ploeg M, Besseling E, Koelmans AA, Geissen V (2016) Microplastics in the terrestrial ecosystem: implications for Lumbricus terrestris (Oligochaeta, Lumbricidae). Environ Sci Technol 50:2685–2691. https://doi.org/10.1021/acs.est.5b05478
64. Bläsing M, Amelung W (2018) Plastics in soil: analytical methods and possible sources. Sci Total Environ 612:422–435. https://doi.org/10.1016/j.scitotenv.2017.08.086

65. Li W, Wufuer R, Duo J, Wang S, Luo Y, Zhang D, Pan X (2020) Microplastics in agricultural soils: extraction and characterization after different periods of polythene film mulching in an arid region. Sci Total Environ 749. https://doi.org/10.1016/j.scitotenv.2020.141420
66. Hurley RR, Lusher AL, Olsen M, Nizzetto L (2018) Validation of a method for extracting microplastics from complex, organic-rich, environmental matrices. Environ Sci Technol 52: 7409–7417. https://doi.org/10.1021/acs.est.8b01517
67. Li L, Zhou Q, Yin N, Tu C, Luo Y (2019) Uptake and accumulation of microplastics in an edible plant | 食用蔬菜能吸收和积累微塑料. Kexue Tongbao/Chin Sci Bull 64:928–934. https://doi.org/10.1360/N972018-00845
68. Wang W, Yuan W, Chen Y, Wang J (2018) Microplastics in surface waters of Dongting Lake and Hong Lake, China. Sci Total Environ 633:539–545. https://doi.org/10.1016/j.scitotenv.2018.03.211
69. Ruggero F, Gori R, Lubello C (2020) Methodologies for microplastics recovery and identification in heterogeneous solid matrices: a review. J Polym Environ 28:739–748. https://doi.org/10.1007/s10924-019-01644-3
70. Qiu Q, Tan Z, Wang J, Peng J, Li M, Zhan Z (2016) Extraction, enumeration and identification methods for monitoring microplastics in the environment. Estuar Coast Shelf Sci 176:102–109. https://doi.org/10.1016/j.ecss.2016.04.012
71. Chai B, Wei Q, She Y, Lu G, Dang Z, Yin H (2020) Soil microplastic pollution in an e-waste dismantling zone of China. Waste Manag 118:291–301. https://doi.org/10.1016/j.wasman.2020.08.048
72. Fei Y, Huang S, Wang J, Luo Y, Zhang H (2021) Microplastics contamination in the protected agricultural soils and its effects on bacterial community diversity. Chin Sci Bull 66:1592–1601. https://doi.org/10.1360/TB-2020-0685
73. Isari EA, Papaioannou D, Kalavrouziotis IK, Karapanagioti HK (2021) Microplastics in agricultural soils: a case study in cultivation of watermelons and canning tomatoes. Water 13. https://doi.org/10.3390/w13162168
74. Yan C, Mei X, He W, Zheng S (2006) Present situation of residue pollution of mulching plastic film and controlling measures. Nongye Gongcheng Xuebao/Trans Chin Soc Agric Eng 22:269–272
75. Nguyen B, Claveau-Mallet D, Hernandez LM, Xu EG, Farner JM, Tufenkji N (2019) Separation and analysis of microplastics and nanoplastics in complex environmental samples. Acc Chem Res 52:858–866. https://doi.org/10.1021/acs.accounts.8b00602
76. Renner G, Schmidt TC, Schram J (2018) Analytical methodologies for monitoring micro (nano)plastics: which are fit for purpose? Curr Opin Environ Sci Heal 1:55–61. https://doi.org/10.1016/j.coesh.2017.11.001
77. Lorenzo-Navarro J, Castrillón-Santana M, Sánchez-Nielsen E, Zarco B, Herrera A, Martínez I, Gómez M (2021) Deep learning approach for automatic microplastics counting and classification. Sci Total Environ 765. https://doi.org/10.1016/j.scitotenv.2020.142728
78. Abu-Hilal AH, Al-Najjar TH (2009) Plastic pellets on the beaches of the Northern Gulf of Aqaba, Red Sea. Aquat Ecosyst Heal Manag 12:461–470. https://doi.org/10.1080/14634980903361200
79. Betzig E, Patterson GH, Sougrat R, Lindwasser OW, Olenych S, Bonifacino JS, Davidson MW, Lippincott-Schwartz J, Hess HF (2006) Imaging intracellular fluorescent proteins at nanometer resolution. Science 313:1642–1645. https://doi.org/10.1126/science.1127344
80. Watteau F, Dignac M-F, Bouchard A, Revallier A, Houot S (2018) Microplastic detection in soil amended with municipal solid waste composts as revealed by transmission electronic microscopy and pyrolysis/GC/MS. Front Sustain Food Syst 2. https://doi.org/10.3389/fsufs.2018.00081
81. Li W, Luo Y, Pan X (2020) Identification and characterization methods for microplastics basing on spatial imaging in micro-/nanoscales. Handb Environ Chem. https://doi.org/10.1007/698_2020_446

82. Shim WJ, Hong SH, Eo SE (2017) Identification methods in microplastic analysis: a review. Anal Methods 9:1384–1391. https://doi.org/10.1039/c6ay02558g
83. Lusher AL, Welden NA, Sobral P, Cole M (2017) Sampling, isolating and identifying microplastics ingested by fish and invertebrates. Anal Methods 9:1346–1360. https://doi.org/10.1039/c6ay02415g
84. Silva AB, Bastos AS, Justino CIL, da Costa JP, Duarte AC, Rocha-Santos TAP (2018) Microplastics in the environment: challenges in analytical chemistry – a review. Anal Chim Acta 1017:1–19. https://doi.org/10.1016/j.aca.2018.02.043
85. Junhao C, Xining Z, Xiaodong G, Li Z, Qi H, Siddique KHM (2021) Extraction and identification methods of microplastics and nanoplastics in agricultural soil: a review. J Environ Manag 294:112997. https://doi.org/10.1016/J.JENVMAN.2021.112997
86. Tagg AS, Sapp M, Harrison JP, Ojeda JJ (2015) Identification and quantification of microplastics in wastewater using focal plane Array-based reflectance micro-FT-IR imaging. Anal Chem 87:6032–6040. https://doi.org/10.1021/acs.analchem.5b00495
87. Bayo J, Olmos S, López-Castellanos J (2020) Microplastics in an urban wastewater treatment plant: the influence of physicochemical parameters and environmental factors. Chemosphere 238. https://doi.org/10.1016/j.chemosphere.2019.124593
88. Huang D, Xu Y, Lei F, Yu X, Ouyang Z, Chen Y, Jia H, Guo X (2021) Degradation of polyethylene plastic in soil and effects on microbial community composition. J Hazard Mater 416. https://doi.org/10.1016/j.jhazmat.2021.126173
89. Möller JN, Heisel I, Satzger A, Vizsolyi EC, Oster SDJ, Agarwal S, Laforsch C, Löder MGJ (2021) Tackling the challenge of extracting microplastics from soils: a protocol to purify soil samples for spectroscopic analysis. Environ Toxicol Chem. https://doi.org/10.1002/etc.5024
90. Wang F, Lai Z, Peng G, Luo L, Liu K, Huang X, Xu Y, Shen Q, Li D (2021) Microplastic abundance and distribution in a central Asian desert. Sci Total Environ 800. https://doi.org/10.1016/j.scitotenv.2021.149529
91. Elert AM, Becker R, Duemichen E, Eisentraut P, Falkenhagen J, Sturm H, Braun U (2017) Comparison of different methods for MP detection: what can we learn from them, and why asking the right question before measurements matters? Environ Pollut 231:1256–1264. https://doi.org/10.1016/j.envpol.2017.08.074
92. Okoffo ED, O'Brien S, Ribeiro F, Burrows SD, Toapanta T, Rauert C, O'Brien JW, Tscharke BJ, Wang X, Thomas KV (2021) Plastic particles in soil: state of the knowledge on sources, occurrence and distribution, analytical methods and ecological impacts. Environ Sci Process Impacts 23:240–274. https://doi.org/10.1039/d0em00312c
93. Cabernard L, Roscher L, Lorenz C, Gerdts G, Primpke S (2018) Comparison of Raman and Fourier transform infrared spectroscopy for the quantification of microplastics in the aquatic environment. Environ Sci Technol 52:13279–13288. https://doi.org/10.1021/acs.est.8b03438
94. Mallikarjunachari G, Ghosh P (2016) Analysis of strength and response of polymer nano thin film interfaces applying nanoindentation and nanoscratch techniques. Polymer (Guildf) 90:53–66. https://doi.org/10.1016/j.polymer.2016.02.042
95. Loeder M, Gerdts G (2015) Methodology used for the detection and identification of microplastics – a critical appraisal. In: Marine anthropogenic litter. Springer, Cham, pp 201–227
96. Gu W, Yang G, Liu Y, Mao Y, Li H, Ai H, He Q (2020) Treatment and detection methods of microplastics from environmental media: a review. J Civ Environ Eng 42:135–143. https://doi.org/10.11835/j.issn.2096-6717.2019.134
97. Renner G, Schmidt TC, Schram J (2017) Characterization and quantification of microplastics by infrared spectroscopy. Compr Anal Chem. https://doi.org/10.1016/bs.coac.2016.10.006
98. Wang W, Ge J, Yu X, Li H (2020) Environmental fate and impacts of microplastics in soil ecosystems: progress and perspective. Sci Total Environ 708. https://doi.org/10.1016/j.scitotenv.2019.134841

99. Sun J, Dai X, Wang Q, van Loosdrecht MCM, Ni B-J (2019) Microplastics in wastewater treatment plants: detection, occurrence and removal. Water Res 152:21–37. https://doi.org/10.1016/j.watres.2018.12.050
100. Lares M, Ncibi MC, Sillanpää M, Sillanpää M (2019) Intercomparison study on commonly used methods to determine microplastics in wastewater and sludge samples. Environ Sci Pollut Res 26:12109–12122. https://doi.org/10.1007/s11356-019-04584-6
101. Xu G, Cheng H, Jones R, Feng Y, Gong K, Li K, Fang X, Tahir MA, Valev VK, Zhang L (2020) Surface-enhanced Raman spectroscopy facilitates the detection of microplastics <1 μm in the environment. Environ Sci Technol 54:15594–15603. https://doi.org/10.1021/acs.est.0c02317
102. Käppler A, Fischer D, Oberbeckmann S, Schernewski G, Labrenz M, Eichhorn K-J, Voit B (2016) Analysis of environmental microplastics by vibrational microspectroscopy: FTIR, Raman or both? Anal Bioanal Chem 408:8377–8391. https://doi.org/10.1007/s00216-016-9956-3
103. Lenz R, Enders K, Stedmon CA, MacKenzie DMA, Nielsen TG (2015) A critical assessment of visual identification of marine microplastic using Raman spectroscopy for analysis improvement. Mar Pollut Bull 100:82–91. https://doi.org/10.1016/j.marpolbul.2015.09.026
104. Zhang B, Yang X, Chen L, Chao J, Teng J, Wang Q (2020) Microplastics in soils: a review of possible sources, analytical methods and ecological impacts. J Chem Technol Biotechnol 95: 2052–2068. https://doi.org/10.1002/jctb.6334
105. Eisentraut P, Dümichen E, Ruhl AS, Jekel M, Albrecht M, Gehde M, Braun U (2018) Two birds with one stone – fast and simultaneous analysis of microplastics: microparticles derived from thermoplastics and tire wear. Environ Sci Technol Lett 5:608–613. https://doi.org/10.1021/acs.estlett.8b00446
106. David J, Steinmetz Z, Kučerík J, Schaumann GE (2018) Quantitative analysis of poly(ethylene terephthalate) microplastics in soil via thermogravimetry-mass spectrometry. Anal Chem 90: 8793–8799. https://doi.org/10.1021/acs.analchem.8b00355
107. Steinmetz Z, Kintzi A, Muñoz K, Schaumann GE (2020) A simple method for the selective quantification of polyethylene, polypropylene, and polystyrene plastic debris in soil by pyrolysis-gas chromatography/mass spectrometry. J Anal Appl Pyrolysis 147. https://doi.org/10.1016/j.jaap.2020.104803
108. Corradini F, Bartholomeus H, Huerta Lwanga E, Gertsen H, Geissen V (2019) Predicting soil microplastic concentration using Vis-NIR spectroscopy. Sci Total Environ 650:922–932. https://doi.org/10.1016/j.scitotenv.2018.09.101
109. Du C, Wu J, Gong J, Liang H, Li Z (2020) ToF-SIMS characterization of microplastics in soils. Surf Interface Anal 52:293–300. https://doi.org/10.1002/sia.6742
110. Dierkes G, Lauschke T, Becher S, Schumacher H, Földi C, Ternes T (2019) Quantification of microplastics in environmental samples via pressurized liquid extraction and pyrolysis-gas chromatography. Anal Bioanal Chem 411:6959–6968. https://doi.org/10.1007/s00216-019-02066-9
111. Paul A, Wander L, Becker R, Goedecke C, Braun U (2019) High-throughput NIR spectroscopic (NIRS) detection of microplastics in soil. Environ Sci Pollut Res 26:7364–7374. https://doi.org/10.1007/s11356-018-2180-2
112. Shan J, Zhao J, Liu L, Zhang Y, Wang X, Wu F (2018) A novel way to rapidly monitor microplastics in soil by hyperspectral imaging technology and chemometrics. Environ Pollut 238:121–129. https://doi.org/10.1016/j.envpol.2018.03.026

Emerging Viruses in Sewage Sludge and Soils

David Polo and Jesús L. Romalde

Contents

Abstract Human pathogenic viruses can be introduced into sewage sludge and soils via fecal material from a variety of human activities. These contaminated matrices can play a substantial role in the dispersion of pathogenic viruses in the environment, constituting a potential public health problem if they enter the water cycle or the food chain. However, the interactions between pathogenic viruses and these matrices have received less attention compared to other environmental compartments. Understanding the presence of viruses, their persistence and fate in solid or semi-solid matrices like sludge and soil is important for the effective control of the infections they may cause. In this chapter, we summarize current knowledge about human pathogenic viruses in sewage sludge and soil, their importance in public health, and the factors that govern their transport and persistence in soil matrices. We also review the occurrence and variety of common and emerging viruses excreted in the feces and their presence in sewage sludge and soil, as well as the potential use of certain viruses as indicators of fecal pollution.

D. Polo (✉) and J. L. Romalde (✉)
Department of Microbiology and Parasitology, CRETUS and CIBUS-Faculty of Biology, Universidade de Santiago de Compostela, Santiago de Compostela, Spain
e-mail: david.polo.montero@usc.es; jesus.romalde@usc.es

Avelino Núñez-Delgado and Manuel Arias-Estévez (eds.),
Emerging Pollutants in Sewage Sludge and Soils,
Hdb Env Chem (2023) 114: 289–306, DOI 10.1007/698_2022_860,

Published online: 12 February 2022

Keywords Emerging viruses, Fecal pollution, SARS-CoV-2, Sewage sludge, Soil

1 Introduction

Transmission of human pathogenic viruses can occur either directly, by close contact between two hosts, or by indirect transmission through the environment. The main source of pathogenic viruses in the environment is the discharge of human and animal fecal matter from infected individuals, being transmitted back to susceptible hosts by direct contact or through consumption of water or foods (shellfish, fruits and vegetables) contaminated with wastewater in their harvesting areas [1]. This mode of transmission is known as the fecal-oral route and viruses transmitted in this manner are commonly referred to as enteric viruses [2]. The potential modes of transmission of enteric viruses through the environment are illustrated in Fig. 1.

Enteric viruses typically replicate in the gut, then are shed in large quantities in the feces of infected individuals, including asymptomatic carriers, and are widely prevalent in the community [3]. An infected person with diarrhea can shed between 10^{10} and 10^{12} viral particles per gram of feces [4]. However, our inability to detect the numerous asymptomatic infections caused by enteric viruses and the fact that environmental transmission is sometimes masked by person-to-person spread has traditionally made difficult the study and the appreciation of the real scope of the environmental transmission of these viruses [5]. In addition, our knowledge of the presence of other viruses, not commonly associated with the fecal-oral route of transmission, like coronaviruses and other respiratory viruses, in wastewater, is largely limited, which, as indicated by Bibby and Peccia [6], would be probably caused by scarce environmental investigations dealing with the fate of these viruses in the urban water cycle. Table 1 includes the main viruses that can be found in environmental samples.

The majority of the viral load associated with human and animal feces enters municipal wastewater through the sewage systems. Their number in raw municipal

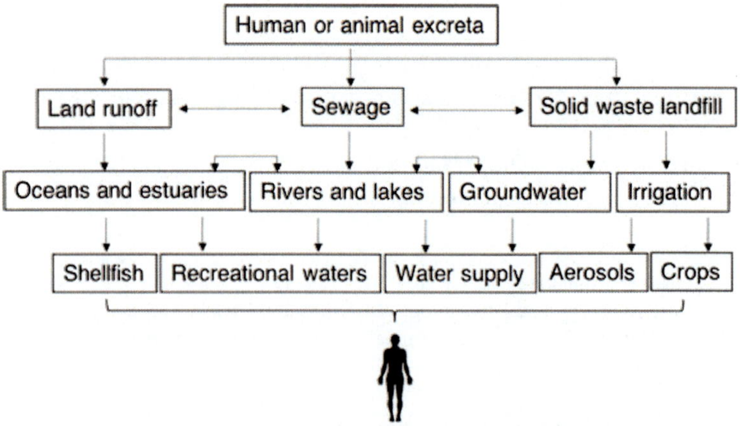

Fig. 1 Possible modes of transmission of enteric viruses in the environment

Table 1 Human pathogenic viruses, taxonomy, genomic characteristic, and disease symptoms

Virus		Genus (family)	Genome	Symptoms
Non-enveloped viruses	Adenovirus (AdV)	*Mastadenovirus* (*Adenoviridae*)	dsDNA	From common cold to severe pneumonia, gastro-enteritis, conjunctivitis, cystitis, and neurological illness
	Aichi virus (AiV)	*Kobuvirus* (*Picornaviridae*)	+ssRNA	Fever, nausea, diarrhea, vomiting, and abdominal pain
	Astrovirus (AsV)	*Mamastrovirus* (*Astroviridae*)	+ssRNA	Mild gastroenteritis (nau-sea dehydration, vomiting, and diarrhea)
	BK polyomavi-rus (BKPyV)	*Betapolyomavirus* (*Polyomaviridae*)	dsDNA	Emerging pathogen in kidney and bone marrow transplant recipients, caus-ing related nephropathy and hemorrhagic cystitis
	Bocavirus (HBoV)	*Bocavirus* (*Parvoviridae*)	ssDNA	Cold with fever, upper respiratory infections, and gastroenteritis
	Coxsackievirus type B (CVB)	*Enterovirus* (Non-polio) *Picornaviridae*	+ssRNA	Myocarditis, chronic dilated cardiomyopathy, diabetes, heart failure, encephalitis, aseptic meningitis
	Hepatitis A virus (HAV)	*Hepatovirus* (*Picornaviridae*)	+ssRNA	Fever, diarrhea, malaise, nausea, jaundice, and dark-colored urine
	Hepatitis E virus (HEV)	*Orthohepevirus* (*Hepeviridae*)	+ssRNA	Mild fever, anorexia, nau-sea, vomiting, abdominal pain, itching, skin rash, joint pain, jaundice, and hepatomegaly
	JC polyomavi-rus (JCV)	*Betapolyomavirus* (*Polyomaviridae*)	dsDNA	Progressive multifocal leukoencephalopathy (PML), an often-lethal ill-ness of the human brain
	Reovirus (ReoV)	*Orthoreovirus* *Reoviridae*	dsRNA	Pneumonia, meningitis, myocarditis, encephalitis, choledochal cysts, biliary atresia
	Norovirus (NoV)	*Norovirus* *Caliciviridae*	+ssRNA	Headaches, fever, vomiting, abdominal cramps, diarrhea, stomach pain
	Sapovirus (SaV)	*Sapovirus* (*Caliciviridae*)	+ssRNA	Vomiting, cramps, diar-rhea, myalgia, headache, chills, nausea, and abdom-inal pain

(continued)

Table 1 (continued)

Virus		Genus (family)	Genome	Symptoms
	Parechovirus (HPeV)	*Parechovirus* (Picornaviridae)	+ssRNA	Fever, diarrhea, cold, sepsis-like syndrome, meningitis, flaccid paralysis, encephalitis, seizures, and hepatitis
	Rotavirus (RoV)	*Rotavirus* *Reoviridae*	dsRNA (segmented)	Abdominal pain, watery diarrhea, vomiting, and fever
	Saffold virus (SAFV)	*Cardiovirus* (*Picornaviridae*)	+ssRNA	High fever, neurological symptoms
	Torque teno virus (TTV)	*Anellovirus* (*Circoviridae*)	ssDNA (circular)	Hepatitis-associated aplastic anemia, liver failure, and cryptogenic cirrhosis
	Human papillomavirus (HPV)	*Papillomavirus* (*Papillomaviridae*)	dsDNA	Warts, particularly genital warts
Enveloped viruses	Severe acute respiratory syndrome (SARS-CoV-1)	*Betacoronavirus* (*Coronaviridae*)	+ssRNA	Fever, fatigue, headache, chills, muscle pains, diarrhea, dry cough, shortness of breath
	Severe acute respiratory syndrome coronavirus 2 (SARS-CoV-2)	*Betacoronavirus* (*Coronaviridae*)	+ssRNA	Headache, loss of taste, sore throat, congestion, runny nose, nausea, vomiting, diarrhea, fever, chills, cough, shortness of breath, fatigue, and body aches

sewage may range from 10^2 to 10^7 genome copies/L [7]. However, sewage treatment processes, when present, are only partially effective in viral removal [8, 9]. Effluents and sludges from wastewater treatment plants (WWTPs) are frequently recycled in agricultural applications [9, 10]. In addition, enteric viruses from animal husbandry, runoff, and other origins can be present in the solid phase of natural environments that receive solid or liquid fecal materials [11]. The exposure of workers to pathogens is also a potential risk due to the generation of aerosols in the aeration facilities of the WWTPs or during the handling or land application of sludge and biosolids [12].

The presence, persistence, and fate of human pathogenic viruses in solid or semisolid matrices like sewage sludge and soils can thus constitute potential repositories of these pathogens in the environment that may come into contact with the population in a variety of ways. However, the interactions between pathogenic viruses and these matrices have received less attention compared to the other compartments, like coastal areas and surface water bodies. Understanding their presence, persistence, and fate in the environment is important for the effective control of these infections.

2 Persistence and Fate of Human Viruses in Sewage Sludge and Soils

Figure 2 shows a summarized graphical representation of interconnected factors that drive the environmental persistence of viruses.

2.1 Viruses in Sewage Sludge

Sewage treatment procedures are not totally effective in virus removal [8, 9] so part of these viruses is merely transferred to sludge. In sludge, viruses tend to be solid associated. Adsorption efficiency varies for different viruses and is influenced by certain solids, water, or virus characteristics such as ionic force, pH, organic matter, dissolved salts, or the size and the isoelectric point (surface charge) of the virus [13, 14]. Viruses adsorbed to solid particles are usually less sensitive to inactivation than those in water suspension, not bond to particles that reduce the area exposed to substances with inactivation potential, also considering the shadowing effect due to solid particles, which could protect viruses from UV radiation [15, 16].

Prado et al. [17] reported that elevated norovirus (NoV) counts were detected in secondary effluents even after activated sludge treatment (10^5–10^3 genome copies/L) and that an additional reduction in their concentrations could be reached using tertiary treatments. Tertiary treatments can include chlorination, and/or sand-anthracite filters, coagulation, and membrane bioreactor (MBR)/reverse osmosis (RO). Prado et al. [17] indicate that "although sand-anthracite filters and chlorination are economically more viable as tertiary effluent treatment, MBR/RO systems are

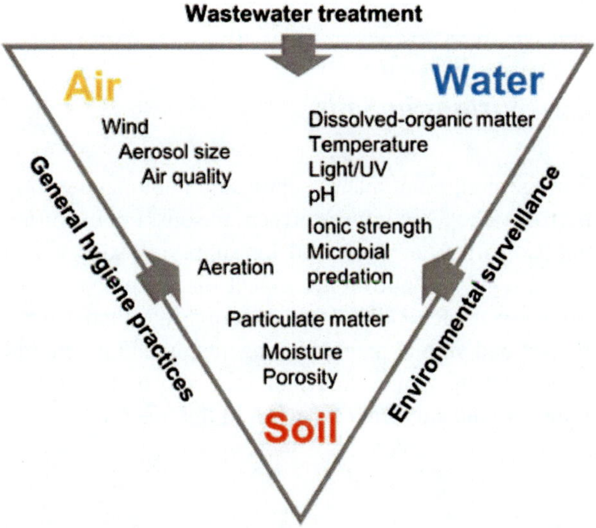

Fig. 2 Summary of the interconnected factors that drive the environmental persistence of viruses

more suitable for the production of reclaimed water where the application requires higher quality criteria." As regards virus removal, MBR and RO are considered high-performance systems, reaching, in the case of NoV, an overall mean reduction of 3.35 LOG [18]. However, Francy et al. [19] and Miura et al. [20] indicate that even with MBR being highly efficient in virus removal, NoV have been detected in effluents from these systems.

The vast majority of research related to virus in biosolids has dealt with non-enveloped viruses. Virus densities vary for specific pathogens and as a function of the procedures employed in their removal. Non-enveloped viruses (such as adenoviruses, polioviruses, or norovirus) are much more robust than those enveloped, like influenza viruses or coronaviruses, due to the higher susceptibility of the outer lipid membrane of enveloped viruses to disinfection and environmental stressors. Overall, the risk of exposure to any pathogenic virus shows a decrease with more intensive treatments applied to fecal waste and biosolids. In fact, the highest risk is related to the spreading of non-treated feces, untreated municipal sludge, class B biosolids, and Class A biosolids.

Thermophilic digestion has been reported to be more effective than mesophilic anaerobic digestion for non-enveloped enteric viruses [21, 22]. Mesophilic anaerobic digestion decreased enteric virus numbers in digested sludge by an average of 1.97 LOG reduction (94.4%), while the reduction obtained by thermophilic anaerobic digestion reached 4.6–7.1 LOGs or to levels below the detection capacity (>2.8–>5.8 LOGs), depending on the virus [21]. Furthermore, Viau and Peccia [23] indicated that "mesophilic anaerobically digested Class B biosolids further treated to Class A with heat pelletization (35° to 37°C for 10 to 20 days, dewatering, followed by a low-pressure oxidation drying system) and composting (agitated windrow method) resulted in even lower virus levels." Other reported LOG reduction ranges for certain sludges treatments were 0.5–2 LOGs for aerobic digestion, 2 to >4 for composting, 0.5 to >4 for air drying, and >4 for lime stabilization [24].

2.2 *Viruses in Soils*

Sewage sludge and manure application onto land is, traditionally, the main source of human pathogenic viruses present in soils. For centuries, untreated wastes have been applied to land as a method for supplying soils with valuable nutrients. Until the 1940s, only the nutritional aspect of this practice was considered, and the health aspects were generally of no concern. Since then, we know that sewage carries many human and animal pathogenic agents, including enteric viruses. The fate of viruses in soil depends on a variety of soil, environmental, and viral factors that affect the transport and survival of viruses in this matrix.

2.2.1 Transport of Viruses in Soils

Key factors with the main effect on virus transport through soils are virus adsorption, pH, ionic strength of the soil solution, presence of soluble organic matter, and the hydraulic flow rate. These factors are interconnected and interact with each other to determine the viral transport in the soil column.

Virus adsorption to soil is principally determined by the soil texture and composition, the ionic strength and the type of virus. It is well known that clay soils show an overall higher virus-retaining capacity than sandy-textured soils, and muck soils have a low affinity for viruses, as indicated by authors such as Scheuerman et al. [25], Sobsey et al. [26], or Meschke and Sobsey [27].

Both dissolved and soil-associated organic matter weakens the electrostatic binding established between soils and viruses. As a general rule, clay minerals are much more stronger sorbents than soils with organic matter contents. The suppression mechanism of these organic compounds on viral adsorption to clay minerals is the competition of proteins with the virus for sites on the minerals [28]. For example, montmorillonite, glauconite, and bituminous shale are less effective than magnetite sand and hematite, both of which are predominantly iron oxides, that have been reported as one of the most effective adsorbers [29–31]. Viral transport is also promoted by soluble organic materials present in sludge and wastewater effluents, and by humic and fulvic acids [31, 32]. As previously noted by Chattopadhyay et al. [33], "domestic wastewater contains surfactants that result from the use of laundry detergents, and these surfactants inhibit virus adsorption to surfaces, thus increasing transport through soils."

It is relevant that hydrophobic as well as electrostatic interactions play a role in virus adsorption to soils. In this way, it is of main importance the fact that, generally, soils have both negatively and positively charged sites (as well as hydrophobic sites), while viruses usually show negative charges, above the isoelectric point. As indicated by Armanious et al. [34], small amounts of hydrophobic material can retard virus transport in porous media. Previous studies, such as those by Lance and Gerba [35], Vaughn et al. [36], and Bales et al. [37], have shown that virus transport is also promoted by increasing hydraulic flow rate, or by increasing pH. To note that pH value influences the charge on both soils and viruses, while the ionic strength and its constituents determine the binding force affecting them. Soils with a low pH are generally good adsorbers for viruses, while viruses adsorb poorly to soils in situations where solutions have low ionic strength. It has been reported that higher ionic strength, due to the presence of cations, reduced the electrokinetic potential of both clay and virus particles, enhancing viral adsorption, with divalent cations like Ca^{2+} and Mg^{2+} being more effective than monovalent cations [34, 38].

Adsorption also varies with the type and strain of the virus. Distinctive adsorption characteristics of four similarly sized and spherical viruses can be related to differences in polarities of the viruses and to surface charges on the capsid. This is important since model viruses do not always simulate the survival and distribution of other viruses of public health importance [32, 34].

2.2.2 Survival of Viruses in Soils

Viral decay in the soil provides a barrier to the transmission of infectious diseases to humans and livestock, but those not yet inactivated can return to surface water as leachate and/or runoff, and to groundwater following subsurface and deep-transport processes. Those released by rainwater may contaminate ground and surface waters through vertical and lateral movement [38]. They can also be incorporated into crops after irrigation with contaminated water. Although the exact mechanism and the extent of this phenomenon are not well known, previous studies reported the internalization of enteric viruses such as human and murine norovirus and hepatitis A virus through the roots of different vegetables [39–41]. In addition, Chandrasekaran and Jiang [39] also demonstrated that viral attachment to soil particles is an important mechanism for viral risk reduction in comparison with hydroponically grown vegetables.

Authors such as Gessel et al. [42] and Campos et al. [43] have indicated that survival times of enteric microorganisms vary widely in top soil, reaching as long as several months. Survival of viruses in soils depends on factors that also govern their transport such as the nature and texture of the soil, the characteristics of the virus; the pH, ionic strength, and the dissolved organic matter of embedding water [13, 44]. However, soil temperature and moisture appear to be the most important factors in viral persistence. Viruses survive better at low than at high temperatures. Virus inactivation is also accelerated by a decrease in soil moisture [45]. Hurst et al. [46] and Sobsey et al. [26] have shown that soil microorganisms may also produce antiviral substances that increase the rate of viral inactivation. The levels of aeration or aerobicity (anaerobiosis), as well as the presence of heavy metals and acid pollutants, can also influence the inactivation of viruses in soils [16, 45–47].

3 Emerging Viruses in Soil and Sewage Sludge

The presence of human pathogenic viruses in sewage reflects part of the viral infections that are transmitted in the population. Viral shedding in wastewater allows tracking of emerging pathogens and the subsequent study of the epidemiology of enteric infectious diseases in communities. Environmental surveillance of urban wastewater can also be useful in tracking emerging viral pathogens and monitoring the changing epidemiology of enteric infectious diseases [48]. Studies, as well as regulations, on the presence and fate of viruses in sewage and aquatic environments, are often limited to enteric viruses, that are typically non-enveloped. However, wastewater can also contain a number of human viruses outside of the common enteric viruses, including enveloped respiratory viruses like influenza virus and coronavirus [6].

3.1 Non-enveloped Viruses

Common enteric viruses present in wastewater are non-enveloped viruses from different genera, most of them belonging to four viral families: *Picornaviridae*, *Caliciviridae*, *Adenoviridae,* or *Reoviridae*. These viruses typically replicate in the gastrointestinal tract and are excreted in large quantities in the feces of symptomatic or asymptomatic infected individuals and for several weeks. They are highly resistant in the environment and transmitted via the fecal-oral route following exposure to drinking water, recreational waters, and foods contaminated by wastewater and effluents from WWTP. Adenovirus (AdV), norovirus (NoV), sapovirus (SaV), rotavirus (RoV), hepatitis A virus (HAV), astrovirus (AsV), and enterovirus (EV), including polio and non-polio EV, are the most common genera found in wastewater and sewage sludge. There is an extensive literature regarding these viruses in wastewater and sewage sludge [6, 8, 22, 49, 50]. Emerging non-enveloped enteric viruses include Aichi virus (AiV), BK polyomavirus (BKPyV), JC polyomavirus (JCV), Human bocaviruses (HBoVs), Human parechovirus (HPeV), Reovirus (ReoV), Human Saffold virus (SAFV), Torque teno virus (TTV), and Hepatitis E virus (HEV).

The case of HEV is of special relevance since it become a worldwide public health concern. HEV is a + ssRNA virus (*Hepeviridae* family), with zoonotic potential, that infects a wide range of mammalian species. HEV is increasingly reported in Europe, attracting the attention of the European Authorities [51]. However, data regarding the burden of HEV infections are limited due to the lack of HEV surveillance in many European countries [52]. HEV transmission normally occurs via contaminated drinking water or contaminated food. The environmental contamination from human/animal sources may have relevance in the dissemination of HEV. In fact, this virus has been detected in urban sewage, being frequent in sewage and sludge from pig slaughterhouses in Europe. As specific details, note that HEV prevalence within herds of domestic swine, as well as in pig production regions in Europe, is often really high (98% in Spain; 55% in the Netherlands), and HEV genotype 3 has been detected in domestic swine in many high-income countries [53]. Shellfish harvested close to land and animal farms can bioaccumulate HEV, including genotypes 3 and 4 that have been isolated from both humans and animals, and are recognized as zoonotic pathogens [54].

3.2 Enveloped Viruses

Enveloped viruses (e.g., coronaviruses, Ebola virus, or influenza viruses) are traditionally considered more susceptible to inactivation in the environment and generally not associated with fecal routes of transmission in humans. Fecal samples are rarely collected for non-enteric illnesses; however, recent wastewater metagenomics studies have revealed a large diversity of human pathogenic viruses in feces and different sewage-related samples, including some enveloped viruses, typically considered respiratory viruses [6, 49]. From the wastewater perspective, coronaviruses (CoVs) and avian influenza viruses are particularly important to consider.

Avian influenza viruses usually cause illness in birds, but occasionally cross over to humans and result in serious illnesses. Unlike human influenza viruses, certain avian influenza strains are transmitted via the fecal-oral route in birds [55]. The highly pathogenic H5N1 virus crossed over from birds to humans in 2004. The virus was present in stool samples [56], infected and replicated in human gut tissues [57] and patients experienced severe diarrhea [56, 58]. Other strains, like the H7N9 virus, which emerged in humans in 2013, or the H1N1 strain that caused illness in humans, were also detected in stool samples and exhibited unusually high gastrointestinal symptoms [59–62].

The knowledge of the presence of CoVs in wastewater is relatively limited due to the lack of previous environmental investigations focusing on these viruses. During the SARS outbreak in 2004 in China, SARS-CoV-1 RNA was detected both in untreated and treated wastewater samples collected from a hospital in Beijing, receiving SARS patients [63]. Wastewater was also believed to be at least partly responsible for a previous SARS outbreak due to a faulty ventilation and plumbing system [64]. However, the recent COVID-19 pandemic has drastically increased interest in the detection of CoVs in wastewater, mainly as a disease surveillance tool, an application known as wastewater-based epidemiology [48]. There have been initial reports of the molecular detection of SARS-CoV-2 in wastewater in the Netherlands, USA, France, Australia, and Spain [65–67]. These studies reported the detection of SARS-CoV-2 RNA in untreated wastewater with maximum concentrations over 10^6 genome copies/L.

Regarding the presence of CoVs in sludge or soils, the literature is scarce. For example, Bibby and Peccia [6] detected the RNA from two CoV strains (HKU1 and 229E) in untreated wastewater sludge and Class B biosolids in 83% of the samples. On the other hand, SARS-CoV-2 RNA was also detected in sludge. Serra-Compte et al. [68] reported that non-treated sludge (from primary and secondary treatments) presented a higher occurrence of SARS-CoV-2 RNA than the corresponding water samples, demonstrating the affinity of virus particles for solids. Furthermore, SARS-CoV-2 RNA was detected in treated sludge after thickening and anaerobic digestion (quantification levels of 10^3 genome copies/L), whereas viral RNA was completely eliminated from sludge only when thermal hydrolysis was applied. Balboa et al. [69] also reported the occurrence of SARS-CoV-2 in thickened sludge and pinpointed it as a hotspot of virus RNA in the plant. SARS-CoV-2 viral RNA was also quantified in solids collected from influent post grit solids (PGS) and primary clarified sludge (PCS), in levels ranging from 10^3 to 10^5 genome copies/L [70]. So far, the presence of infectious SARS-CoV-2 in wastewater or biosolids has not been documented. However, Zhang et al. [71] reported the presence of SARS-CoV-2 RNA in hospital outdoor soil (205–550 genome copies/g), pointing to respiratory droplets from asymptomatic COVID-19 patients, contamination from medical wastes or deposition from aerosols as the potential routes of soil contamination.

4 Emerging Viral Indicators of Fecal Pollution

Fecal indicator bacteria (FIB) such as *E. coli* and enterococci are typically used as indicators of fecal pollution in water. However, FIB differ from viral pathogens in terms of persistence in water. To overcome this limitation viral indicators have been proposed to represent viruses in sewage-contaminated water better than FIB [72].

Previous studies have investigated human pathogenic viruses as possible viral indicators. Human AdVs, PyVs, and AiVs are frequently found in wastewater and other polluted environments, being suggested as potential fecal indicators [50, 73–75]. However, and despite being human-specific, human pathogenic viruses can also show distinct seasonality or variable prevalence by region [76]. Bacteriophages infecting bacteria associated with the human gut are also common in wastewater. Somatic coliphages (phages infecting *E. coli*) and F-specific RNA bacteriophages (phages infecting bacteria through the F-pili) are commonly used to assess wastewater contamination [50].

Recently, two novel viruses have been reported as promising viral indicators: a cross-assembly phage (crAssphage) and the pepper mild mottle virus (PMMoV). CrAssphage is a bacteriophage, discovered in 2014 by computational analysis of human fecal metagenomes [77]. This bacteriophage, that infects the human gut symbiont *Bacteroides intestinalis*, is highly abundant in the human microbiome on a global scale [78] and its usefulness as an indicator of human fecal pollution for impacted urban watersheds was demonstrated [79, 80].

Interestingly, a plant virus, the pepper mild mottle virus (PMMoV), has also been shown to be associated with human wastewater and suggested to be a useful indicator for wastewater contamination [81, 82]. The primary source of PMMoV in human excreta is through consumption of peppers (*Capsicum* spp.) [83]. However, its shape and size differ from other pathogenic viruses with icosahedral capsids and hence its fate and behavior in the environment may be different.

5 Conclusions

Understanding the presence, persistence, fate, and diversity of pathogenic viruses in biosolids and manure and their introduction into the soils, are crucial aspects to providing better control and management strategies designed to mitigate their potential microbiological health risks associated with common agriculture practices, like land application of raw or treated sewage sludge and crop irrigation with reclaimed water. This information would also assist in making rational regulatory and treatment design decisions in the assessment of human exposure routes to viruses, wastewater source tracking applications, and the efficacy of virus removal from a variety of wastewater treatment schemes.

Acknowledgments Prof. Jesuús L. Romalde acknowledges the support of Consellería de Educación, Universidade e Formacioón Profesional, Xunta de Galicia, Spain (Grants 2014-PG110 and ED431C 2018/18), that funded the research that generated part of the results presented here.

References

1. Bosch A, Pintó RM, Abad FX (2006) Survival and transport of enteric viruses in the environment. In: Goyal SM (ed) Viruses in foods. Food microbiology and food safety. Springer, Boston. https://doi.org/10.1007/0-387-29251-9_6
2. Symonds EM, Griffin DW, Breitbart M (2009) Eukaryotic viruses in wastewater samples from the United States. Appl Environ Microbiol 75(5):1402–1409. https://doi.org/10.1128/AEM.01899-08
3. Bishop RF, Kirkwood CD (2008) Enteric viruses. Encycl Virol 116–123. https://doi.org/10.1016/B978-012374410-4.00386-1
4. Haas CH, Rose JB, Gerba CP (eds) (2014) Quantitative microbial risk assessment. John Wiley & Sons
5. Bartsch SM, Lopman BA, Ozawa S, Hall AJ, Lee BY (2016) Global economic burden of norovirus gastroenteritis. PLoS One 11(4):e0151219. https://doi.org/10.1371/journal.pone.0151219
6. Bibby K, Peccia J (2013) Identification of viral pathogen diversity in sewage sludge by metagenome analysis. Environ Sci Technol 47(4):1945–1951. https://doi.org/10.1021/es305181x
7. Bisseux M, Colombet J, Mirand A, Roque-Afonso AM, Abravanel F, Izopet J, Archimbaud C, Peigue-Lafeuille H, Debroas D, Bailly JL, Henquell C (2018) Monitoring human enteric viruses in wastewater and relevance to infections encountered in the clinical setting: a one-year experiment in Central France, 2014 to 2015. Euro Surveill 23(7):17-00237. https://doi.org/10.2807/1560-7917.ES.2018.23.7.17-00237
8. Campos C, Avant J, Lowther J, Till D, Lees DN (2016) Human norovirus in untreated sewage and effluents from primary, secondary and tertiary treatment processes. Water Res 103:224–232. https://doi.org/10.1016/j.watres.2016.07.045
9. Gerba CP, Betancourt WQ, Kitajima M (2017) How much reduction of virus is needed for recycled water: a continuous changing need for assessment? Water Res 108:25–31. https://doi.org/10.1016/j.watres.2016.11.020
10. Grobelak A, Czerwińska K, Murtaś A (2019) General considerations on sludge disposal, industrial and municipal sludge. In: Industrial and municipal sludge. Elsevier, pp 135–153. https://doi.org/10.1016/B978-0-12-815907-1.00007-6
11. Corsi SR, Borchardt MA, Spencer SK, Hughes PE, Baldwin AK (2014) Human and bovine viruses in the Milwaukee River watershed: hydrologically relevant representation and relations with environmental variables. Sci Total Environ 490:849–860. https://doi.org/10.1016/j.scitotenv.2014.05.072
12. Pasalari H, Ataei-Pirkooh A, Aminikhah M, Jafari AJ, Farzadkia M (2019) Assessmentof airborne enteric viruses emitted from wastewater treatment plant: atmospheric dispersion model, quantitative microbial risk assessment, disease burden. Environ Pollut 253:464–473. https://doi.org/10.1016/j.envpol.2019.07.010
13. Fongaro G, García-González MC, Hernández M, Kunz A, Barardi CR, Rodríguez-Lázaro D (2017) Different behavior of enteric bacteria and viruses in clay and sandy soils after biofertilization with swine digestate. Front Microbiol 8:74. https://doi.org/10.3389/fmicb.2017.00074

14. Kimura M, Jia ZJ, Nakayama N, Asakawa S (2008) Ecology of viroses in soils: past, present and future perspectives. Soil Sci Plant Nutr 54:1–32. https://doi.org/10.1111/j.1747-0765.2007.00197.x
15. Hassard F, Gwyther CL, Farkas K, Andrews A, Jones V, Cox B, Brett H, Jones DL, McDonald JE, Malham SK (2016) Abundance and distribution of enteric bacteria and viruses in coastal and estuarine sediments-a review. Front Microbiol 7:1692. https://doi.org/10.3389/fmicb.2016.01692
16. Zhao B, Zhang H, Zhang J, Jin Y (2008) Virus adsorption and inactivation in soil as influenced by autochthonous microorganisms and water content. Soil Biol Biochem 40:649–659. https://doi.org/10.1016/j.soilbio.2007.09.020
17. Prado T, de Castro Bruni A, Barbosa M, Garcia SC, Moreno LZ, Sato M (2019) Noroviruses in raw sewage, secondary effluents and reclaimed water produced by sand-anthracite filters and membrane bioreactor/reverse osmosis system. Sci Total Environ 646:427–437. https://doi.org/10.1016/j.scitotenv.2018.07.301
18. Sano D, Amarasiri M, Hata A, Watanabe T, Katayama H (2016) Risk management of viral infectious diseases in wastewater reclamation and reuse: review. Environ Int 91:220–229. https://doi.org/10.1016/j.envint.2016.03.001
19. Francy DS, Stelzer EA, Bushon RN, Brady AM, Williston AG, Riddell KR, Borchardt MA, Spencer SK, Gellner TM (2012) Comparative effectiveness of membrane bioreactors, conventional secondary treatment, and chlorine and UV disinfection to remove microorganisms from municipal wastewaters. Water Res 46(13):4164–4178. https://doi.org/10.1016/j.watres.2012.04.044
20. Miura T, Okabe S, Nakahara Y, Sano D (2015) Removal properties of human enteric viruses in a pilot-scale membrane bioreactor (MBR) process. Water Res 75:282–291. https://doi.org/10.1016/j.watres.2015.02.046
21. Sassi HP, Ikner LA, Abd-Elmaksoud S, Gerba CP, Pepper IL (2018) Comparative survival of viruses during thermophilic and mesophilic anaerobic digestion. Sci Total Environ 615:15–19. https://doi.org/10.1016/j.scitotenv.2017.09.205
22. Wong K, Onan B, Xagoraraki I (2010) Quantification of enteric viruses, pathogen indicators, and salmonella bacteria in class B anaerobically digested biosolids by culture and molecular methods. Appl Environ Microbiol 76:6441–6448. https://doi.org/10.1128/AEM.02685-09
23. Viau E, Peccia J (2009) Survey of wastewater indicators and human pathogen genomes in biosolids produced by class A and class B stabilization treatments. Appl Environ Microbiol 75(1):164–174. https://doi.org/10.1128/AEM.01331-08
24. Godfree A (2003) Health constraints on the agricultural recycling of wastewater sludges. In: Mara D, Horan NJ (eds) Handbook of water and wastewater microbiology. Elsevier, London
25. Scheuerman PR, Bitton G, Overman AR, Gifford GE (1979) Transport of viruses through organic soils and sediments. J Environ Eng Div ASCE 105:629–640
26. Sobsey MD, Dean CW, Knuckles ME, Wagner RA (1980) Interactions and survival of enteric viruses in soil materials. Appl Environ Microbiol 40:92–101
27. Meschke JS, Sobsey MD (2003) Comparative reduction of Norwalk virus, poliovirus type 1, F +RNA coliphage MS2 and Escherichia coli in miniature soil columns. Water Sci Technol 47(3):85–90
28. Carlson Jr GF, Woodard FE, Wentworth DF, Sproul OJ (1968) Virus inactivation on clay particles in natural waters. J Water Pollut Control Fed 40:R98 R106
29. Lipson SM, Stotzky G (1984) Effect of proteins on reovirus adsorption to clay minerals. Appl Environ Microbiol 48:525–530
30. Moore RS, Taylor DH, Sturman LS, Reddy MM, Fuhs GW (1981) Poliovirus adsorption by 34 minerals and soils. Appl Environ Microbiol 42:963–975
31. Zhuang J, Jin Y (2003) Virus retention and transport as influenced by different forms of soil organic matter. J Environ Qual 32:816–823

32. Dizer H, Nasser A, Lopez JM (1984) Penetration of different human pathogenic viruses into sand columns percolated with distilled water, groundwater or wastewater. Appl Environ Microbiol 47:409–415
33. Chattopadhyay D, Chattopadhyay S, Lyon WG, Wilson JT (2002) Effect of surfactants on the survival and sorption of viruses. Environ Sci Technol 36:4017–4024
34. Armanious A, Aeppli M, Jacak R, Refardt D, Sigstam T, Kohn T, Sander M (2016) Viruses at solid-water interfaces: a systematic assessment of interactions driving adsorption. Environ Sci Technol 50(2):732–743. https://doi.org/10.1021/acs.est.5b04644
35. Lance JC, Gerba CP (1980) Poliovirus movement during high rate land filtration of sewage water. J Environ Qual 9:31–34
36. Vaughn JM, Landry EF, Beckwith CA, Thomas MZ (1981) Virus removal during groundwater recharge: effects of infiltration rate on adsorption of poliovirus to soil. Appl Environ Microbiol 41:139–143
37. Bales RC, Li S, Maguire KM, Yahya MT, Gerba CP, Harvey RW (1995) Virus and bacteria transport in a sandy aquifer, Cape Cod, MA. Ground Water 33:653–881
38. Wong K, Mukherjee B, Kahler AM, Zepp R, Molina M (2012) Influence of inorganic ions on aggregation and adsorption behaviors of human adenovirus. Environ Sci Technol 46(20): 11145–11153. https://doi.org/10.1021/es3028764
39. Chandrasekaran S, Jiang SC (2018) A dynamic transport model for quantification of norovirus internalization in lettuce from irrigation water and associated health risk. Sci Total Environ 643: 751–761. https://doi.org/10.1016/j.scitotenv.2018.06.158
40. Hirneisen KA, Kniel KE (2013) Comparative uptake of enteric viruses into spinach and green onions. Food Environ Virol 5(1):24–34. https://doi.org/10.1007/s12560-012-9093-x
41. Yang Z, Chambers H, DiCaprio E, Gao G, Li J (2018) Internalization and dissemination of human norovirus and Tulane virus in fresh produce is plant dependent. Food Microbiol 69:25–32. https://doi.org/10.1016/j.fm.2017.07.015
42. Gessel PD, Hansen NC, Goyal SM, Johnston LJ, Webb J (2004) Persistence of zoonotic pathogens in surface soil treated with different rates of liquid pig manure. Appl Soil Ecol 25: 237–243. https://doi.org/10.1016/j.apsoil.2003.09.008
43. Campos C, Beltran M, Duarte M, Medina L, Lucena F, Jofre J (2013) Abatement of helminth eggs and bacterial and viral indicators in soil after land application of treated sludges. J Water Resour Prot 5:1155–1164. https://doi.org/10.4236/jwarp.2013.512122
44. Bradford SA, Morales VL, Zhang W, Harvey RW, Packman AI, Mohanram A, Welty C (2013) Transport and fate of microbial pathogens in agricultural settings. Crit Rev Environ Sci Technol 43:775–893. https://doi.org/10.1080/10643389.2012.710449
45. Straub TM, Pepper IL, Gerba CP (1992) Persistence of viruses in desert soils amended with anaerobically digested sewage sludge. Appl Environ Microbiol 58:636–641
46. Hurst CJ, Gerba CP, Cech I (1980) Effects of environmental variables and soil characteristics on virus survival. Appl Environ Microbiol 40:1067–1079
47. Song I, Choi C, O'Shaughnessy Y, Gerba S (2005) Effects of temperature and moisture on coliphage PRD-1 survival in soil. J Food Prot 68:2118–2122. https://doi.org/10.4315/0362-028X-68.10.2118
48. Polo D, Quintela-Baluja M, Corbishley A, Jones DL, Singer AC, Graham DW, Romalde JL (2020) Making waves: wastewater-based epidemiology for COVID-19 - approaches and challenges for surveillance and prediction. Water Res 186:116404. https://doi.org/10.1016/j.watres.2020.116404
49. Cantalupo PG, Calgua B, Zhao G, Hundesa A, Wier AD, Katz JP, Grabe M, Hendrix RW, Girones R, Wang D, Pipas JM (2011) Raw sewage harbors diverse viral populations. mBio 2(5):e00180–e00111. https://doi.org/10.1128/mBio.00180-11
50. Farkas K, Walker DI, Adriaenssens EM, McDonald JE, Hillary LS, Malham SK, Jones DL (2020) Viral indicators for tracking domestic wastewater contamination in the aquatic environment. Water Res 181:115926. https://doi.org/10.1016/j.watres.2020.115926

51. ECDC (2017) Hepatitis E in the EU/EEA, 2005–2015. Stockholm, June 2017. https://doi.org/10.2900/059144
52. Horn J, Hoodgarzadeh M, Klett-Tammen CJ, Mikolajczyk RT, Krause G, Ott JJ (2018) Epidemiologic estimates of hepatitis E virus infection in European countries. J Infect 77(6): 544–552. https://doi.org/10.1016/j.jinf.2018.09.012
53. Pavio N, Meng XJ, Renou C (2010) Zoonotic hepatitis E: animal reservoirs and emerging risks. Vet Res 41(6):46. https://doi.org/10.1051/vetres/2010018
54. Rivadulla E, Varela MF, Mesquita JR, Nascimento M, Romalde JL (2019) Detection of hepatitis E virus in shellfish harvesting areas from Galicia (Northwestern Spain). Viruses 11(7):618. https://doi.org/10.3390/v11070618
55. WHO (2006) Review of latest available evidence on risks to human health through potential transmission of avian influenza (H5N1) through water and sewage. http://www.who.int/water_sanitation_health/emerging/h5n1background.pdf
56. de Jong MD, Bach VC, Phan TQ, Vo MH, Tran TT, Nguyen BH, Beld M, Le TP, Truong HK, Nguyen VV, Tran TH, Do QH, Farrar J (2005) Fatal avian influenza A (H5N1) in a child presenting with diarrhea followed by coma. N Engl J Med 352(7):686–691. https://doi.org/10.1056/NEJMoa044307
57. Shu Y, Li CK, Li Z, Gao R, Liang Q, Zhang Y, Dong L, Zhou J, Dong J, Wang D, Wen L, Wang M, Bai T, Li D, Dong X, Yu H, Yang W, Wang Y, Feng Z, McMichael AJ, Xu XN (2010) Avian influenza A (H5N1) viruses can directly infect and replicate in human gut tissues. J Infect Dis 201(8):1173–1177. https://doi.org/10.1086/651457
58. Beigel JH, Farrar J, Han AM, Hayden FG, Hyer R, de Jong MD, Lochindarat S, Nguyen TK, Nguyen TH, Tran TH, Nicoll A, Touch S, Yuen KY, Writing Committee of the World Health Organization (WHO) Consultation on Human Influenza A/H5 (2005) Avian influenza A (H5N1) infection in humans. N Engl J Med 353(13):1374–1385. https://doi.org/10.1056/NEJMra052211
59. Hu Y, Lu S, Song Z, Wang W, Hao P, Li J, Zhang X, Yen HL, Shi B, Li T, Guan W, Xu L, Liu Y, Wang S, Zhang X, Tian D, Zhu Z, He J, Huang K, Chen H, Yuan Z (2013) Association between adverse clinical outcome in human disease caused by novel influenza A H7N9 virus and sustained viral shedding and emergence of antiviral resistance. Lancet 381(9885): 2273–2279. https://doi.org/10.1016/S0140-6736(13)61125-3
60. To KK, Chan KH, Li IW, Tsang TY, Tse H, Chan JF, Hung IF, Lai ST, Leung CW, Kwan YW, Lau YL, Ng TK, Cheng VC, Peiris JS, Yuen KY (2010) Viral load in patients infected with pandemic H1N1 2009 influenza A virus. J Med Virol 82(1):1–7. https://doi.org/10.1002/jmv.21664
61. Yoo SJ, Moon SJ, Kuak EY, Yoo HM, Kim CK, Chey MJ, Shin BM (2010) Frequent detection of pandemic (H1N1) 2009 virus in stools of hospitalized patients. J Clin Microbiol 48(6): 2314–2315. https://doi.org/10.1128/JCM.00325-10
62. Zeng H, Pappas C, Katz JM, Tumpey TM (2011) The 2009 pandemic H1N1 and triple-reassortant swine H1N1 influenza viruses replicate efficiently but elicit an attenuated inflammatory response in polarized human bronchial epithelial cells. J Virol 85(2):686–696. https://doi.org/10.1128/JVI.01568-10
63. Wang XW, Li J, Guo T, Zhen B, Kong Q, Yi B, Li Z, Song N, Jin M, Xiao W, Zhu X, Gu C, Yin J, Wei W, Yao W, Liu C, Li J, Ou G, Wang M, Fang T, Li J (2005) Concentration and detection of SARS coronavirus in sewage from Xiao Tang Shan Hospital and the 309th Hospital of the Chinese People's Liberation Army. Water Sci Technol 52(8):213–221
64. McKinney KR, Gong YY, Lewis TG (2006) Environmental transmission of SARS at Amoy Gardens. J Environ Health 68(9):26–52
65. Ahmed W, Angel N, Edson J, Bibby K, Bivins A, O'Brien JW, Choi PM, Kitajima M, Simpson SL, Li J, Tscharke B, Verhagen R, Smith W, Zaugg J, Dierens L, Hugenholtz P, Thomas KV, Mueller JF (2020) First confirmed detection of SARS-CoV-2 in untreated wastewater in Australia: a proof of concept for the wastewater surveillance of COVID-19 in the community. Sci Total Environ 728:138764. https://doi.org/10.1016/j.scitotenv.2020.138764

66. Lodder W, de Roda Husman AM (2020) SARS-CoV-2 in wastewater: potential health risk, but also data source. Lancet Gastroenterol Hepatol 5(6):533–534. https://doi.org/10.1016/S2468-1253(20)30087-X
67. Randazzo W, Truchado P, Cuevas-Ferrando E, Simón P, Allende A, Sánchez G (2020) SARS-CoV-2 RNA in wastewater anticipated COVID-19 occurrence in a low prevalence area. Water Res 181:115942. https://doi.org/10.1016/j.watres.2020.115942
68. Serra-Compte A, González S, Arnaldos M, Berlendis S, Courtois S, Loret JF, Schlosser O, Yáñez AM, Soria-Soria E, Fittipaldi M, Saucedo G, Pinar-Méndez A, Paraira M, Galofré B, Lema JM, Balboa S, Mauricio-Iglesias M, Bosch A, Pintó RM, Bertrand I, Litrico X (2021) Elimination of SARS-CoV-2 along wastewater and sludge treatment processes. Water Res 202: 117435. https://doi.org/10.1016/j.watres.2021.117435
69. Balboa S, Mauricio-Iglesias M, Rodriguez S, Martínez-Lamas L, Vasallo FJ, Regueiro B, Lema JM (2021) The fate of SARS-COV-2 in WWTPS points out the sludge line as a suitable spot for detection of COVID-19. Sci Total Environ 772:145268. https://doi.org/10.1016/j.scitotenv.2021.145268
70. D'Aoust PM, Mercier E, Montpetit D, Jia JJ, Alexandrov I, Neault N, Baig AT, Mayne J, Zhang X, Alain T, Langlois MA, Servos MR, MacKenzie M, Figeys D, MacKenzie AE, Graber TE, Delatolla R (2021) Quantitative analysis of SARS-CoV-2 RNA from wastewater solids in communities with low COVID-19 incidence and prevalence. Water Res 188:116560. https://doi.org/10.1016/j.watres.2020.116560
71. Zhang D, Zhang X, Yang Y, Huang X, Jiang J, Li M, Ling H, Li J, Liu Y, Li G, Li W, Yi C, Zhang T, Jiang Y, Xiong Y, He Z, Wang X, Deng S, Zhao P, Qu J (2021) SARS-CoV-2 spillover into hospital outdoor environments. J Hazard Mater Lett 2:100027. https://doi.org/10.1016/j.hazl.2021.100027
72. Bibby K, Crank K, Greaves J, Li X, Wu Z, Hamza IA, Stachler E (2019) Meta-genomics and the development of viral water quality tools. NPJ Clean Water 2(1):1–13
73. Kitajima M, Gerba CP (2015) Aichi virus 1: environmental occurrence and behavior. Pathogens 4(2):256–268. https://doi.org/10.3390/pathogens4020256
74. Rachmadi AT, Torrey JR, Kitajima M (2016) Human polyomavirus: advantages and limitations as a human-specific viral marker in aquatic environments. Water Res 105:456–469. https://doi.org/10.1016/j.watres.2016.09.010
75. Rames E, Roiko A, Stratton H, Macdonald J (2016) Technical aspects of using human adenovirus as a viral water quality indicator. Water Res 96:308–326. https://doi.org/10.1016/j.watres.2016.03.042
76. Harwood VJ, Boehm AB, Sassoubre LM, Vijayavel K, Stewart JR, Fong TT, Caprais MP, Converse RR, Diston D, Ebdon J, Fuhrman JA, Gourmelon M, Gentry-Shields J, Griffith JF, Kashian DR, Noble RT, Taylor H, Wicki M (2013) Performance of viruses and bacteriophages for fecal source determination in a multi-laboratory, comparative study. Water Res 47(18): 6929–6943. https://doi.org/10.1016/j.watres.2013.04.064
77. Dutilh BE, Cassman N, McNair K, Sanchez SE, Silva GG, Boling L, Barr JJ, Speth DR, Seguritan V, Aziz RK, Felts B, Dinsdale EA, Mokili JL, Edwards RA (2014) A highly abundant bacteriophage discovered in the unknown sequences of human faecal metagenomes. Nat Commun 5:4498. https://doi.org/10.1038/ncomms5498
78. Guerin E, Shkoporov A, Stockdale SR, Clooney AG, Ryan FJ, Sutton T, Draper LA, Gonzalez-Tortuero E, Ross RP, Hill C (2018) Biology and taxonomy of crAss-like bacteriophages, the most abundant virus in the human gut. Cell Host Microbe 24(5):653–664.e6. https://doi.org/10.1016/j.chom.2018.10.002
79. Polo D, Lois M, Fernández-Núñez MT, Romalde JL (2021) Detection of SARS-CoV-2 RNA in bivalve mollusks and marine sediments. Sci Total Environ 786:147534. https://doi.org/10.1016/j.scitotenv.2021.147534

80. Stachler E, Akyon B, de Carvalho NA, Ference C, Bibby K (2018) Correlation of crAssphage qPCR markers with culturable and molecular indicators of human Fecal pollution in an impacted urban watershed. Environ Sci Technol 52(13):7505–7512. https://doi.org/10.1021/acs.est.8b00638
81. Kitajima M, Sassi HP, Torrey JR (2018) Pepper mild mottle virus as a water quality indicator. NPJ Clean Water 1. https://doi.org/10.1038/s41545-018-0019-5
82. Symonds EM, Nguyen KH, Harwood VJ, Breitbart M (2018) Pepper mild mottle virus: a plant pathogen with a greater purpose in (waste)water treatment development and public health management. Water Res 144:1–12. https://doi.org/10.1016/j.watres.2018.06.066
83. Zhang T, Breitbart M, Lee WH, Run JQ, Wei CL, Soh SW, Hibberd ML, Liu ET, Rohwer F, Ruan Y (2006) RNA viral community in human feces: prevalence of plant pathogenic viruses. PLoS Biol 4(1):e3. https://doi.org/10.1371/journal.pbio.0040003